Understanding Essential Chemistry

Understanding Essential Chemistry

Max Diem
Professor Emeritus
Department of Chemistry and Chemical Biology
Northeastern University, Boston, MA, USA

This edition first published 2025

© 2025 John Wiley & Sons, Inc.

All rights reserved, including rights for text and data mining and training of artificial intelligence technologies or similar technologies. No part of this publication may be reproduced, stored in a retrieval system, or transmitted, in any form or by any means, electronic, mechanical, photocopying, recording or otherwise, except as permitted by law. Advice on how to obtain permission to reuse material from this title is available at http://www.wiley.com/go/permissions.

The right of Max Diem to be identified as the author(s) of this work has been asserted in accordance with law.

Registered Office(s)
John Wiley & Sons, Inc., 111 River Street, Hoboken, NJ 07030, USA
For details of our global editorial offices, customer services, and more information about Wiley products visit us at www.wiley.com.

The manufacturer's authorized representative according to the EU General Product Safety Regulation is Wiley-VCH GmbH, Boschstr. 12, 69469 Weinheim, Germany, e-mail: Product_Safety@wiley.com.

Wiley also publishes its books in a variety of electronic formats and by print-on-demand. Some content that appears in standard print versions of this book may not be available in other formats.

Trademarks: Wiley and the Wiley logo are trademarks or registered trademarks of John Wiley & Sons, Inc. and/or its affiliates in the United States and other countries and may not be used without written permission. All other trademarks are the property of their respective owners. John Wiley & Sons, Inc. is not associated with any product or vendor mentioned in this book.

Limit of Liability/Disclaimer of Warranty
In view of ongoing research, equipment modifications, changes in governmental regulations, and the constant flow of information relating to the use of experimental reagents, equipment, and devices, the reader is urged to review and evaluate the information provided in the package insert or instructions for each chemical, piece of equipment, reagent, or device for, among other things, any changes in the instructions or indication of usage and for added warnings and precautions. While the publisher and authors have used their best efforts in preparing this work, they make no representations or warranties with respect to the accuracy or completeness of the contents of this work and specifically disclaim all warranties, including without limitation any implied warranties of merchantability or fitness for a particular purpose. No warranty may be created or extended by sales representatives, written sales materials or promotional statements for this work. The fact that an organization, website, or product is referred to in this work as a citation and/or potential source of further information does not mean that the publisher and authors endorse the information or services the organization, website, or product may provide or recommendations it may make. This work is sold with the understanding that the publisher is not engaged in rendering professional services. The advice and strategies contained herein may not be suitable for your situation. You should consult with a specialist where appropriate. Further, readers should be aware that websites listed in this work may have changed or disappeared between when this work was written and when it is read. Neither the publisher nor authors shall be liable for any loss of profit or any other commercial damages, including but not limited to special, incidental, consequential, or other damages.

Library of Congress Cataloging-in-Publication Data

Names: Diem, Max, 1947- author.
Title: Understanding essential chemistry / Max Diem, Professor emeritus, Boston, MA, US.
Description: Hoboken, NJ : Wiley 2025. | Includes bibliographical references and index.
Identifiers: LCCN 2025009592 | ISBN 9781394321193 (paperback) | ISBN 9781394321230 (epdf) | ISBN 9781394321223 (epub)
Subjects: LCSH: Chemistry—Textbooks. | LCGFT: Textbooks.
Classification: LCC QD31.3 .D54 2025 | DDC 540—dc23/eng/20250326
LC record available at https://lccn.loc.gov/2025009592

Cover Design: Wiley
Cover Image: © Fresh photos from all over the world/Getty Images

SKY10104636_043025

Understanding Essential Chemistry

Other Books by the Author

M. Diem, *Introduction to Modern Vibrational Spectroscopy*, J. Wiley Interscience, 1993 (ISBN-13: 978-0-471-59584-7, ISBN-10: 0-471-59584-5)

"*Vibrational Spectroscopy for Medical Diagnosis*," M. Diem, P. Griffiths and J. Chalmers, Editors, J. Wiley Interscience, 2008 (ISBN-13: 978-0-470-01214-7, ISBN-10: 0-470-01214-5)

M. Diem, *Modern Vibrational Spectroscopy and Micro-spectroscopy: Theory, Instrumentation and Biomedical Applications*, J. Wiley, 2015 (ISBN-13: 978-1-118-82486-3, ISBN-10: 1-118-82486-5)

M. Diem, *Quantum Mechanical Foundation of Molecular Spectroscopy*, Wiley-VCH, 2021 (ISBN-13: 978-3527347926, ISBN-10: 3527347925)

M. Diem, *Quantenmechanische Grundlagen der Molekülspektroskopie*, Wiley-VCH, 2021 (ISBN-13: 978-3527347902, ISBN-10: 3527347909)

Contents

Preface *xi*
About the Companion Website *xiii*

1 The Metric System and Mathematical Tools *1*
1.1 Scientific Notation and Significant Figures *1*
1.2 The Metric System *2*
1.3 Manipulations of Exponential Expressions *4*
1.4 Equations, Proportionality, and Graphs *5*
1.5 Quadratic, Cubic, and Quartic Equations *7*
1.6 Exponential Functions and Logarithms *8*
1.7 Radial and Spherical Polar Coordinates *11*
1.8 Differential and Integral Calculus *12*
1.9 Differential Equations *14*
1.10 Complex Numbers *15*

2 Atoms, Elements, and the Periodic System *17*
2.1 Subatomic Particles and Atoms *17*
2.2 Elements, Isotopes, and Ions *18*
2.3 The Periodic Chart and Periodic Properties of the Elements *21*
2.4 Definition of Atomic Masses, Avogadro's Number, and the Mole *26*
Further Reading *28*

3 Molecules, Compounds, Bonding, and Percent Composition *29*
3.1 Ionic Compounds *29*
3.2 Molecules with Covalent Bonds *30*
3.3 Molecules with Polar Covalent Bonds and Lewis Structures *31*
3.4 Molecular Compounds and the (Gram) Molecular Mass *37*
3.5 Percent Composition and Empirical Formulae *38*
Further Reading *39*

4 Chemical Reactions *41*
4.1 Chemical Reaction and Stoichiometry *41*
4.2 Limiting Reagents, Theoretical Yield, and Percent Yield *42*
4.3 Solutions: General Aspects *43*
4.4 Solution Stoichiometry: Molarity, Molality, Mole Fraction, Dilutions *45*
4.5 Precipitation Reactions *47*
Further Reading *49*

5 Electronic Structure of Atoms *51*
5.1 Description of Light as an Electromagnetic Wave *51*
5.2 Particle Properties of Light and Wave-particle Duality *52*

5.3	The Hydrogen Atom Emission Spectrum: Stationary Atomic States *55*
5.4	Hydrogen Atom Orbitals *58*
5.5	Atoms with Multiple Electrons: The Aufbau Principle Revisited *62*
	Further Reading *67*

6 Chemical Bonding: Covalent Bonding, Molecular Geometries, and Polarity *69*
- 6.1 General Aspects of Covalent Bonding *69*
- 6.2 Lewis and VB Theory *69*
- 6.3 Hybridization and Multiple Bonding *70*
- 6.4 VSEPR Model *73*
- 6.5 Molecular Polarity *75*
- 6.6 MO Theory *76*
- Further Reading *81*

7 Solids and Liquids: Bonding and Characteristics *83*
- 7.1 Metals and Semiconductors *83*
- 7.2 Ionic Solids *85*
- 7.3 Covalent Solids *86*
- 7.4 Intermolecular Forces *87*
- 7.4.1 Hydrogen Bonding *87*
- 7.4.2 Dipole–Dipole Interactions *88*
- 7.4.3 London Dispersion Forces (Induced Dipole Forces) *89*
- 7.5 Macromolecular Solids *89*
- 7.6 Liquids and Solutions *90*
- 7.6.1 General Aspects of Solutions and Solvation *90*
- 7.6.2 Colligative Properties *90*
- Further Reading *93*

8 The Gaseous State *95*
- 8.1 General Properties of Gases *95*
- 8.2 Empirical Gas Laws *97*
- 8.3 The Ideal Gas Law *99*
- 8.4 Real Gases *100*
- 8.5 Gaseous Mixtures and Partial Pressures *101*
- 8.6 Kinetic Theory of Gases *102*
- 8.7 Diffusion and Effusion of Gases *104*
- Further Reading *106*

9 Chemical Equilibrium *107*
- 9.1 What Is a System "at Equilibrium"? *107*
- 9.2 Liquid–Vapor Phase Equilibrium: Vapor Pressure *108*
- 9.3 Temperature Dependence of Vapor Pressure *110*
- 9.4 Chemical Equilibrium and the Equilibrium Constant *113*
- 9.5 Equilibrium Calculations *115*
- 9.6 Direction of a Chemical Reaction and the Concentration Quotient Q *119*
- 9.7 Numerical Determination of Equilibrium Constants from Experimental Data *120*
- 9.8 Perturbations of Equilibria: Le Chatelier's Principle *120*
- 9.9 Solubility and Solubility Product *122*
- 9.9.1 The Solubility Product Constant, K_{sp} *123*
- 9.9.2 Solubility Calculations *123*
- 9.9.3 Common Ion Effect *124*
- 9.9.4 Experimental Determination of K_{sp} *125*

9.9.5	Precipitation Reactions	*126*
	Further Reading *127*	

10 Acids and Bases *129*
- 10.1 What Are Acids/Bases? *129*
- 10.2 Strong Acids and Bases; Definition of pH and pOH *130*
- 10.3 Weak Acids/Bases *131*
- 10.4 The Relationship Between pH and pOH: Self-dissociation of Water *134*
- 10.5 Common Ion Effect *135*
- 10.6 Acidic and Basic Salts *136*
- 10.7 Buffers *138*
- 10.8 Acid–Base Titrations *140*
- 10.8.1 Titration of a Strong Acid with a Strong Base *141*
- 10.8.2 Titration of a Weak Acid with a Strong Base *143*
- 10.8.3 Acid–Base Indicators *145*
  - Further Reading *146*

11 Thermodynamics: Energy, Energy Conversions, and Spontaneity *147*
- 11.1 Energetics of Chemical Reactions *147*
- 11.2 Thermochemistry *147*
- 11.2.1 Definition of Energy, Work, and Heat *147*
- 11.2.2 Calorimetry: Measurement of Heat Flow *150*
- 11.3 The First Law of Thermodynamics *152*
- 11.4 State Functions *153*
- 11.5 Definition of Enthalpy *153*
- 11.6 Hess' Law and Reaction Enthalpies *154*
- 11.6.1 Enthalpy of Crystal Formation: Lattice Energy of MgO *156*
- 11.7 Enthalpy of Phase Transitions *157*
- 11.8 Entropy *158*
- 11.8.1 Entropy and Probability *161*
- 11.8.2 Entropy and Heat Flow *162*
- 11.8.3 Entropy as an Indicator of Energy Exhaustion *163*
- 11.9 Free Enthalpy *164*
- 11.10 Free Enthalpy and Equilibrium *165*
  - Further Reading *168*

12 Reduction–Oxidation (Redox) Reactions and Electrochemistry *169*
- 12.1 Oxidation State and Oxidation Numbers: Balancing Redox Equations *170*
- 12.2 Galvanic Cells, Electric Work, and Electromotive Force *173*
- 12.3 Batteries *177*
- 12.3.1 Alkaline Dry Cell (AA Battery) *177*
- 12.3.2 Lead–Acid Battery *178*
- 12.3.3 Lithium-ion Battery *180*
- 12.4 Relationship Between Cell Potential and Free Enthalpy *181*
- 12.5 Concentration and Temperature Dependence of EMF *181*
  - Further Reading *183*

13 Chemical Kinetics: Rates of Reactions and Reaction Mechanisms *185*
- 13.1 Scope of Kinetics Discussion *185*
- 13.2 Elementary Steps and Chemical Reactions *185*
- 13.2.1 Kinetic Model of Chemical Reactions *185*
- 13.2.2 Basics of Chemical Kinetics: Rate Law and Rate Constant *187*

13.2.3	Time Dependence of the Reaction Rate *188*
13.2.4	Integrated Rate Law *189*
13.3	Rates of Multistep Reactions, and Equilibria *191*
13.4	Reaction Rates for Reactions That Are Nonlinear in Concentrations *194*
13.5	Reaction Path and Catalysis *195*
	Further Reading *198*

14 Nuclear Reactions *199*

14.1	Nuclear Reactions and Transmutations *199*
14.2	The Structure of Atomic Nuclei *199*
14.3	Radioactive Decay and Decay Chains *200*
14.3.1	α-Decay *200*
14.3.2	β-Decay *200*
14.3.3	γ-Emission (γ-Decay) *201*
14.3.4	Positron Emission *202*
14.3.5	Nuclear Decay Chains *202*
14.3.6	Nuclear Dating *203*
14.4	Nuclear Fission and Nuclear Fusion *204*
14.4.1	Nuclear Binding Energy *205*
14.4.2	Nuclear Fusion *205*
14.4.3	Nuclear Fission *206*
	Further Reading *207*

15 Fundamentals of Quantum Chemistry, Spectroscopy, and Structural Chemistry *209*

15.1	Wavefunctions and the 1D and 2D Particle in a Box *209*
15.2	Spherical Harmonics, Hydrogen Atom Wavefunctions, and Hydrogen Atomic Orbitals *213*
15.3	Atomic Energy Levels and Atomic Emission Spectroscopy *217*
15.4	Molecular Energy Levels, Spectroscopy, and Structural Methods *219*
15.4.1	Electronic Energy Levels and UV-vis Absorption Spectroscopy *219*
15.4.2	Vibrational Energy Levels and Infrared Spectroscopy *221*
15.4.3	Rotational Energy Levels and Microwave Spectroscopy *225*
15.4.4	Nuclear Magnetic Resonance Spectroscopy *226*
15.4.5	X-ray Diffraction *228*
15.5	Mass Spectrometry *230*
	Further Reading *230*

Epilogue *231*

Appendix *233*

List of Constants *233*

List of Abbreviations and Symbols *234*

Index *235*

Preface

Understanding Essential Chemistry (UEC) is written with your future career in mind. Whether you pursue a pure science field, a medical field, an environmental field, an engineering field, or others, you will appreciate having a deep understanding of essential chemical principles. That is because chemistry can be viewed as the central science, located at the interface between physics, biology, biochemistry, medicine, pharmacology, engineering, environmental sciences, geology, etc. A deep understanding of chemistry will allow you to apply and transfer this knowledge to whatever field you choose for your future.

UEC is designed as a low-cost supplement to accompany any one of the expensive textbooks your professor may require you to purchase or rent for the particular general chemistry course in which you are enrolled. It is unique from all other textbooks in several ways. First, the narrative focuses on the "big picture," helping you to see the beauty and interrelationship among chemical principles. It helps you to see the historical use of the scientific method in the elucidation of the body of chemical principles we study today. It also provides a *path to deep understanding* by directly linking the custom diagrams, charts, and tables to the narrative. This approach enables the narrative to be more interactive, concise, and simply more readable and memorable.

Second, ***UEC*** embraces a different approach to problem-solving. It again focuses on the integral relationship between the narrative and the actual solving of each example problem. In other words, the narrative introduces new concepts and then asks you to apply and transfer this knowledge to solve related problems. Short answers to the example problems are given with the problem, but you must use the narrative to help you arrive at the given answer. If you are unable to solve a particular problem in this way, an answer booklet found online can be consulted for extra help and detailed methods to answer the example problems. This approach to solving problems helps you to enhance and expand your understanding of the concepts.

Third, the narrative assumes you have had a high school level algebra course and that you retained this knowledge. This implies that you can manipulate algebraic expressions, rearrange equations to isolate unknowns, graph functions, and solve quadratic equations. If you cannot comfortably perform these steps, a review of the necessary concepts is presented in Chapter 1. More advanced mathematics is necessary in the discussions in Chapter 5 (Electronic Structure of Atoms), Chapter 13 (Chemical Kinetics), and Chapter 15 (Fundamentals of Quantum Theory), including some differential and integral calculus and complex numbers. Exposure to these more complicated branches of mathematics is necessary to show that chemistry and all the sciences and engineering are governed by a common language, mathematics. These advanced mathematical concepts are also presented in Chapter 1. If you are familiar with most of the mathematical principles, feel free to skip over Chapter 1, and refer to it only when you feel the need to do so.

Finally, since ***UEC*** is designed as a supplementary textbook for all major general chemistry textbooks, you may rearrange the order in which you study the chapters. This is possible as each chapter is self-contained. This enables you to select the chapter that corresponds to the chapter your professor has chosen to cover.

I hope that ***UEC*** helps you to form a deep and lasting understanding of chemistry, as it relates to the world of science and to the natural world as a whole. I hope it empowers you to pass, with flying colors, the tests and exams that accompany your studies. Finally, I hope that you will keep this book as a personal, professional resource library once your general chemistry class is over and refer to it when you need to brush up on essential concepts.

Boston, 2024
Max Diem, Professor of Chemistry (Emeritus)

About the Companion Website

This book is accompanied by a companion website.

www.wiley.com\go\Diem\EssentialChemistry

This website includes:

- Answer Booklet

1

The Metric System and Mathematical Tools

This chapter presents a review of mathematical concepts and procedures, as well as a concise summary of the metric system. In contrast to all other chapters in this book, which should be read in the order they are presented, this chapter can be skipped if you are familiar with the concepts or can be used to refresh certain aspects you may not have been exposed to. It is not recommended that you go through the entirety of Chapter 1 before you start working on the Chemistry sections, since many aspects of Chapter 1 are not needed until much later in the discussions.

1.1 Scientific Notation and Significant Figures

In science, one often has to deal with very large or very small numbers. Avogadro's number, given the symbol N_A in this text (see Section 2.4), is $N_A = 602\,214\,000\,000\,000\,000\,000\,000$. This representation is rather inconvenient and is written in scientific notation as $N_A = 6.02214 \cdot 10^{23}$. In general, numbers are written in scientific notation as

$$m \cdot 10^n \tag{1.1}$$

where n, the exponent, is a positive or negative integer number, and m, the mantissa, is $|-10| < m < 10$. Remember that $10^0 = 1$, $10^1 = 10$, $10^2 = 100$, $10^3 = 1000$, etc., and $10^{-1} = 0.1$, $10^{-2} = 0.01$, $10^{-3} = 0.001$, etc. Thus, in very small numbers, the exponent n is negative. The mass of a hydrogen atom, for example,

$$m_H = 0.00000000000000000000000016735575 \text{ [g]} \tag{1.2}$$

is presented in scientific notation as

$$m_H = 1.6735575 \cdot 10^{-24} \text{ [g]} = 1.6735575 \cdot 10^{-27} \text{ [kg]} \tag{1.3}$$

since 1 [g] = 10^{-3} [kg]. Eq. 1.3 introduced two further aspects: the metric system, to be reviewed in Section 1.2, and the manipulation of exponential expressions, discussed in Section 1.3.

Most scientific calculators abbreviate the exponential part, 10^n, as "En," for example, 10^2 appears as E2, or worse, as e+2. Do not confuse this with the number "e," Euler's constant (2.71828...), the basis of the natural logarithm that we will encounter in Section 1.3. So, when your calculator (or the calculator function in Windows) gives you a result "2.13158e-5," it is implied that it means $2.13158 \cdot 10^{-5}$. Also beware of a strange inconsistency in putting an exponential expression into the Windows calculator: the number 1 000 000, 1 million, or 10^6, must be entered as 1 followed by the "exp" key, followed by the number 6. If you enter 10 exp 6, it will read the number as 10 000 000. So BEWARE!!!

When you perform a numerical calculation, particularly with a calculator, you may be tempted to copy the result from the calculator without any further scrutiny. For example, if you are computing the circumference c of a circle with a radius r of 1.5 [cm], according to $c = 2\pi r$, your calculator will display 9.4247779607693797153879301498385 or something similar as an answer, if you use the value of π programmed into the calculator (you can, however, specify a different format). This answer implies that it is known to 32 significant figures. However, unless the radius is known to the same number of significant figures, the answer is worthless and misleading. When the radius is reported as 1.5 [cm], it implies that the next digit after the 5 is not known for sure: it could be 1.51 [cm], or 1.499 [cm], rounded down or up to 1.5 [cm]. This uncertainty results from the uncertainty in the measurement. If you use a simple ruler, 1.5 [cm] may be the best answer, given the graduation of the ruler. If you use a better measuring device, you may find that the radius is 1.514 [cm]. The accuracy

Understanding Essential Chemistry, First Edition. Max Diem.
© 2025 John Wiley & Sons, Inc. Published 2025 by John Wiley & Sons, Inc.

of a measurement implies how close the measurement is to the true value. To indicate the improved accuracy of your measurement, you should report the radius as 1.514 [cm], with four significant figures.

If the calculation you are performing requires the radius to be entered in units of meters (see Section 1.2), with

$$1 \text{ [cm]} = 10^{-2} \text{ [m]} \tag{1.4}$$

you retain the number of significant figures as follows:

$$r = 1.514 \cdot 10^{-2} \text{ [m]} \tag{1.5}$$

where as

$$r = 1.5 \cdot 10^{-2} \text{ [m]} \tag{1.6}$$

would imply the former situation with two significant figures. In any calculations, the input with the fewest significant figures determines with how many significant figures the results should be reported. Thus, the circumference should be reported as 9.4 [cm] in the first case above, and as 9.513 (9.5127 rounded to four significant figures) in the second case. Your main text most likely will have a number of examples. Most constants reported in this book are given with three or four significant digits although many are known to higher accuracy. As indicated, the number of significant figures is an indicator of the accuracy of a measurement.

When dealing with significant figures, notice that there are cases where a number occurs in an expression that has more significant figures than there are written. In the above example, $c = 2\pi r$, the number 2 is considered to be an exact number that is known to as many significant figures as necessary. This is, since the diameter and the radius of a circle are related by a factor exactly equal to 2.

1.2 The Metric System

The metric system of measurements is used in science, rather than the "imperial system" used predominantly in the English-speaking world. The metric system was introduced in France at the beginning of the nineteenth century with the aim of standardizing all measurements and providing reproducible standards. The unit of **length** in the metric system is the meter, abbreviated as [m], which is about 1.1 [yard]. The principle of the metric system is that all larger and smaller units (of length, in this case) are related to the meter by powers of 10. Thus, a kilometer [km] is 1000 [m], and a centimeter is 0.01 or 10^{-2} [m], unlike in the imperial system, where the conversion from miles to yards involves the factor of 1760, and from inches to yards by a factor of 36.

In the metric system, the powers of 10 for the conversion from larger to smaller (or vice versa) units are abbreviated as shown in Table 1.1 (notice that these abbreviations are case sensitive):

Table 1.1 Prefixes and abbreviations used in the metric system.

Factor	Prefix	Symbol
10^{15}	Peta	P
10^{12}	Tera	T
10^{9}	Giga	G
10^{6}	Mega	M
10^{3}	Kilo	k
10^{-2}	Centi	c
10^{-3}	Milli	m
10^{-6}	Micro	μ
10^{-9}	Nano	n
10^{-12}	Pico	p
10^{-15}	Femto	f

The conversion from centimeter to meter we encountered in Section 1.1

$$1.5 \,[cm] = 1.5 \cdot 10^{-2} \,[m]$$

follows directly from the information in Table 1.1, since $1\,[cm] = 10^{-2}\,[m]$. For the reverse conversion, from meter to centimeter, we multiply both sides of Eq. 1.4 by 100, and obtain

$$1\,[cm] \cdot 10^2 = 10^{-2}\,[m] \cdot 10^2 \tag{1.4}$$

$$100\,[cm] = 1\,[m] \tag{1.7}$$

Area in the metric system is measured in units of $[m^2]$. Due to the ease of manipulating exponential expressions (see Section 1.3), the metric system affords easy conversion between, for example, $[km^2]$ to $[m^2]$. Since $1\,[km] = 10^3\,[m]$, $1\,[km^2] = (10^3)^2\,[m^2] = 10^6\,[m^2] = 1\,000\,000\,[m^2]$. It is easy to see why the metric system is superior to the imperial system, where a similar conversion, say from square mile to square feet, would require paper and pencil (or a calculator), since

$$1\,[\text{mile}] = 1760\,[\text{yards}] = 5280\,[\text{ft}] \tag{1.8}$$

$$1\,[\text{mile}^2] = 5280^2\,[\text{ft}^2] = 27\,878\,400\,[\text{ft}^2] \tag{1.9}$$

By the same logic, we can define **volumes** in the metric system: the volume is measured in units of $[m^3]$. The conversion factors shown in Table 1.1 have to be raised to the third power when converting volumes, for example, $1\,[cm^3] = (10^{-2})^3\,[m^3] = 10^{-6}\,[m^3] = 0.000\,001\,[m^3]$. The volume V of a cube 10 [cm] on edge would be

$$V = (10\,[cm])^3 = 10^3\,[cm^3] = 1000\,[cm^3] \tag{1.10}$$

is also called 1 liter [L]. Consequently,

$$1\,[mL] = 1\,[cm^3] \tag{1.11}$$

Mass originally was defined in units of gram [g], where the mass of $1\,[cm^3] = 1\,[mL]$ of water is 1 [g], or the mass of 1 [L] of water is 1 [kg]. This was based on the definition of the density (see next paragraph) of water to be 1 [g/mL]. The units discussed so far have been redefined to much higher accuracy and reproducibility, but in the context of this book, the old definitions are sufficiently accurate.

Many units are derived from these basic units. **Density d**, for example, is defined as the ratio of mass over volume:

$$d = m/V \tag{1.12}$$

Water, as mentioned above, has a density of 1 [g/mL]. Mercury, a very dense (and liquid) metal, has a density of 13.6 [g/mL] or 13.6 [kg/L]. So, if you are asked what volume of mercury will have a mass of 1.00 [g], you would solve the definition of density for the volume,

$$V = m/d = 1.00/13.6 = 0.0735 = 7.34 \cdot 10^{-2} \left[\frac{g}{\frac{g}{mL}} = mL \right] \tag{1.13}$$

Force (mass · acceleration) has units of $[kg\,m/s^2] = 1[N]$ (Newton), and **energy** has units of force · distance or $[Nm] = [kg\,m^2/s^2] = [J]$ (Joule, see below).

Temperature, in the metric system, is reported on the Celsius scale in degrees centigrade [°C]. It was originally defined by two limiting temperatures: the ice/water equilibrium mixture was defined to have a temperature of 0 [°C], and the water boiling temperature was defined as 100 [°C] (at an ambient pressure of 1 [atm], see below). These two experimental conditions can be readily reproduced allowing easy calibration of thermometers worldwide. However, as shown in the discussion of the gaseous state (Chapter 8), it becomes clear that the centigrade scale is as arbitrary as other temperature scales (for example, the Fahrenheit scale used in the United States), and another temperature scale needed to be defined. In this new temperature scale, the zero point is not defined arbitrarily, but based on the concept that an ideal gas approaches zero pressure (or zero volume) when the temperature is zero degrees on a new scale. This is due to the fact that the pressure p depends on the temperature as given by the ideal gas law

$$p = \frac{nR}{V} T \tag{1.14}$$

(see Eq. 8.14). Thus, the temperature has to be defined such that p is zero as T is zero. The temperature zero point can be obtained by extrapolating the p,T diagram shown in Figure 1.1 to zero pressure. In this temperature scale, known as the Kelvin (K) temperature, the size of the increment is the same in the centigrade and the Kelvin scale, $1\,[K] = 1\,[°C]$. Hence,

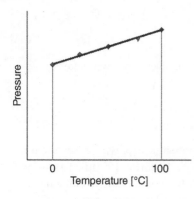

Figure 1.1 Plot of ideal gas pressure vs. temperature in [°C].

we are allowed to use [°C] units in calorimeter (Section 11.2.2) since we discuss temperature changes in this section.

The zero point of the Kelvin temperature scale is found to be $T_K = -273.16$ [°C]. For temperature conversions, the following equations are useful:

$$T_F = 1.8\ T_C + 32 \tag{1.15}$$

$$T_K = T_C + 273 \tag{1.16}$$

Here, T_C denotes temperature in centigrade units, T_F in Fahrenheit units, and T_K in Kelvin units. In section 1.4, a few examples of temperature conversions will be discussed.

Pressure is defined as force per unit area, and thus, the pressure units in the metric system are Newton per square meter $[N/m^2] = 1$ [Pa] (Pascal). However, in this book, and in most Chemistry texts, the older definition, based on the atmospheric pressure of the earth's atmosphere, is used. Here,

$$1\ [atm] = 760\ [mm\ Hg] = 101{,}325\ [Pa] \tag{1.17}$$

As we shall see in the section on Thermochemistry (Section 11.2), **heat** is a form of energy, and consequently, has units of energy. Originally the unit of heat was defined in the metric system as the amount of heat required to raise the temperature of 1 [g] = 1 [mL] of water by 1 [°C]. This amount of heat was called 1 calorie [cal]. In modern scientific work, the calorie has been replaced by the metric unit of energy: 1 [cal] = 4.18 [J]. This unit of heat, and its multiple, 1 [kJ] = 1000 [J] is used throughout this book. Interestingly, the unit [cal] is still used in nutritional science and sports medicine; however, it is used incorrectly. What a nutritionist calls 1 [cal] is actually 1 [kcal]; thus, if a paper in sports medicine reports the base metabolic rate of an adults as 2000 [cal], it actually implies 2000 [kcal].

Time, the unit of which in both the metric and imperial system is the second [s], uses the same prefixes for fractions of a second: 1 [ms] = 10^{-3} [s], 1 [ns] = 10^{-9} [s], etc. However, for time periods larger than 1 [s], we use minutes and hours, rather than kiloseconds (you see here that there are a few inconsistencies in the metric system as well). This has mostly historical reasons, since we are accustomed to a day's length as 24 [h], with 1 [h] = 3600 [s], and one year as 365 days. The latter time units (days, years) are relevant in Chapters 13 and 14, where we deal with the half-lives of very slow reactions. By the way, ignore the occurrence of leap years (with 366 days/year) in these calculations, since the few extra days do not matter, within the significant figures, if a half-life is given as 240 000 years (see Figure 14.1).

1.3 Manipulations of Exponential Expressions

Addition/subtraction of exponential expressions: ascertain that the exponents are the same, then proceed by adding the mantissa.

Examples:

$$1.20 \cdot 10^{12} + 2.3 \cdot 10^{11} = 1.20 \cdot 10^{12} + 0.23 \cdot 10^{12} = 1.43 \cdot 10^{12} \quad \text{(three significant figures)} \tag{1.18}$$

$$1.20 \cdot 10^{-12} - 2.3 \cdot 10^{-11} = 0.12 \cdot 10^{-11} - 2.3 \cdot 10^{-11} = -2.2 \cdot 10^{-11} \quad \text{(two significant figures)} \tag{1.19}$$

Multiplication/division of exponential expressions:

$$a^n \cdot a^m = a^{(n+m)};\quad a^n/a^m = a^{(n-m)};\quad (a^n)^2 = a^{2n} \tag{1.20}$$

Examples:

$$10^{20} \cdot 10^3 = 10^{23};\ 10^{20}/10^3 = 10^{17} \tag{1.21}$$

$$6.3 \cdot 10^3 \cdot 4.5 \cdot 10^5 = 6.3 \cdot 4.5 \cdot 10^8 = 28.35 \cdot 10^8 = 2.8 \cdot 10^9 \quad \text{(two significant figures)} \tag{1.22}$$

To experience a typical computation involving exponential expressions, we turn to a calculation from Chapter 15 (Example 15.3):

$$\frac{(6.6 \cdot 10^{-34})^2}{8 \cdot 9.1 \cdot 10^{-31} \cdot (10^{-9})^2} = \frac{(6.6)^2}{8 \cdot 9.1} \frac{10^{-68}}{10^{-49}} = \frac{43.56}{72.8} \cdot 10^{-19} = 0.598 \cdot 10^{-19} = 6.0 \cdot 10^{-20} \quad \text{(two significant figures)} \quad (1.23)$$

When using an electronic calculator, not all intermediate steps shown above are necessary, and the fraction $\frac{(6.6 \cdot 10^{-34})^2}{8 \cdot 9.1 \cdot 10^{-31} \cdot (10^{-9})^2}$ could have been entered directly into the calculator. However, the step $\frac{43.56}{72.8} \cdot 10^{-19}$ was added since it allows you to estimate, in your head, the order of magnitude of the result. Thus, if you enter any number incorrectly into your calculator, this step allows you to verify the result you get from the calculator. Notice also that the number "8" in the denominator is an exact number, so it does not influence the number of significant figures.

1.4 Equations, Proportionality, and Graphs

The **proportionality** expresses the dependence of a variable (the "dependent" variable) on another variable referred to as the independent variable. The independent can assume any value, whereas the dependent variable can assume only values given by the proportionality. The circumference C of a circle, for example, is proportional to the diameter d of the circle according to:

$$C \propto d \qquad (1.24)$$

Any proportionality can be written as an equation by inserting a proportionality constant:

$$C = \pi \cdot d \qquad (1.25)$$

where the proportionality constant is π. A plot of c, the dependent variable, vs. d (the independent variable), gives a straight line with the slope of 3.14159, as shown in Figure 1.2.

In general, any proportionality can be written as an equation, and we have made use of this statement many times in the text of the book. An equation can be solved for either of the variables by straightforward algebraic manipulations, as long as the manipulation is applied to both sides of the equation (with the exception of division by zero, which is a no-no). Let us take the relationship between temperature, expressed in degrees Fahrenheit (T_F), and in centigrade scale (T_C) for an example. This relationship is

$$T_F = 1.8 \, T_C + 32 \qquad (1.15)$$

What does an equation like Eq. 1.15 tell us? First of all, it tells us that the dependence of T_F is linear with T_C. You may ascertain that this equation is correct by inserting temperature values with known outcome: the boiling temperature of water in the centigrade scale, $T_C = 100$ [°C], when inserted into Eq. 1.15, correctly results in the boiling temperature of water as $T_F = 212$ [°F]. Now, if a temperature is given in [°F] and you want to express it in [°C], use a little algebra and get

$$1.8 \, T_C = T_F - 32 \quad \text{or} \quad T_C = (T_F - 32)/1.8 \qquad (1.26)$$

For example, a cold morning with a temperature of -4 [°F] would have registered as -20 [°C] on a thermometer calibrated in centigrade units. I would recommend not to memorize both Eqs. 1.15 and 1.26 (because the storage space in your brain is limited) but rather, memorize one, say Eq. 1.15, and derive the other form as needed.

This last statement is even more important when you have an equation that contains more than two variables. Take the ideal gas law (see Chapter 8),

$$p \, V = n \, R \, T \qquad (8.14)$$

Here, p, V, n, and T are all variables, and only R is a constant (hence the name "gas constant"). If you wish to visualize how the volume of a gas depends on its pressure (keeping the temperature and amount of gas constant), you would write

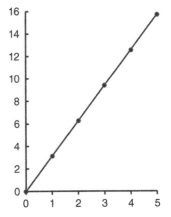

Figure 1.2 Dependence of a circle's circumference on diameter.

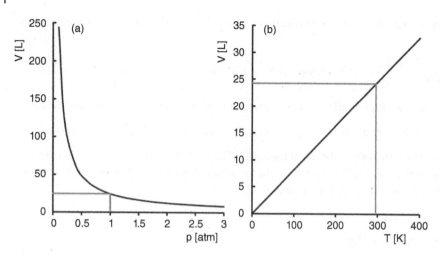

Figure 1.3 (a) Plot of gas volume vs. pressure at constant temperature, 298 [K]. The gray lines denote the volume of 1 mole of gas, V = 24.2 [L] at 298 [K]. (b) Plot of gas volume vs. temperature at constant pressure, 1 [atm]. The gray lines denote the volume of 1 mole of gas, V = 24.2 [L] at 1 [atm] pressure.

$$V = nRT \frac{1}{p} \tag{1.27}$$

$$\text{which implies} \quad V \propto 1/p \tag{1.28}$$

If you use 1.00 [mol] of an ideal gas at 298 [K] (do not forget, always use Kelvin temperatures), you get a plot shown in Figure 1.3a. The shape of this curve is a hyperbola, as expected from the inverse proportionality. If you wish to visualize how the volume of a gas depends on temperature (keeping the pressure and amount of gas constant), you would write

$$V = \frac{nR}{p} T \tag{1.29}$$

$$\text{or} \quad V \propto T \tag{1.30}$$

which is shown in Figure 1.3b.

The take home point from this discussion is that in a scientific equation such as the ideal gas law, the same rules apply that you have learned in algebra classes, and that simply solving the equation for the unknown (by algebraic rules) allows you to predict the functionality, and the shape of plots you obtain. When any of the variables appear in the second or third power, we have to revert to different methods to solve for the allowed numerical values. This will be discussed in Section 1.5.

Eq. 1.26 was written as a function that relates two variables – depending which way they are written – either temperature can be the dependent or independent variable. Let us now look at another situation (a common question in general chemistry exams), namely "is there a temperature where the numeric values of the Fahrenheit and centigrade scale are equal?" Since the relationship of the two scales is given by

$$T_F = 1.8 \, T_C + 32 \tag{1.15}$$

we need to find a temperature for which

$$T_F = T_C \tag{1.31}$$

that is, we have two equations that need to be solved together. In algebra class of earlier years, you would have encountered the same problem written as

$$y = ax + b \tag{1.32}$$

$$\text{and} \quad y = x. \tag{1.33}$$

Then, you would have substituted Eq. 1.33 into Eq. 1.32 to obtain $x = ax + b$, and solved for x

$$\text{to get} \quad x = b/(1 - a) \tag{1.34}$$

Applying the same approach to the pair of Eqs. 1.15 and 1.31 gives the temperature for which Eq. 1.31 holds as

$$T = -40 \, [°F] \text{ or } [°C] \tag{1.35}$$

where the numerical values of T_F and T_C are the same. You also could graph the two functions and search for their intersect. This is shown in Figure 1.4. Here, the black line shows the Fahrenheit temperature as a function of the centigrade temperature. As you can see, the slope is larger than 1 (in fact, 1.8), and it intersects the y-axis at 32, as expected. The gray line is a plot of the centigrade temperature vs. itself; therefore, we get a line with a slope of one and a zero intersect. Finally, we see that the lines intersect at −40 [°C], as calculated by Eq. 1.35.] By the arguments shown above, think about the following two questions:

(1) At what temperature does the Fahrenheit temperature intersect the abscissa? (In other words, at what centigrade temperature is the Fahrenheit temperature 0?)
(2) Is there a temperature where the centigrade and Kelvin scale have the same numerical value?

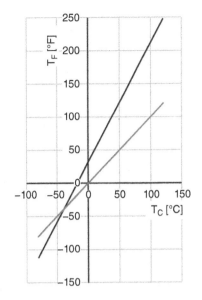

Figure 1.4 Plot of centigrade and Fahrenheit temperature scales.

1.5 Quadratic, Cubic, and Quartic Equations

Quadratic equations are equations in which the variable occurs to the second power, or the first and second power. Here, we consider two cases.
Case 1: the variable occurs to the second power only:

$$x^2 - a = 0, \quad \text{or} \quad x^2 = a; \tag{1.36}$$

The equation has two solutions x_+ and x_- given by

$$x_{+,-} = \pm \sqrt{a} \tag{1.37}$$

Example: The area A of a circle is given by $A = \pi r^2$. Let the area of a circle be 200 [cm]². What is the radius of the circle? $A = \pi r^2$; $r^2 = A/\pi$; $r = \pm \sqrt{200/\pi} = \pm 7.98$ [cm]. Here, only the positive answer makes sense.
Case 2: the variable occurs to the second and first power:

$$ax^2 + bx + c = 0 \tag{1.38}$$

This equation has two solutions x_+ and x_- given by

$$x_+ = \frac{-b + \sqrt{b^2 - 4ac}}{2a} \quad \text{and} \quad x_+ = \frac{-b - \sqrt{b^2 - 4ac}}{2a} \tag{1.39}$$

This is generally written as $x_{+,-} = \dfrac{-b \pm \sqrt{b^2 - 4ac}}{2a}$

Example: (from Chapter 10, Eq. 10.25)

$$x^2 + 7.2 \cdot 10^{-4} \times -7.2 \cdot 10^{-6} = 0$$

Here, $a = 1$, $b = 7.2 \cdot 10^{-4}$, $c = -7.2 \cdot 10^{-6}$

$$x_{+,-} = \frac{-b \pm \sqrt{b^2 - 4ac}}{2a} = \frac{-7.2 \cdot 10^{-4} \pm \sqrt{(7.2 \cdot 10^{-4})^2 - 4 \cdot 1 \cdot (-7.2 \cdot 10^{-6})}}{2} \tag{1.40}$$

Notice that in real applications such as general equilibrium calculations, the radical (the part under the square root symbol) must be positive. If it is not, you – most likely – have made a mistake with the sign (remember, it is −4ac)

$$x_{+,-} = \frac{-7.2 \cdot 10^{-4} \pm 5.415 \cdot 10^{-3}}{2} \tag{1.41}$$

$$x_+ = 2.34703 \cdot 10^{-3} \tag{1.42}$$

$$x_- < 0 \tag{1.43}$$

Which of the two solutions of a quadratic equation is physically meaningful often cannot be determined from the mathematical view but has to be decided based on the conditions. The result for x_- (Eq. 1.43), for example, is negative, but a negative concentration is physically not meaningful.

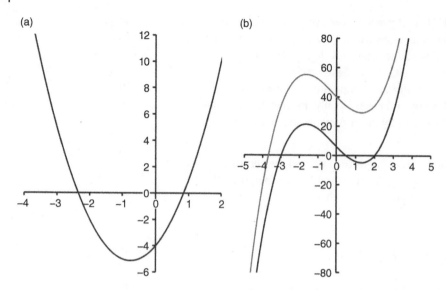

Figure 1.5 Graphic solution of (a) a quadratic equation and (b) a cubic equation.

The expression $f(x) = ax^2 + bx + c$ that we encountered in Eq. 1.38 actually is what we call a polynomial (see Eq. 5.21) in the single variable x. Such polynomials also can be solved for x graphically by plotting the function $f(x)$ and searching for the intersections of the functions with the abscissa, at which $f(x)$ is zero. Consider the function $f(x) = 2x^2 + 3x - 4$. The plot of this function is shown in Figure 1.5a with the two abscissa intersects at -2.35 and $+0.85$, corresponding to the numerical values you would obtain if you substituted $a = 2$, $b = 3$, and $c = -4$ into Eq. 1.39.

The graphical method outlined above can be used to solve cubic (or higher) equations as well. For cubic equations (polynomials) of the form $f(x) = ax^3 + bx^2 + cx + d$, there is no simple solution corresponding to Eq. 1.39, and the graphical approach can show how many real solutions exist. This is shown in Figure 1.5b (black trace), which depicts the function $f(x) = 2x^3 + x^2 - 13x + 6$. This function has roots (solutions) at $x = 2$, $x = 0.5$, and $x = -3$. Once one root is known, for example $x = 2$, we can divide the original function $f(x)$ by $(x - 2)$, if we exclude $x = 2$ in the division (since that would be a disallowed division by zero). This step reduces the cubic equation to a quadratic equation, since

$$(2x^3 + x^2 - 13x + 6)/(x - 2) = 2x^2 + 5x - 3 \quad (\text{for } x \neq 2) \tag{1.44}$$

The quadratic equation $2x^2 + 5x - 3 = 0$ has roots of 0.5 and -3, as shown in Figure 1.5b. Notice that the same function offset along the y-axis, that is $f(x) = 2x^3 + x^2 - 13x + 40$ (Figure 1.5b, gray trace) has only one solution.

Quartic equations have the variable at up to the fourth power. In the context of this book, we encounter quartic equations in equilibrium calculations. Fortunately, they contained only even powers of x, such as

$$ax^4 + bx^2 + c = 0 \tag{1.45}$$

which can be solved by successive application of the quadratic equation approach.

1.6 Exponential Functions and Logarithms

In Section 1.3, we have dealt with numbers that have exponential expressions and have learned how to manipulate them. Next, we will take a look at functions that have the variable x in the exponent, such as the functions

$$y = 10^x \quad \text{or} \quad y = e^x \tag{1.46}$$

You may or may not have encountered functions of this kind in high school algebra; thus, it is necessary to spend some time here and explain where these functions are coming from. The most tangible way to explain these "growth functions" is as follows. Imagine you put US$1.00 into a bank account that rewards you with 5% interest, compounded yearly. So, after one year, you find US$1.05 in your account. After two years, the interest was calculated for the new starting amount in your account, so the interest earned in the second year is $1.05 \cdot 0.05 = 0.0525$, which is added to your balance after the second year. Your balance after the second year, therefore, is US$1.1025. If you keep the funds in the account, they will grow according to

$$(1 + 1 \cdot 0.05)^x \tag{1.47}$$

where x is the number of years. There are several interesting aspects to Eq. 1.47. First, the growth function given by Eq. 1.47 is shown by the gray trace in Figure 1.6 and indicates that after 60 years, your US$1 investment would have grown to 18 bucks. Not bad for doing nothing (imagine what would have happened if you had invested US$10 000?). Second, the factor 0.05, of course, is the interest rate of 5%. Even at a moderately larger interest rate of 7.5%, your capital would have grown much faster, to US$64.5 in 60 years. Finally, if the interest was not compounded yearly, but monthly, daily, or even hourly, your capital would have grown slightly faster than at the yearly rate. This is shown by the black trace in Figure 1.6, which is for infinitesimally short time increments, or a growth function where the compounding period approaches zero. This is known as the natural growth function, $y = e^{ax}$. Here, e (represented by a script letter "e" to distinguish it from the electronic charge, e) is Euler's number, which, like π, is an irrational number with a numerical value

$$e \approx 2.71828 \ldots \tag{1.48}$$

and is known as the basis of natural growth, as we have seen in the example of our investment scheme above. The value of e itself is defined as a growth function:

$$e = \lim_{n \to \infty} \left(1 + \frac{1}{n}\right)^n \tag{1.49}$$

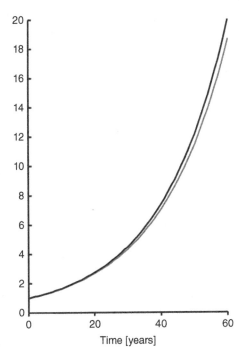

Figure 1.6 Growth functions. Black trace: $e^{0.05x}$; gray trace $(1 + 1 \cdot 0.05)^x$.

Since it is easier for us to use powers of 10, rather than powers of e (after all, we all know what 10^3 is, but who knows what e^3 is?), we often convert powers of e to powers of 10 according to

$$10^x = e^{2.303x} \tag{1.50}$$

$$\text{since} \quad e^{2.303} \approx 10 \tag{1.51}$$

Although the conversion from powers of e to powers of 10 is useful, I suggest that you get familiar with the e^{ax} and e^{-ax} functions on your calculator, since they (and their inverse function, the logarithm functions) occur frequently in Chapters 9 and 14.

While we call the function $y = e^{ax}$ the natural growth function, the function $y = e^{-ax}$ is referred to as the natural decay function. A plot of both exponential functions is presented in Figure 1.7. The natural decay function predicts, for example, the decay in concentration of medication in the bloodstream with time, or the amount of an element left over after it undergoes radioactive decay. Examples of such calculations are found after the discussion of logarithms.

The inverse mathematical operations of exponential functions are logarithmic functions. Logarithms occur in the text of this book in several places (Chapter 7 in the discussion of vapor pressure, Chapter 10 in the discussion of pH and pOH, and in Chapter 14 in the discussion of reaction rates). A logarithm is the exponent to which a base must be raised to yield a given number. For example, the equation

$$10^x = 2 \tag{1.52}$$

has a solution

$$x = \log 2 \approx 0.3010\ldots \tag{1.53}$$

In other words, taking the logarithm is the inverse operation to exponentiation. Log 2 could have been written as $\log_{10} 2$ to indicate the base 10. However, in this book, the symbol "log" always will indicate base 10.

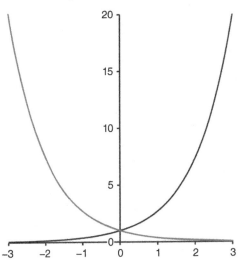

Figure 1.7 Plot of $y = e^x$ (black trace) and $y = e^{-x}$ (gray trace).

Remembering Eq. 1.20:
$10^n \cdot 10^m = 10^{(n+m)}$ and $\log 2 = 0.3010$, we see that the logarithm of

$$\log 20 = \log(10 \cdot 2) = \log 10 + \log 2 = 1.3010 \text{ and}$$
$$\log 200 = \log 100 + \log 2 = 2.3010. \quad (1.54)$$

It also follows that

$$\log(0.2) = \log(10^{-1}) + \log 2 = -0.6990 \quad (1.55)$$

When taking the logarithm of an exponential expression, such as 2^x, the exponent becomes a factor:

$$\log(2^x) = x \log(2) \quad (1.56)$$

In many calculations in the text, the natural logarithm, rather than the decadic logarithm, is used. The natural logarithm (called the "ln" function in this book) is the exponent to which the number e must be raised to yield a given number. For example, the equation

$$e^x = 2 \quad (1.57)$$

with $x = \ln 2 \approx 0.693$

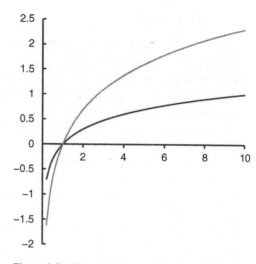

Figure 1.8 Plot of the log x (black) and ln x (gray) functions.

$$(1.58)$$

The prevalence of natural logarithm and natural growth functions in Chemistry arises because the integral

$$\int_{x_a}^{x_b} \frac{1}{x} dx = \ln \frac{x_b}{x_a} = \ln(x_b) - \ln(x_a) \quad (1.59)$$

is given by the ratio of the natural logarithm of the two values between which the integration is carried out, x_b and x_b. This integral occurs frequently in the derivation of equations such as the first-order rate law in kinetics, and other places, and will be discussed in more detail in Section 1.8. A graph of the log x and ln x functions is shown in Figure 1.8.

We now turn to a number of examples from the main text to practice exponential and logarithmic functions.

Example 10.3 Calculate the pH of a 1.5 [M] aqueous solution of HCl.

Answer: $\text{pH} = -\log[H^+] = -\log(1.5) = -(0.1761) = -0.18 \quad (1.60)$

This example is discussed here, although it seems very easy and straightforward, to indicate an issue with significant figures in exponential calculations. The result in Eq. 1.60 was rounded to two significant figures, which is appropriate for the problem as stated. If we want to check out our calculation, we would use the reverse of the logarithm function, the "anti-log" function. The anti-log $(0.18) = 10^{0.18} = 1.513$. Here, we see a feature commonly encountered in exponential and logarithmic manipulations: the result in Eq. 1.60 was recorded to the correct number of significant figures. The reverse calculation, however, gave a different result. Therefore, it is advisable to perform intermediate calculations with more significant figures in logarithmic calculations. In the example above, the difference between the original concentration (1.5[M]) and the recalculated concentration of 1.513 [M] is due to truncation of the pH to two significant figures. If the pH is reported as pH = -0.1761, then $10^{0.1761}$ gives a more satisfactory result of 1.5000.

Example 9.2 Use the enthalpy of vaporization of butane, 22.4 [kJ/mol], and its vapor pressure at 298 [K], 2.41 [atm], to calculate the vapor pressure of butane at 473 [K] (this problem is solved in Example 9.2 using the logarithmic approach. Here, it is solved using the corresponding exponential equations):

$$\frac{p_{473}}{p_{298}} = e^{-\frac{\Delta H_{vap}}{R}\left(\frac{1}{298} - \frac{1}{478}\right)}, \text{ or } p_{473} = p_{298}\, e^{-\frac{22{,}400}{R}\left(\frac{1}{298} - \frac{1}{473}\right)} \text{ with } R = 8.3[\text{J/(K mol)}]$$

$$p_{473} = 2.41\, e^{2698.8\,(0.0033557 - 0.0021142)}$$

$$p_{473} = 2.41\, e^{2698.8\,(0.0012415)} = 2.41\, e^{3.35066} = 2.41 \cdot 28.52 = 68.7 \text{ [atm]}.$$

Example 13.2 Consider an elementary step A → C with a rate constant k = 0.3 [s^{-1}]. Calculate the ratio [A]$_0$/[A]$_t$ after 10 s.

Answer: $\log \frac{[A]_0}{[A]_t} = k\, t/2.303 = 0.3 \cdot 10/2.303\,[s^{-1} s] = 1.3026$;

$$\frac{[A]_0}{[A]_{10}} = \text{antilog}\,(1.3026) = 20.1;\quad [A]_{10} \approx \frac{[A]_0}{20}$$

1.7 Radial and Spherical Polar Coordinates

When plotting functions as discussed in Sections 1.4–1.6, we define a point in a plane by its x and y coordinates, as shown in Figure 1.9a. The same point in a plane can also be described in polar coordinates by its distance from the origin, given by the vector **r**, and the angle φ between the vector **r** and the positive X-axis, with $0 < \varphi < 2\pi$, or $0 < \varphi < 360°$ (Figure 1.9b).

The decision on which of the two coordinate systems to use often is dictated by the problem itself: the particle-in-a-square-box is described previously in Cartesian coordinates (see Figure 15.3 in Chapter 15), whereas the particle-in-a-circular-box is presented in polar coordinates (see Figure 15.4 in Chapter 15) due to the different shapes and symmetries of the confinements.

Using our knowledge of trigonometry, we find that the two coordinate systems can be related to each other by

$$x = r \cos \varphi \qquad (1.61)$$
$$y = r \sin \varphi \qquad (1.62)$$

Since $x^2 = r^2 \cos^2 \varphi$ and $y^2 = r^2 \sin^2 \varphi$, it follows that

$$x^2 + y^2 = r^2 \cos^2 \varphi + r^2 \sin^2 \varphi = r^2\left(\cos^2 \varphi + \sin^2 \varphi\right) = r^2 \quad \text{since} \quad \cos^2 \varphi + \sin^2 \varphi = 1$$

The equation $x^2 + y^2 = r^2$ \hfill (1.63)

describes a circle with radius r, located at the center of the coordinate system. This leads to the polar equation r(φ) of a circle at the center of the coordinate system:

$$r(\varphi) = a \qquad (1.64)$$

where the constant *a* is just the radius of the circle. This is an important aspect, as we shall see below (Eq. 1.69).

In three-dimensional space, we define the spherical polar coordinates (rather than polar coordinates introduced above) for a point with Cartesian coordinates x, y, and z in terms of the variables r, θ, and φ as shown in Figure 1.10. The angle θ is measured between the line OP and the positive z-axis and runs from +90 to −90 degrees {−π/2 to π/2}, just as the latitude of a point on earth. The other angle, φ, is defined between the positive x-axis and the projection of OP onto the x–y plane, and runs between 0 and 2π. This angle corresponds to the longitude of a point on earth. The coordinates x, y, z of an arbitrary point in space are described in polar coordinates as

$$x = r \sin \theta \cos \varphi \qquad (1.65)$$
$$y = r \sin \theta \sin \varphi \qquad (1.66)$$
$$z = r \cos \theta \qquad (1.67)$$

In analogy to Eq. 1.63, the equation

$$x^2 + y^2 + z^2 = r^2 \qquad (1.68)$$

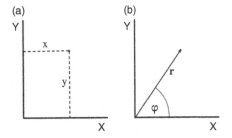

Figure 1.9 Cartesian (a) and polar (b) coordinates.

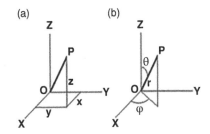

Figure 1.10 Cartesian (a) and spherical polar (b) coordinates.

describes a sphere with radius r, located at the center of the coordinate system. In polar coordinates φ and θ, the equation of a sphere is

$$r(\varphi, \theta) = a \tag{1.69}$$

in analogy to Eq. 1.64, where *a* is the radius of the sphere. This explains a statement in Section 15.2 that the spherical harmonic function $Y_0^0 = Y(\varphi, \theta) = \frac{1}{2}\sqrt{\frac{1}{\pi}}$ (Eq. 15.18) represents a sphere.

Section 15.2 also pointed out that the second derivative of a function, given in Cartesian coordinates, $\frac{\partial^2}{\partial x^2} + \frac{\partial^2}{\partial y^2} + \frac{\partial^2}{\partial z^2}$, becomes a scary-looking

$$\frac{1}{r^2}\frac{\partial}{\partial r}\left(r^2\frac{\partial}{\partial r}\right) + \frac{1}{r^2 \sin^2\theta}\frac{\partial^2}{\partial \phi^2} + \frac{1}{r^2 \sin\theta}\frac{\partial}{\partial \theta}\left(\sin\theta \frac{\partial}{\partial \theta}\right) \tag{15.13}$$

in spherical polar coordinates.

1.8 Differential and Integral Calculus

Differential calculus is a branch of mathematics dealing with the slope, or the rate of change, of a function. This is understood best by considering the graph in Figure 1.11. Here, we have a function, $f(x) = x^2$, plotted vs. x. The slope of the function f(x) at x = 1 (heavy black line in Figure 1.11) can be determined as follows. If we draw a secant from the point (x = 4, y = 16) to the point (x = 1, y = 1), we get a slope S_a of this line as $S_a = \Delta y/\Delta x = 15/3 = 5$. The secant from point x = 3, y = 9 to the point (x = 1, y = 1) has a slope of $S_b = 8/2 = 4$. Finally, the slope of secant "c" is $S_c = 3/1 = 3$. If we let Δx get infinitesimally small, the slope at point x=1, y=1 (which is a tangent, rather than a secant) becomes 2. If we repeat the same approach for other points along the $f(x) = x^2$ curve, we find that the slope of this function is 2x at any point. We then define the slope of the function f(x) for infinitesimally small values of dx, as the derivative, $\frac{df(x)}{dx} = \frac{dx^2}{dx} = 2x$. For any function of the form $f(x) = a x^b$, the derivative is given by

$$\frac{df(x)}{dx} = (ab)x^{b-1} \tag{1.70}$$

Thus, the derivative of the function $f(x) = x^2$ is $\frac{d}{dx}f(x) = \frac{d}{dx}(x^2) = 2x$, and the derivative of the function $f(x) = 4x^3$ is $\frac{d}{dx}f(x) = \frac{d}{dx}(4x^3) = 12x^2$.

The derivatives of trigonometric functions are very important for our discussion:

$$\frac{d \sin(ax)}{dx} = a \cos(ax) \quad \text{and} \quad \frac{d \cos(ax)}{dx} = -a \sin(ax) \tag{1.71}$$

Furthermore, $\frac{d e^{ax}}{dx} = a e^{ax}$ \hfill (1.72)

Taken derivatives of a function successively leads to the second derivative, $\frac{d^2 f(x)}{dx^2}$. In physics, you may have encountered that velocity is defined as the derivative of the position with respect to time,

$$v = \frac{dx}{dt} \tag{1.73}$$

and acceleration a as the second derivative of position with respect to time,

$$a = \frac{d^2 x}{dt^2} \tag{1.74}$$

We shall encounter this equation in Section 1.9 again. We also shall find that the second derivatives of trigonometric functions are quite important:

$$\frac{d^2 \sin(ax)}{dx^2} = \frac{d}{dx}\{a \cos(ax)\} = -a^2 \sin(ax) \tag{1.75}$$

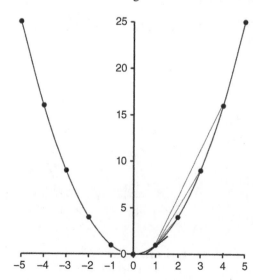

Figure 1.11 Slope of an algebraic function.

In other words, the second derivative of the sine function is just the negative sine function, multiplied by a constant. Notice that the derivative of a constant is zero, according to Eq. 1.70. Thus, the derivative of the function $f(x) = 2x^2 + 3x - 4$ (see Figure 1.5a) is

$$\frac{df(x)}{dx} = 4x + 3 \tag{1.76}$$

One of the important features of differential calculus is the fact that the derivative of a function easily lets us determine where maxima or minima of functions occur. Take the derivative shown in Eq. 1.76 as an example. At the minimum of the original function $f(x) = 2x^2 + 3x - 4$, its slope will be zero. Thus, we set

$$\frac{df(x)}{dx} = 4x + 3 = 0 \tag{1.77}$$

and obtain $x = -0.75$. Inspection of Figure 1.5a reveals that the minimum of the function $f(x) = 2x^2 + 3x - 4$, indeed, occurs at this abscissa value.

When taking the derivatives of products or quotients of functions, special rules (the product and quotient rules) of differentiation apply. Similarly, when the function to be differentiated involves nested functions, the chain rule of differentiation applies. Since these cases do not arise in the context of this book, these rules of differentiation are not discussed any further.

Integration is the inverse operation to differentiation, just like division is the inverse of multiplication, or taking the logarithm is the inverse to exponentiation. Since differentiation of a function $f(x)$ was written as $\frac{df(x)}{dx}$, we define the inverse function as $\int f(x)dx$. This is called the indefinite integral, and, since it is the inverse of differentiation, we can write

for $f(x) = 1$, the integral is $\int 1\, dx = x$ and

for $f(x) = x$, the integral is $\int x\, dx = x^2/2$ (take these two integrals with a grain of salt, for the moment)

You can convince yourself that the derivative of the function $x^2/2$ is, indeed, x and demonstrate hereby that differentiation and integration are inverse functions.

Now to the grain of salt. You can easily convince yourself that the functions $\frac{1}{2}x^2$, $\frac{1}{2}x^2 + 2$ and $\frac{1}{2}x^2 + a$ all have the same derivative, x, since the constants in the functions disappear upon differentiation. Consequently, integration of the function $f(x) = x$ (i.e. performing the reverse of differentiation) cannot determine which of the three functions $\frac{1}{2}x^2$, $\frac{1}{2}x^2 + 2$, and $\frac{1}{2}x^2 + a$ was the original function. Therefore, we write the integral as

$$\int x\, dx = \frac{1}{2}x^2 + C \tag{1.78}$$

where C is a constant that can be 0, 2, or "a" in the examples given above. Therefore, we call the integral in Eq. 1.78 as the indefinite integral. The general rule for simple integration is

$$\int x^n\, dx = \frac{x^{n+1}}{n+1} + C \tag{1.79}$$

$$\int \frac{1}{x}\, dx = \ln x + C \tag{1.80}$$

$$\int e^x\, dx = e^x + C \tag{1.81}$$

$$\int \sin(x)\, dx = -\cos(x) + C \tag{1.82}$$

$$\int \cos(x)\, dx = \sin(x) + C \tag{1.83}$$

The ambiguity in the indefinite integral may be avoided by the use of the definite integral which may be interpreted as the area under the integrand $f(x)$, delimited within the interval $[a, b]$:

$$\int_a^b f(x)dx = [F(x)]_a^b = F(b) - F(a) \tag{1.84}$$

Here, $F(x)$ is the anti-derivative, or the inverse derivative of $f(x)$, as defined in Eqs. 1.80 through Eq. 1.83, evaluated at the values "a" and "b". The area under the curve $f(x) = x^2$ shown in Figure 1.11 in the interval from 1 to 3 would be

$$\int_1^3 x^2\, dx = \left[\frac{1}{3}x^3\right]_1^3 = \frac{1}{3}27 - \frac{1}{3} = 8\frac{2}{3} \tag{1.85}$$

where the super- and subscripts at the square bracket denote that $F(x)$ has to be evaluated for the values "3" and "1."

1.9 Differential Equations

We encounter a simple differential equation in Chapter 13, when we found that the rate of a first-order reaction, expressed as the change in the concentration of A, $-\frac{d[A]}{dt}$, is proportional to the momentary concentration of A:

$$-\frac{d[A]}{dt} = k\,[A] \tag{13.20}$$

which can be rewritten as

$$\frac{d[A]}{[A]} = -k\,dt$$

Taking the definite integral on both sides between time zero and a later time "t", with the condition that the concentration of A at t=0 was $[A]_0$, gives

$$\int_{t=0}^{t=t'} \frac{d[A]}{[A]} = -k \int_{t=0}^{t=t'} dt \tag{1.86}$$

$$\ln \frac{[A]_{t'}}{[A]_0} = -k(t') \quad \text{or} \tag{1.87}$$

$$\ln \frac{[A]_0}{[A]_{t'}} = k(t') \tag{13.22}$$

This is one of the easiest differential equations that could be solved by just taking the definite integral on both sides of the equation. A slightly more complicated differential equation is encountered in Chapter 15, during the discussion of the vibration of a diatomic molecule. There, we discuss that the restoring force for the elongation of the bond (spring) between two masses is

$$F = k\,x \tag{15.33}$$

With Newton's second law of motion, which states that force equals mass times acceleration, we may write this force as

$$F = m_R \frac{d^2 x}{dt^2} \tag{1.88}$$

where m_R is the reduced mass of the system defined in Eq. 15.35. Thus, we may write the equation of motion for a diatomic molecule as

$$\frac{d^2 x}{dt^2} + \frac{k}{m_R} x = 0 \tag{1.89}$$

Here, we need a solution of the form x(t) such that the second derivative of this function, $\frac{d^2 x(t)}{dt^2}$ is equal to this function, multiplied by a constant. We have encountered functions that fulfill this condition before:

$$\frac{d^2 \sin(ax)}{dx^2} = -a^2 \sin(ax) \tag{1.75}$$

Similarly, the cosine or exponential functions could be used as a trial function. We use

$$x(t) = A \sin(\omega t) \tag{1.90}$$

as a trial function where ω is the angular frequency, $\omega = 2\pi \nu$ and A is an amplitude factor. Then

$$\frac{d^2 x(t)}{dt^2} = -A\,\omega^2 \sin(\omega t) \tag{1.91}$$

Substituting this result back into Eq. 1.89 yields

$$-A\,\omega^2 \sin(\omega t) + \frac{k}{m_R} A \sin(\omega t) = 0 \tag{1.92}$$

from which we obtain (by cancelling $A\,\sin(\omega t)$ on both sides of the equation)

$$\omega^2 = \frac{k}{m_R} \quad \text{or} \quad \nu = \frac{1}{2\pi} \sqrt{\frac{k}{m_R}} \tag{15.34}$$

(neglecting the negative root in the quadratic equation above).

1.10 Complex Numbers

Complex numbers are an extension of real numbers that consist of a real part and an imaginary part that contains the imaginary unit i, defined as $i = \sqrt{-1}$. Complex numbers are necessary to solve equations such as $x^2 = -9$, which has no solution in real number space but has solutions in complex number space, $x_\pm = \pm(3i)$. Complex numbers generally are represented as

$$z = a + b\,i \tag{1.93}$$

where "a" is the real part and "b i" the imaginary part. They are represented graphically in an x–y plane where x-axis represents the real part and y-axis the imaginary part.

In the context of the discussions in this book, complex numbers are encountered only once, in the discussion of the wave functions of p-orbitals (Eqs. 15.24 and 15.25) that contain an expression $e^{\pm i\varphi}$. At this point, it is important to realize that the functions $e^{\pm i\varphi}$ are not exponential growth or decay function that were described by $e^{\pm x}$ (see Figure 1.7), but represent periodic functions in complex space. This is indicated by Euler's formula,

$$e^{i\varphi} = \cos\varphi + i\sin\varphi \tag{15.26}$$

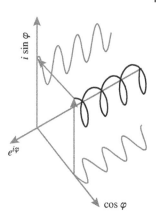

Figure 1.12 Visualization of the function $e^{i\varphi}$ as a periodical function in complex space. *Source:* Adapted from https://en.wikipedia.org/wiki/Euler%27s_formula.

which depicts a helical path for $e^{i\varphi}$ when plotted in a complex representation: $\cos\varphi$ being associated with the real, and $i\sin\varphi$ with the imaginary axis. A graphical depiction of this function is shown in Figure 1.12, which is adapted from https://en.wikipedia.org/wiki/Euler%27s_formula under "Visualization of Euler's formula as a helix in three-dimensional space." Although the concept of complex numbers may seem somewhat absurd at first, they appear frequently in physics, engineering, signal processing and mathematics.

2

Atoms, Elements, and the Periodic System

2.1 Subatomic Particles and Atoms

From a chemist's viewpoint, matter consists of atoms, and atoms themselves consist of protons, neutrons, and electrons. Atoms can form molecules, or exist on their own. This viewpoint is substantially different from that a physicist may have since several branches of modern physics deal with the composition of protons and neutrons in terms of other, more basic elementary particles such as quarks and even smaller ("more elementary") particles. The existence of different viewpoints does not imply that one is wrong and the other correct. It is a matter, as we shall see throughout this book, of the scientific method which requires us to build models that explain experimental observations. A simplistic model will have limitations, and a more sophisticated model is constructed to explain observations that cannot be explained with the simplistic model. Such is the case of the different viewpoints of atomic structures: at the energies of chemical and biochemical reactions, and in everyday life, the view of atoms consisting of protons, neutrons, and electrons is perfectly adequate. In nuclear chemistry (see Chapter 14), one reaches the limits of the simplistic chemist's view, and at enormously increased energies, for example in the center of a star, or high energy physics (such as the large hadron collider) the refined model of particle physics needs to be invoked.

So, for the remainder of this book, we shall view atoms as being composed of protons and neutrons in the nucleus of an atom, and electrons. The pertinent properties of the three relevant elementary particles are presented in Table 2.1.

Table 2.1 Absolute masses and charges of elementary particles.

	Proton (p)	Neutron (n)	Electron (e)
Mass [kg]	$1.6726 \cdot 10^{-27}$	$1.6749 \cdot 10^{-27}$	$9.1094 \cdot 10^{-31}$
Charge [C]	$1.602 \cdot 10^{-19}$	0	$-1.602 \cdot 10^{-19}$

Here, masses are expressed in [kg] and charges in Coulomb, abbreviated [C]. Notice that the mass of the neutron is slightly larger than that of the proton. The mass of the electron is about 1/1836 of the proton mass and can be ignored for chemical calculations involving masses, such as in stoichiometry (see Section 4.1). The charges of the proton and electron are numerically equal, but opposite in sign. Thus, we express their charge in terms of the "elementary charge," $e = 1.602 \cdot 10^{-19}$ [C], and assign a value of +1 to the proton and −1 to the electron. For simplicity's sake, we express the masses of both the proton and the neutron in terms of the atomic mass unit (amu), with 1 [amu] = $1.66 \cdot 10^{-27}$ [kg] as shown in Table 2.1. (The reason the atomic mass unit is not exactly $1.6726 \cdot 10^{-27}$ kg will be elaborated upon in Section 4 of this chapter). We then obtain a simplified Table 2.2 as follows:

Table 2.2 Relative masses and charges of elementary particles.

	Proton (p)	Neutron (n)	Electron (e)
Mass [amu]	~1	~1	1/1830
Charge [e]	+1	0	−1

Notice that in this discussion, and in the remainder of the early chapters in this book, the electron is described as a particle with mass and charge. In the discussion of atomic structure and quantum chemistry in Chapters 5 and 15, respectively, an alternate description of the electron, namely as a (standing) wave, will be introduced. The electron exhibits what

Understanding Essential Chemistry, First Edition. Max Diem.
© 2025 John Wiley & Sons, Inc. Published 2025 by John Wiley & Sons, Inc.

is called the "particle-wave duality" which implies that it can exhibit wave-like and particle-like properties, depending on the context. See more on this in Chapter 5.

In an atom, the heavy particles, i.e. the protons and neutrons, make up the nucleus of an atom. They are therefore referred to collectively as "nucleons." In an oxygen atom, for example, there are eight protons and eight neutrons in the nucleus. The electrons form an "electron cloud" around the nucleus that will be the subject of many pages later in this book. The size of this electron cloud determines the atomic radius of an atom. To present a comparison of nuclear vs. atomic size, we pick the oxygen atom as an example. The experimentally determined radius of an oxygen atom is 152 [pm] = $1.52 \cdot 10^{-10}$ [m] (for a review of the metric system, and the manipulation of exponential expression, see Chapter 1). The nucleus of an oxygen atom is very much smaller; its radius is about $2.8 \cdot 10^{-15}$ [m]. Thus, we see that the nucleus of an oxygen atom is about 50000 smaller than the oxygen atom, but it contains nearly all the mass of the oxygen atom. The density of an atomic nucleus is enormous, as the following example calculation will demonstrate.

Example 2.1 (a) Estimate the density of an oxygen nucleus, in [kg/m^3] from the data of Table 2.1 and the discussion in the previous paragraph.
(b) Compare this density with the density of the oxygen atom.

Answer: (a) d $\approx 3 \cdot 10^{17}$ [kg/m^3]; (b) d $\approx 2 \cdot 10^3$ [kg/m^3]

The sample calculations in Example 2.1 demonstrate that an atom has densities comparable to the densities of materials in everyday life. Liquid water, for example, has a density of about 10^3 [kg/m^3] or about 1 [g/cm^3], and solid iron has a density of 7.87 [g/cm^3]. The nucleus of oxygen has a density 10^{14} times larger, or a hundred trillion times larger. This demonstrates that an atom is mostly "empty space" with a very small, enormously dense nucleus, and a large volume occupied by very light electrons.

Example 2.2 Verify the density unit conversion of the previous paragraph, 10^3 [kg/m^3] = 1[g/cm^3].

Answer: See answer booklet

The eight positively charged protons in the oxygen atom, confined in the size of the nucleus, exert very strong repulsive forces toward each other. These repulsive forces are counteracted by the neutrons in what is called the "strong force" in the standard model of nuclear physics. The presence of neutrons is required for any nucleus that contains more than one proton: the simplest stable atomic nucleus with more than one proton is helium (He) that has two protons and two neutrons "gluing" the protons together. Going back to the oxygen atom, we now need to consider the number of electrons to form a neutral atom (and atoms are neutral by definition). We need to balance the positive charge of the eight protons in the nucleus by an equal number of electrons. Thus, for the oxygen atom to be electrically neutral, it must have eight electrons. We conclude this discussion of the oxygen atom with the statement that it has eight protons, eight neutrons, and eight electrons.

2.2 Elements, Isotopes, and Ions

We have seen above that atoms are composed of protons and neutrons, collectively referred to as nucleons, which make up an atomic nucleus, and electrons that occupy space pretty "far" away from the nucleus. The term "far" needs to be seen in the context of atomic dimensions: a nucleus may have a radius of about 10^{-15} [m], whereas the atom may have a radius of about 10^{-10} [m]. Translating this to dimensions of our everyday life reveals that if the atomic nuclear size were that of a softball (diameter of 10 cm), the size of the corresponding atom would be 10^5 times larger, or about 10 km in diameter. So, atoms are mostly empty space.

Example 2.3 Verify that an atom would have a diameter of about 10 [km] if the nucleus had a diameter of 10 [cm].

Answer: See answer booklet

Different atoms have different numbers of protons and neutrons in their nuclei. We define a quantity known as the atomic number "Z," which is always an integer, as the number of protons in the nucleus of an atom. In any neutral atom, the

Transition metals

1A	2A											3A	4A	5A	6A	7A	8A
1 H 1.008																	2 He 4.003
3 Li 6.94	4 Be 9.01											5 B 10.81	6 C 12.01	7 N 14.01	8 O 16.0	9 F 19.0	10 Ne 20.18
11 Na 22.99	12 Mg 24.31											13 Al 26.98	14 Si 28.09	15 P 30.97	16 S 32.06	17 Cl 35.45	18 Ar 39.95
19 K 39.1	20 Ca 40.08	21 Sc 44.96	22 Ti 47.90	23 V 50.94	24 Cr 52.0	25 Mn 54.94	26 Fe 55.85	27 Co 58.93	28 Ni 58.7	29 Cu 63.55	30 Zn 65.38	31 Ga 69.72	32 Ge 72.59	33 As 74.92	34 Se 78.96	35 Br 79.9	36 Kr 83.8
37 Rb 85.47	38 Sr 87.62	39 Y 88.91	40 Zr 91.22	41 Nb 92.91	42 Mo 95.94	43 Tc (98)	44 Ru 101.1	45 Rh 102.9	46 Pd 106.4	47 Ag 107.9	48 Cd 112.4	49 In 114.8	50 Sn 118.7	51 Sb 121.8	52 Te 127.6	53 I 126.9	54 Xe 131.3
55 Cs 132.9	56 Ba 137.3	57 La* 138.9	72 Hf 178.5	73 Ta 180.9	74 W 183.9	75 Re 186.2	76 Os 190.2	77 Ir 192.2	78 Pt 195.1	79 Au 197.0	80 Hg 200.6	81 Tl 204.4	82 Pb 207.2	83 Bi 209.0	84 Po (209)	85 At (210)	86 Rn (222)
87 Fr (223)	88 Ra 226	89 Ac† (227)	104 Rf	105 Ha													

*	58 Ce 140.1	59 Pr 140.9	60 Nd 144.2	61 Pm (145)	62 Sm 150.4	63 Eu 152.0	64 Gd 157.3	65 Tb 158.9	66 Dy 162.5	67 Ho 164.9	68 Er 167.3	69 Tm 168.9	70 Yb 173.0	71 Lu 175.0
†	90 Th 232.0	91 Pa (231)	92 U 238.0	93 Np (237)	94 Pu (244)	95 Am (243)	96 Cm (247)	97 Bk (247)	98 Cf (251)	99 Es (252)	100 Fm (257)	101 Md (258)	102 No (259)	103 Lr (260)

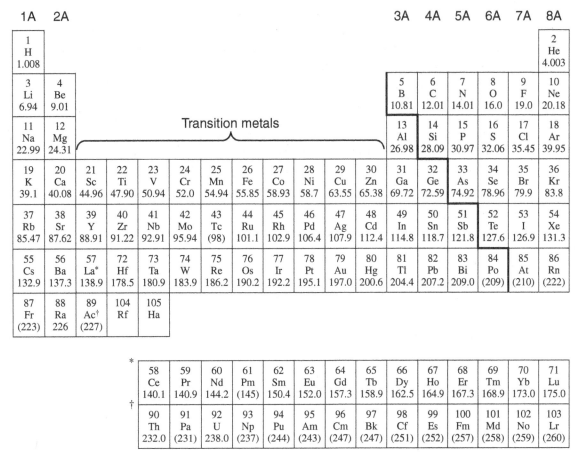

Figure 2.1 Periodic chart of the elements. Notice that elements 58–71 and 90–103 should be inserted at the symbols * and †.

number of electrons orbiting the nucleus must therefore also be "Z." We define an element to be an ensemble of atoms with the same number "Z," and this number defines an element. There are just under 120 elements we know of; the exact numbers may vary because more elements are being discovered. Some of them are very unstable, and might exist for a few microseconds only, whereas others are stable for millennia. The number of stable elements is thought to be about 80, but even this number is not exactly defined.

9 F 19.0

Figure 2.2 Section of periodic chart: the element fluorine.

The elements – stable or not – are summarized in the "Periodic Chart of Elements" (Figure 2.1), which will be the subject of many paragraphs of discussion in this and later chapters. In this table, elements are listed in order of increasing number of protons in the nucleus, or in order of increasing Z. This number is listed above the element name abbreviated by a one or two-letter symbol. Figure 2.2 shows a small portion of Figure 2.1 of an element with the atomic number 9 and the name "fluorine".

The other numerical value that appears in Figure 2.2, and, indeed in the periodic chart under the symbol for the element, is the so-called "mass number" of an element. This is the mass of an atom, expressed in [amu], as defined in Table 2.2. Since we have assumed that both protons and neutrons have a mass of about 1 [amu], we conclude that a neutral fluorine atom has 9 protons, 10 neutrons (and therefore, a mass number of 19), and 9 electrons. The equal number of protons and electrons ascertains that the atom is neutral since the nine positive charges of the protons exactly balance the nine negative charges of the electrons. The 10 neutrons are required to confine the 9 positively charged protons to the tiny nuclear space. Without the neutrons, the positively charged protons would repel each other so strongly that the nucleus would disintegrate immediately. In the periodic table, we find that the number of neutrons in stable nuclei either equals or exceeds the number of protons (more on this below).

Upon inspection of the periodic chart, we find that the mass number generally is not an integer. This is because many elements are mixtures of isotopes. Isotopes are atoms of the same element (that is, with the same Z-value) that differ in the number of neutrons. Carbon, for example, has an isotope with six protons and six neutrons. Carbon also has an isotope with six protons and seven neutrons. Both isotopes have six electrons. The former one is referred to as the $^{12}_{6}C$ isotope and the latter as the $^{13}_{6}C$ isotope. The left subscript here indicates the number of protons in the nucleus (or Z), whereas the left superscript indicates the mass number. The abundance of the $^{13}_{6}C$ isotope is just over 1%; notice that both isotopes are "stable," which indicate that they do not undergo radioactive decay. Another isotope of carbon is $^{14}_{6}C$, that is, a carbon atom with eight neutrons. This isotope is not stable, and decays slowly by radioactive decay into another element. More on this later (see Chapter 14).

When elements exist as mixtures of different isotopes, their average mass number is determined by the relative abundance of the isotopes and the mass numbers of the individual isotopes. Let us consider the element Li, for example. It exists in nature as a mixture of approximately 93% $^{7}_{3}Li$ and 7% of $^{6}_{3}Li$. Therefore, its average mass number is obtained as

$$(0.93 \cdot 7 + 0.07 \cdot 6) = 6.93 \tag{2.1}$$

in close agreement with the number shown for Li in Figure 2.1. In the calculation above, we have assumed that the atomic mass numbers are integers, which is important for the consideration of significant figures. When an element, such as chlorine, is listed with an average mass number of ca. 35.5, and we know that naturally occurring chlorine is composed of the isotopic species $^{37}_{17}Cl$ and $^{35}_{17}Cl$ only, we can solve for the relative abundances of both isotopes by proceeding as follows. Let x be the fraction of the $^{37}_{17}Cl$ isotope and y the fraction of the $^{35}_{17}Cl$ isotope. Then, x + y = 1, since there are only two isotopic species. By the rules of computing averages, the observed average mass is determined by 35.5 = 37x + 35y.

Example 2.4 Using the two equations in the last two sentences, calculate the abundance of the isotopes $^{37}_{17}Cl$ and $^{35}_{17}Cl$.

Answer: x = 0.25, y = 0.75 or 25% and 75%

Example 2.5 Calculate the average atomic mass number of iron, consisting of a mixture of the following isotopes: 5.8% ^{54}Fe, 91.7% ^{56}Fe, and 2.5% ^{57}Fe.

Answer: 55.9 [amu]

Many atoms in the periodic chart, particularly those in Columns 1A and 2A, and in Columns 6A and 7A have the tendency to form ions. Ions are atoms that have lost or gained electrons to form cations and anions, respectively. In this process, they have lost their electric neutrality. Let us look at one of the most abundant ionic compounds, table salt or sodium chloride, NaCl. For sodium, Z = 11; thus, the neutral sodium atom has 11 protons and 11 electrons. It readily gives up one electron according to

$$Na \rightarrow Na^+ + e^- \tag{2.2}$$

where e^- represents the electron, and Na^+ is the sodium cation. It still has 11 protons (11 positive charges) and only 10 electrons (negative charges), since it gave up one electron according to Eq. 2.2. Thus, the sodium ion, Na^+, has a net charge of +1.

On the other hand, chlorine, with Z = 17, has 17 protons and 17 electrons. It readily accepts an electron according to

$$Cl + e^- \rightarrow Cl^- \tag{2.3}$$

and turns into a negatively charged anion, since the "chloride ion" has 17 protons and 18 electrons. It should be noted that the processes shown in Eqs. 2.2 and 2.3 change the nature of the species enormously: whereas the element sodium, consisting of sodium atoms, reacts nearly explosively with water, the sodium ion dissolves in water peacefully. Similarly, atomic chlorine is a very reactive agent, the chloride ion is chemically much more inert. The formation of NaCl (= Na^+ Cl^-) from the elements can be summarized as an exchange of an electron between the elements.

When elements form anions, their names are derived from the element name by appending "ide" to the root of the element name. The cation name is unchanged, and the species Na^+ is simply referred to as the sodium ion. This explains the chemical name of NaCl, sodium chloride. Other atoms form doubly charged ions, such as Mg^{2+}, Ca^{+2}, O^{2-}, S^{2-}, N^{3-}, etc.

Example 2.6 Name the five species listed in the last line of the last paragraph.

Answer: See answer booklet

2.3 The Periodic Chart and Periodic Properties of the Elements

The shape of the periodic chart, as shown in Figure 2.1, needs to be discussed further. Early versions of the periodic chart were based strictly on empirical observations, namely an apparent periodicity of chemical properties of elements. For example, the elements in the rightmost column are gases at room temperature (the "noble gases") that do not readily undergo chemical reactions. The elements in Column 7A tend to form anions with a single negative charge, such as the fluoride, chloride or bromide ions. The elements in the leftmost column (except for hydrogen), marked as column 1A are known as the alkali metals that react quite violently with water, resulting in strong bases. Similar trends were found to hold for most elements in a given column of the periodic chart. What could be the reason for the similarity in chemical behavior of some elements? The answer to this question led early chemists to realize that the electronic configuration, or the distribution of electrons, must be responsible for the chemical properties of elements. This fact, of course, is well-known today, but was an enormous breakthrough in the beginning of the twentieth century. The early models for the electronic structure were based on the assumption that there are distinct "shells" around an atom that can accommodate a fixed number of electrons. These shells were designated the K, L, M... shells and could accommodate 2, 8, 8 (or 18) electrons, respectively. Since the number of electrons in an atom equals the number of protons (given by Z), a conclusion was reached that chemical similarity was caused by the number of electrons in the outermost shell. The element Li would have one electron in the L shell, and the element Na would have one electron in the M shell, and so on. This was a step in the right direction for the understanding of chemical properties of the elements.

Beginning in the third decade of the twentieth century, with the advent of quantum mechanics, a more refined view of the elements and the periodic chart emerged. As it turns out, the "electron cloud" mentioned earlier in the discussion of the atomic radii follows certain rules of quantum mechanics that define (energy) levels and sublevels that determine the way electrons are arranged around the nucleus. These levels and sublevels are associated with spatial regions in which the electrons are able to exist, and are referred to as **orbitals**. These orbitals are also associated with the energy of interaction between the electrons and the nucleus. The earlier classification scheme, using the K, L, M... notation, however, was not thrown out completely, but refined, as is happens frequently in science. These refinements showed that each "shell" had "subshells," and a new nomenclature for these subshells, or orbitals, was devised. Instead of referring to the shells by the letters K, L, M..., they are now referred to by the main quantum number "n," with $n = 1, 2, 3...$ The subshells are defined by two new quantum numbers, "ℓ" and "m" (see below) such that each orbital has a unique "address" of three quantum numbers n, ℓ, and m. This notation shows the influence of the "new" way of thinking due to quantum mechanics.

The energies of the atomic orbitals are shown in Figure 2.3 for atoms with more than one electron. An orbital is represented in this scheme by a horizontal line and is associated with the three quantum numbers, n, ℓ, and m, as pointed out above. "n" is referred to as the main or principal quantum number that previously was referred to as the "shell." "ℓ" is a new designation of the sublevels resulting from quantum mechanics and is referred to as the orbital quantum number that determines the shape of the orbitals, or defines the spatial arrangement of electrons. "m," the magnetic quantum number, indicates the orientation of the orbital in space. There are certain rules, imposed by the mathematics of quantum mechanics, and briefly referred to in Chapter 5, that relate the quantum numbers to each other. The allowed values of ℓ are

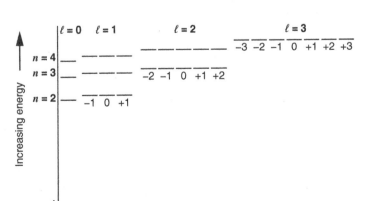

Figure 2.3 Orbitals and approximate orbital energies for atoms in the periodic chart. The designations −1, 0, 1 or −2, −1, 0, 1, 2, etc. of orbitals shown refer to the quantum numbers m.

from 0 to $(n - 1)$. Thus, for $n = 1$, the only allowed value for the quantum number "ℓ" is zero. Consequently, there is only one orbital at the $n = 1$ energy level (see Figure 2.3).

For $n = 2$, "ℓ" can have values of 0 or 1, as shown in Figure 2.3. When $\ell = 0$, m must be zero. However, when $\ell = 1$, m can have values of −1, 0, or 1 (see caption of Figure 2.3). Thus, there are four orbitals at the $n = 2$ energy level, and nine orbitals for the $n = 3$ energy level. At this level, "ℓ" can have values of 0, 1, or 2, and the allowed m-values are indicated in Figure 2.3.

The rules governing quantum numbers must be accepted for the time being without an understanding of their origins. The existence of these rules, which govern the entire periodic system, and thereby, just about all of chemistry, is a direct consequence of the mathematics of quantum mechanics, and can be traced back to the brilliant early quantum physicists and chemists who are responsible for putting chemistry onto a solid, theoretical foundation. This will be discussed in more detail in Chapter 5.

To simplify bookkeeping, the orbitals are referred to as shown in Table 2.3. Notice that the quantum number m is referred to as m_l in many texts. Each orbital shown in Figure 2.3 can accommodate two electrons (see below).

Example 2.7 Why cannot there be 2d orbitals?

Answer: For $n = 2$, the maximum allowed value of ℓ is 1.

In a hydrogen atom (Z = 1), there is only one electron, which occupies the lowest energy orbital with $n = 1$ and $\ell = 0$ (the 1s orbital, see Table 2.3). This is shown in Figure 2.4a by an arrow representing the electron. In Figure 2.4, only the lowest energy orbitals with $n = 1$ and $n = 2$ are shown. The element helium (Z = 2) has two electrons that can, in principle, occupy the $n = 1$, $\ell = 0$ (or 1s) orbital, since, as stated above, all orbitals can accommodate two electrons. However, one more condition has to be fulfilled for two electrons occupying the same orbital. This condition depends on a property of electrons known as their "spins." The spin is an inherent property of many nuclear particles and may be visualized (although not really correctly)

Table 2.3 Nomenclature of atomic orbitals.

Main quantum number	Orbital quantum number	Possible magnetic quantum number	Orbital name
$n = 1$	$\ell = 0$	$m = 0$	1s
$n = 2$	$\ell = 0$	$m = 0$	2s
$n = 2$	$\ell = 1$	$m = -1, 0, 1$	2p
$n = 3$	$\ell = 0$	$m = 0$	3s
$n = 3$	$\ell = 1$	$m = -1, 0, 1$	3p
$n = 3$	$\ell = 2$	$m = -2, -1, 0, 1, 2$	3d
$n = 4$	$\ell = 0$	$m = 0$	4s
$n = 4$	$\ell = 1$	$m = -1, 0, 1$	4p
$n = 4$	$\ell = 2$	$m = -2, -1, 0, 1, 2$	4d
$n = 4$	$\ell = 3$	$m = -3, -2, -1, 0, 1, 2, 3$	4f
etc.			

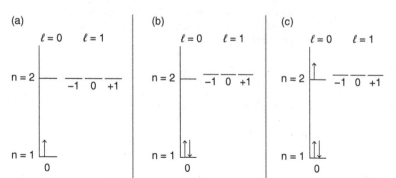

Figure 2.4 (a) Schematic of the only electron in an H atom. (b) Schematic of two electrons with opposite spins in the He atom. (c) Schematic of the three electrons in a Li atom. Notice that for atoms with more than two electrons, the s and p orbitals are no longer degenerate.

as a particle spinning around an axis. More on this is given in Chapter 5. For two electrons to share an orbital, their spins must be opposite, as indicated by up- and down arrows in Figure 2.4b for the helium atom.

Going back to Figure 2.1, the two elements discussed so far, H and He, constitute the first row in the periodic chart and have their electrons in the $n = 1, \ell = 0$ orbital. The element Li, with Z = 3, fills two of its three electrons into the lowest energy 1s orbital (with $n = 1, \ell = 0$), and the third one into the energetically next lowest orbital, the $n = 2, \ell = 0$ (2s) orbital, as shown in Figure 2.4c. In the periodic chart in Figure 2.1, this corresponds to the first position in the second row. The orbitals identified by the "main" quantum number $n = 2$ can accommodate eight electrons, two in the orbital identified by $n = 2, \ell = 0$ (2s), and six in the three orbitals with $n = 2, \ell = 1$ (2p). The 2p orbitals are further identified by the magnetic quantum number m which can have values of $-1, 0,$ and $+1$ (see Table 2.3). The eight elements from Li to Ne, representing the second row of the periodic chart, fill in these orbitals with electrons, but with one more caveat. This is known as Hund's rule, which implies that orbitals with the same energy (degenerate orbitals, such as the 2p orbitals) are filled with electrons of the same spin to maximize the number of unpaired spins when going from one element to the next. This is shown schematically in Figure 2.5a for the carbon atom, in Figure 2.5b for the nitrogen atom, and in Figure 2.5c for the oxygen atom. Here, the two 2p electrons are in orbitals with different magnetic quantum numbers and equal spin. Whether the spins are drawn using up- or down-arrows is immaterial; what is important is that the spins are the same. The elements Na to Ar (Z = 11 to Z = 18) fill in orbitals identified by $n = 3, \ell = 0$ (the 3s orbitals) and $n = 3, \ell = 1$ (the 3p orbitals) and are represented in the third row of the periodic chart.

Example 2.8 Draw an orbital occupancy diagram for the element Ne.

Answer: See answer booklet

At this point, it is advantageous to pause with this scheme of build-up (German: Aufbau) of the periodic chart, and discuss the consequences and insights gained. First of all, these rules and the schemes presented so far may appear arbitrary and a bit of "black magic." However, these rules are based on experimental results and experimental results reign supreme. As R. Feynman, the Nobel laureate in physics (1965) quoted "It doesn't matter how beautiful a theory is, ... if it doesn't agree with experiment, it's wrong." In the case of the discussion so far, the "Aufbau" principle can be verified experimentally via the measurement of such quantities as the atomic radius, ionization energy, electron affinity, and spin multiplicity, i.e. the observed number of unpaired spins to be discussed in more detail in Chapter 5. The quantum mechanical model, based on the observed results, explains the origin of the quantum numbers and the rules connecting them. For the discussion presented so far, the experiments and the theoretical framework to explain them agree perfectly. However, the mathematics in this field is quite complicated; thus, at this point, we have to accept the result sight unseen.

One of the fascinating aspects in the "Aufbau" scheme discussed so far is that atoms in the second and third row of the periodic chart prefer to have their orbitals filled to capacity (namely eight electrons) by either accepting electrons from other atoms, donating electrons to other atoms, or sharing electrons with other atoms. This is the gist of all chemistry: atoms form

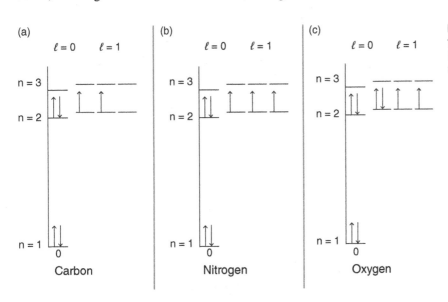

Figure 2.5 Orbital occupancy diagram for (a) carbon, (b) nitrogen, and (c) oxygen, illustrating Hund's rule of maximum spin multiplicity.

ions (positively or negatively charged atoms) or form chemical bonds by sharing electrons, to reach an "octet" of electrons. In the following paragraphs, we introduce a simple picture how octet structures can be achieved in ionic compounds via the formation of ions. The subject of (covalent) chemical bonds will follow in Chapter 6.

In Section 2.2 (Eqs. 2.2 and 2.3) we already discussed that the ionic compound NaCl is formed from by the atoms giving up (the sodium atom) or accepting (the chlorine atom) an electron to form ions that are held together in a crystal by electrostatic forces. With the information we have gained since Section 2.2, we can visualize these two steps of forming ions in the energy diagrams shown in Figure 2.6. The sodium atom ($Z = 11$) has one electron in the $n = 3$, $\ell = 0$ orbital, and eight electrons in the $n = 2$ orbitals (two in the $n = 2$, $\ell = 0$ and six in the $n = 2$, $\ell = 1$ orbitals) as shown in Figure 2.6a. When forming the Na$^+$ ion, it gives up the electron in the 3s orbital, and achieves a configuration with an octet of electrons in the $n = 2$ shell, shown in Figure 2.6b.

Chlorine ($Z = 17$), on the other hand, has seven electrons in the n = 3 orbitals (two in the $n = 3$, $\ell = 0$ and 5 in the $n = 3$, $\ell = 1$ orbitals), see Figure 2.6c. It achieves an octet configuration by accepting the electron released by sodium to become a chloride ion, Cl$^-$, shown in Figure 2.6d. Arguments similar to those described by Eqs. 2.2 and 2.3 can be made for the formation of ionic compounds from the second column, and third to last column of the periodic chart. Magnesium (Mg) and oxygen (O) form a compound magnesium oxide, MgO, that is composed of Mg^{2+} and O^{2-} ions.

Example 2.9 Write the equations for the formation of the magnesium and oxygen ion corresponding to Eqs. 2.2 and 2.3, and account for the numbers of protons and electrons in each species.

Answer: Mg $\rightarrow$ Mg^{+2} + 2e$^-$; O + 2e$^-$ $\rightarrow$ O^{2-}; Mg: 12 p, 12 e$^-$; Mg^{+2}: 12 p, 10 e$^-$; O: 8 p, 8 e; O^{2-}: 8 p 10 e$^-$

Elements in the periodic chart are grouped by similar chemical behavior, which is caused by similar electronic configurations. Li, Na, and K, have a single electron in their outermost "shell" (2s, 3s, 4s, respectively) and therefore, have the tendency to become cations with the charge +1 to achieve an octet configuration. Elements in column 2A (see Figure 2.1) all have the tendency to form cations with 2+ charges, whereas elements in column 7A fill form anions with charge −1. This is, of course, due to the fact that elements in a column have the same outer electron arrangement, albeit with different main quantum numbers "n". This will be discussed further in Section 2.3.

When monovalent ions – those with one positive or negative charge – combine with divalent ions to form ionic compounds, the resulting compound's electrical neutrality must be maintained. Thus, magnesium chloride will have the formula MgCl$_2$, where the two positive charges on the magnesium ion are balanced by two negative charges from each of the chloride ions. We shall discuss ionic compounds in much more detail in later chapters. This will include the discussion of polyatomic ions such as the carbonate ion, CO$_3^{2-}$, and the ammonium ion, NH$_4^+$, and many others, the naming of ionic compounds, and the kind of forces that hold the ions in place in an ionic compound.

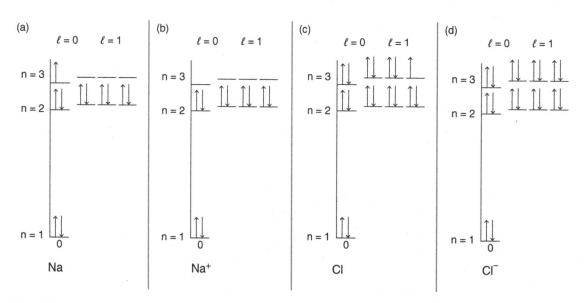

Figure 2.6 Orbital occupancy in (a) the sodium atom, (b) the sodium ion, (c) the chlorine atom, and (d) the chloride ion.

Example 2.10 Write the expected formula for calcium phosphate. The phosphate ion is PO_4^{3-}.

Answer: $Ca_3(PO_4)_2$

As mentioned before, the original periodic chart of elements was established before the results from quantum mechanics were known and was based on the fact that certain elements exhibit similar chemical behavior. All elements in the rightmost column of the periodic chart, for example, are the noble gases: they are monatomic gases (which implies that they exist as individual atoms), that do not readily undergo chemical reactions, or do not undergo chemical reactions at all. The first element in the rightmost column, He, has the $n = 1$, $\ell = 0$ filled to the maximum of two electrons (see Figure 2.4b), and therefore, is not likely to undergo any chemical reaction. Similarly, the second and third elements in the rightmost column of the periodic chart, Ne and Ar, have their n = 2 and n = 3 orbitals, respectively, filled with eight electrons and, therefore, are chemically inert.

The elements in columns 1A through 8A are called the "main group" elements. Among these, the elements in column 1A (with the exception of H) are called the alkali metals, and the elements in column 2A the alkaline earth metals. In column 8A, we find the noble cases, in column 7A the halogens ("salt formers"), and in column 6A the chalcogens ("ore formers"). When going down in one column of the periodic system, the size of the atom, as expressed as the atomic radius, always increases. This is because a new electron "shell" is being filled, or the main quantum number of the outermost electrons increases by one for each element when going down a column. When proceeding from left to right in a row of the periodic chart, the atomic radii generally decrease, since each successive element has one more proton in the nucleus which exerts additional attractive forces to the electrons that all have the same main quantum number.

The elements in the column next to the noble gases (column 7A), the "halogens," comprise the elements F to At. They all have five electrons in the orbitals with $\ell = 1$ and therefore, readily form negatively charged anions by accepting an electron, as shown in Eq. 2.3 for chlorine. In elementary form, they exist as diatomic molecules (much more on this later) F_2, Cl_2, Br_2, etc., since sharing of electrons is one way of obtaining an octet structure of electrons. Whereas F_2 and Cl_2 are gases, Br_2 is a liquid, and I_2 and At_2 are solids at room temperature. This is a general trend observed in the periodic table that heavier elements (those with larger mass numbers) are solids, whereas the light ones are gases.

Column of elements 6A, the chalcogens, includes the elements from O to Po (polonium). They readily form anions with a -2 charge in order to fill their $\ell = 1$ orbitals, as discussed above in the example of MgO. In elementary form, they occur as molecules: in the case of oxygen as the diatomic gas O_2, and in the case of sulfur, mostly as S_8 molecules. This is in analogy to what we found in the halogens: in order to reach octet structures, they may form covalent bonds. These bonds may be with other atoms of the same element, or with other elements. The elements in column 6A all form compounds with hydrogen of the form H_2O, H_2S, etc. These latter compounds are no longer ionic (salt-like) in nature, but involve what is referred to as polar covalent bonds formed by unequal sharing of electrons to achieve octet structures around the central atom (see Chapter 3).

The elements in the first and second column (columns 1A and 2A) are known as the alkali and alkaline earth elements, respectively. They all are metallic (except for hydrogen), and all give up electrons to form positively charged cations. Thus, the equivalent of Eq. 2.2 holds for all elements in the first (leftmost) column. Similarly, the elements in the second column, Be through Ra (Radium) readily form cations with a charge of +2. In particular, the Mg and Ca cations are among the most common ions on earth, and, along with silicon and oxygen, among the most abundant elements on earth.

The elements toward the middle of the periodic chart (columns 3A to 5A) still exhibit similar chemical properties within each column of elements. However, the compounds they form generally are no longer described by ionic structures. Al and Si both form oxides Al_2O_3 and SiO_2 (alumina and quartz, respectively) whose stoichiometry could be attributed to Al^{3+} and Si^{+4} ions paired with O^{2-} ions; however, these compounds have chemical bonds best described as polar covalent bonds.

Notice that the elements that arise from filling the $\ell = 2$ ("d") orbitals all are metals and are referred to as transition metals. Elements Sc(21) to Zn(30) fill in their 3d orbitals, Y(39) to Cd(48) their 4d orbitals, and La(57) to Hg(80) their 5d orbitals. Transition metals tend to form cations by first shedding their outer "s" electrons. In addition, they form polar covalent bonds. Their chemistry is discussed mostly in the context of inorganic chemistry.

The elements 58 (cerium, Ce, through 71, lutetium, Lu) fill the $n = 4$, $\ell = 3$ (4f) orbitals. They are a group of metals known as "rare earth metals" with rather similar chemical properties. Some of them have become rather important in the semiconductor industry. The color "dots" on your cell phone screen are actually tiny semiconductor (see Chapter 7) devices that contain rare earth oxides that emit light of different colors when electric currents pass through them. The actinides, starting with element 89 (Actinium) fill the $n = 5$, $\ell = 3$ (5f) orbitals, although the rules of which orbitals are filled (5f, 7s, or 6d) are no longer as strictly obeyed as in the previous elements. They all are radioactive metals, and starting from element 95 (Americium) are synthetic elements, that is, they do not occur naturally.

Inspection of Figure 2.1 reveals a zigzag line that extends from element 5 (B) to element 85 (At). This line represents a boundary between metals and nonmetals: elements to the left of this line are metals, and to the right are nonmetals. Elements directly adjacent to the zigzag line may exist in metallic or nonmetallic forms and are referred to as metalloids. Some of them are semiconductors. This distinction is based on the fact that metals generally are good conductors of electricity, and nonmetals generally are poor conductors, or insulators. Semiconductors have conductivities somewhat in between those of metals and insulators but differ from metals in the temperature dependence of conductivity (more on this in Chapter 7). Metallic elements do not exist as single atoms (like the noble gases) or small molecules (such as oxygen, fluorine, etc.) but exhibit "metallic bonding" that will be discussed in Chapter 7.

2.4 Definition of Atomic Masses, Avogadro's Number, and the Mole

We now turn to the discussion of one of the most useful concepts in chemistry, the mole. This concept often is hard to grasp for students, so in this discussion, we will approach it from the view of everyday life. Before doing so, we need to revisit briefly the definition of atomic masses, introduced early in this chapter. In Table 2.2, the masses of protons and neutrons were given as approximately 1 [amu]. According to the latest definition, their masses are 1.00728 and 1.00866 unified atomic mass units, corresponding to $1.67262 \cdot 10^{-27}$ and $1.67493 \cdot 10^{-27}$ [kg], respectively. This is based on the international standard that defines the unified atomic mass unit as exactly 1/12 of the mass of the $^{12}_{6}C$ isotope, or $1.66054 \cdot 10^{-27}$ [kg]. For most practical calculations in the early chapters, we can use the approximation shown in Table 2.2.

Let us now consider a sample that consists of 1.00 [g] = $1.00 \cdot 10^{-3}$ [kg] of hydrogen atoms. For the time being, let us assume that this sample consists of hydrogen atoms (as we have seen before, hydrogen exists as a diatomic gas, but forget this aspect for the time being. We will get back to that in the next chapter). Since we know that the hydrogen atom, consisting of one proton, has a mass of $1.67 \cdot 10^{-27}$ [kg] (notice that we are using here the mass to two significant figures, which allows us to ignore the mass of the electron), we can calculate how many hydrogen atoms are in 1 [g] of hydrogen. In order to do so, let us consider a simplified version of the problem:

Example 2.11 In a grocery store, you buy a "box" of eggs that has a mass of 0.84 [kg]. Each egg has a mass of 0.070 [kg] (let us assume hens lay eggs of standard mass!). How many eggs are in the box?

Answer: 12

Hopefully, you solved Example 2.9 by arguing that 0.84 [kg/box of eggs]/0.070 [kg/egg] resulted in 12 [eggs/box of eggs], or a dozen of eggs, since eggs are sold for some archaic reason in units of dozens or half-dozens. Let us do the equivalent calculation for a "box of hydrogen atoms" that has a mass of 1.00×10^{-3} [kg]. The mass of one hydrogen atom is $1.67 \cdot 10^{-27}$ [kg]. Thus, the number of hydrogen atoms in our "box" of hydrogen atoms is, in complete analogy to Example 2.9.

$$N = \frac{1.0 \cdot 10^{-3} \left[\frac{kg}{box}\right]}{1.67 \cdot 10^{-27} \left[\frac{kg}{atom}\right]} = 6.0 \cdot 10^{23} \left[\frac{atoms}{box}\right] \tag{2.4}$$

So, our "1 g box" of hydrogen atoms contains $6.0 \cdot 10^{23}$ atoms. Now let us repeat these calculations for a 4.00 [g] box of helium atoms. Notice, we have used 4[g] of He, since its mass number is 4.0. We therefore know that the mass of one He atom is $4 \cdot 1.67 \cdot 10^{-27}$ [kg], and

$$N = \frac{4.0 \cdot 10^{-3} \left[\frac{kg}{box}\right]}{4 \cdot 1.67 \cdot 10^{-27} \left[\frac{kg}{atom}\right]} = 6.0 \cdot 10^{23} \left[\frac{atoms}{box}\right] \tag{2.5}$$

We can perform this calculation for any element in the periodic chart and realize that an amount of an element given by its mass number expressed in grams, contains the same number of atoms, namely about $6.0 \cdot 10^{23}$ atoms. A more accurate value of this number N is $6.022 \cdot 10^{23}$ and is known by the name of Avogadro's constant and has been given the symbol N_A. The quantity that we called "box" in the equations above is what chemists call "the mole" [mol], or in the case of the eggs, a "dozen". This is so important that we must emphasize it once more:

1 mole of an element is the amount of the element, in grams, that contains $6.022 \cdot 10^{23}$ atoms.

This amount, expressed in grams, formerly was referred to as "gram atomic mass" (GAM), but is now generally called atomic mass and given the symbol $\mathcal{A}$. Thus, $\mathcal{A}$ is the amount of an element given by its mass number, expressed in grams and thus, has the units of [g/mol = g mol^{-1}]. For hydrogen, the atomic mass is 1 which implies that 1 [g] of hydrogen atoms is 1 mole of hydrogen atoms, and contains Avogadro's number of hydrogen atoms. For carbon, the atomic mass is 12 which implies that 12 [g] of carbon are 1 mole of carbon atoms, containing N_A carbon atoms. The last two sentences can be summarized in the form of an equation: the number of moles of an element (which we call "n") is related to the mass of an element ("m") and the atomic mass $\mathcal{A}$ by[1]

$$n = \frac{m}{\mathcal{A}} \left[\frac{g}{\frac{g}{mol}} = \frac{g}{g\, mol^{-1}} = mol \right] \tag{2.6}$$

So, if we have 2.00 [g] or carbon, this amounts to

$$n = 2.00/12.0 = 0.167\,[mol] \tag{2.7}$$

If the question is asked how many grams of Fe (element 26, mass number 55.9) are needed to constitute 0.10 [mol], we solve Eq. 2.6 for m and obtain

$$m = n\,\mathcal{A} = 0.10 \cdot 55.9 \left[\frac{mol \cdot g}{mol} \right] \approx 5.6\,[g] \tag{2.8}$$

Furthermore, we can define the number of moles n of an element in terms of the number of atoms N of the element according to

$$n = N/N_A \tag{2.9}$$

just as one would define three eggs as ¼ dozen eggs. An example combining Eqs. 2.6 and 2.9 is given below. We shall encounter equivalent problems in the next chapter when we discuss molecules.

Example 2.12 Given a typical small iron nail with a mass of 1.50 [g], (a) how many moles of iron, and (b) how many iron atoms does it contain?

Answer: (a) $2.68 \cdot 10^{-2}$ [mol]; (b) $1.6 \cdot 10^{22}$ [atoms]

Example 2.13 (a) Calculate the mass of a sample of 3.0 [fmol] of sodium atoms. (b) How many sodium atoms are there in this sample?

Answer: (a) $m = 6.9 \cdot 10^{-14}$ [g]; (b) $N = 1.8 \cdot 10^9$ [atoms]

One quick comment on the magnitude of Avogadro's number. It is, indeed, a staggeringly large number. For example, the earth – which you hopefully agree is a pretty hefty lump of matter – has a mass of $6 \cdot 10^{24}$ [kg], or about 10 moles of 1 [kg] weight pieces. The total volume of the combined oceans on earth is about $1.4 \cdot 10^{21}$ [L] of water, or about 2 mmol of 1 [L] water bottles!! Or look at the mole another way: the world's richest person is worth about 200 billion dollars (US$$2 \cdot 10^{11}$), which is a bit of dough. If every person in the world, all 8 billion people, were as rich, the total combined wealth of all people would be US$2.7 [mmol] dollars. Avogadro's number really is quite large.

---

[1] Notice that we used italic symbols *n* and *m* for quantum numbers, and n and m for number of moles and masses.

Example 2.14 Which of the following statements is true?

(a) Isotopes of the same element differ in the number of protons in the nucleus
(b) The number of protons equals the number of neutrons in all nuclei
(c) When an anion is formed from an atom, its number of electrons decreases
(d) None of the above is true

Answer: (d) Details in answer booklet

Example 2.15 The element Fermium, $^{257}_{101}$Fm, does not occur naturally but is one of the elements produced by nuclear reactions. Only about $2 \cdot 10^8$ atoms were ever produced. The mass of these $2 \cdot 10^8$ atoms is approximately

(a) $9 \cdot 10^{-14}$ [g]
(b) $3 \cdot 10^{-14}$ [g]
(c) $5 \cdot 10^{10}$ [g]
(d) None of the above

Answer: (a) Details in answer booklet

Further Reading

An excellent summary of atomic structure and the periodic chart can be found at https://en.wikipedia.org/wiki/Periodic_table. This article also contains information that will be discussed in the following chapters in this book.

3

Molecules, Compounds, Bonding, and Percent Composition

3.1 Ionic Compounds

As pointed out earlier, chemistry deals with the electronic configuration around atoms or changes therein during a chemical reaction. This can happen either by transfer of electrons between different atoms, or sharing of electrons, to reach what is energetically the most favorable state. For elements in the first row of the periodic system, this most favorable state is either zero electrons, such as in the H^+ (hydronium or hydrogen) ion, or two electrons, such as in the He atom. Hydrogen can also form a H^- ion (the hydride ion) that, like He, has two electrons.

Elements in the second row of the periodic chart, Li through Ne, have between one and eight valence electrons, as they fill their $n = 2$ (2s and 2p) shells. For these elements, again, it is energetically most favorable to achieve a full octet by either becoming a negatively charged anion, for example, F^-, O^{2-}, N^{3-} (fluoride, oxide, or nitride ion, respectively), or shedding electrons to become Li^+ or Be^{2+} cations with two electrons in the $n = 1$ (1s) orbital and no electrons at all in the $n = 2$ shell. Thus, elements from the first and second columns preferentially form ionic compounds[1] with anions such as the ones elaborated upon earlier. In ionic compounds (salts), the positively and negatively charged ions form crystal lattices, where the ions are held together by electrostatic forces. In such lattices, depending on the size of the ions, the attractive forces between oppositely charged ions are maximized, whereas the repulsive forces due to equally charged ions are minimized. A picture of a simple ionic lattice is shown in Figure 3.1. The energetics of crystal formation will be further discussed in Section 11.6. As one can clearly see, there are no individual NaCl "molecules" in an ionic solid such as NaCl; rather, one deals with a lattice of many ions held together by electrostatic forces. In the gaseous phase, however, that is, after heating solid NaCl to very high temperatures, individual NaCl molecules do exist.

Ionic compounds are very common in nature, because nearly all metals tend to occur as positively charged cations, with various anions. This is because all metals give up their outer electrons rather easily, as will be discussed in Chapter 5 under the subject of ionization energies. The earth crust contains many ionic compounds, such as $CaCO_3$, $CaSO_4$, Ca_2SiO_4, (calcium carbonate, sulfate, and silicate, respectively) and many more complicated compounds. Many commercially important metals occur as ores as mixtures of oxides, sulfides, carbonates, etc. Iron, for example, is found mostly as a mixture of FeO and Fe_2O_3 that may be considered salt-like. This last example shows that many metals can exist as cations of different charges: the iron atom in FeO has a charge of 2+, whereas it has a charge of 3+ in Fe_2O_3. This charge is expressed as Roman numerals when naming ionic compounds. $FeCl_2$ is referred to as iron (II) chloride and $FeCl_3$ as iron (III) chloride.

Even in biological systems, ionic compounds play a major role: the shells of clams and oysters contain mostly $CaCO_3$, and even the bones of mammals are a very sophisticated "composite" material consisting of an ionic compound, calcium hydroxyapatite, $Ca_5(PO_4)_3(OH)$, and collagen, a protein. In bones, the mineral phase (the ionic compound) provides the hardness, whereas the collagen fibers provide the elasticity and resilience to

Figure 3.1 Example of a crystal lattice: NaCl. Small spheres: sodium ions, large spheres: chloride ions. *Source:* Image from Benjah/Wikimedia/Public Domain.

---

1 The term "compound" is used here in the sense of ionic substances formed from cations and anions. A definition of 'molecular compounds' will be given in Section 3.4 of this chapter.

Understanding Essential Chemistry, First Edition. Max Diem.
© 2025 John Wiley & Sons, Inc. Published 2025 by John Wiley & Sons, Inc.

fracture. The structure of the polyatomic anions mentioned in the last paragraphs, e.g. CO_3^{2-}, SO_4^{2-}, PO_4^{3-}, will be discussed with the covalently bonded molecules later.

Example 3.1 Based on the discussion in Section 2.2 and the charges of cations and anions discussed here, write the formula for the ionic compounds cesium nitrate, lithium phosphate, iron (III) sulfate, calcium phosphate, sodium carbonate, and aluminum oxide.

Answers: $CsNO_3$, Li_3PO_4, $Fe_2(SO_4)_3$, $Ca_3(PO_4)_2$, Na_2CO_3, Al_2O_3

3.2 Molecules with Covalent Bonds

As it was pointed out in Chapter 2, atoms also can achieve a "filled shell" situation by sharing electrons to form covalent chemical bonds. The simplest description of such a covalent bond is that the atomic orbitals of two different atoms combine to form a "molecular orbital" that can accommodate two electrons with opposing spins, just as atomic orbital can accommodate two electrons with opposite spins. We call an electron pair in a bonding molecular orbital a "covalent bond," and designate it, in the simple *Lewis* (see below) model, by a line connecting the two atoms, as shown in Eq. 3.1:

$$H \cdot + \cdot H \rightarrow H-H \tag{3.1}$$

Here, the symbol H· represents a hydrogen nucleus with a single electron. In H—H, the dash represents a covalent chemical bond, as pointed out above. We can now argue that formally there are two electrons around each nucleus, and the world is a happy place. We refer to a species such as H_2 as a (diatomic) molecule with a nonpolar covalent bond, and an ensemble of like molecules, such as H_2 molecules, as a chemical compound. This is equivalent to the definition of an element, which is an ensemble of like atoms.

In general, we write the process of two hydrogen atoms to form diatomic H_2 molecules as a chemical equation

$$H + H \rightarrow H_2 \tag{3.2}$$

In a chemical equation, the left- and right-hand sides must contain the same number of atoms. They can be in the form of molecules, or atoms and molecules, as in Eq. 3.2, but the total number of atoms, as well as all charges, must be equal on both sides of the equation. Thus, a chemical equation is like a mathematical equation in that the two sides really must be equal. Much more on this will be explored in our discussions on stoichiometry later.

We can describe the formation of the F_2 molecule in a similar way. As we saw before, the fluorine atom has seven electrons in its outer shell: two in the $n = 2$, $\ell = 0$ (or 2s) orbital, and five in the $n = 2$, $\ell = 1$ (the 2p) orbital. When drawing a bonding scheme similar to Eq. 3.2, we need to include all "valence" electrons in our drawing, where the valence electrons are all electrons in orbitals identified by the main quantum number (in this case, all seven "valence electrons"):

$$|\underline{F}\cdot + \cdot \underline{F}| \rightarrow |\underline{F}-\underline{F}| \tag{3.3a}$$
$$F + F \rightarrow F_2 \tag{3.3b}$$

Here, the line between the F symbols again represents the chemical bond formed by the two unpaired electrons at the fluorine atoms, and the other lines around each fluorine atom represent "nonbonding" electron pairs. Eq. 3.3b again represents the chemical equation for the formation of diatomic fluorine from the atoms. The drawing $|\underline{F}-\underline{F}|$ shown in Figure 3.3a is called a *Lewis structure*, named after Gilbert N. Lewis, one of the early pioneers in chemical bond theory, and the originator of the idea that chemical bonds are formed by paired electrons. Notice that a valid Lewis structure always must include the bonding and nonbonding electrons. We shall see in Chapter 6 how these nonbonding electron pairs, along with the chemical bonds between atoms, determine the geometric shapes of molecules.

The view presented here, the equal sharing of an electron pair between atoms, adequately describes diatomic molecules such as H_2, F_2, Cl_2, etc. The examples shown so far describe bonds where the electron pairs are shared equally, and the bond formed is called a covalent bond. However, in chemistry (as well as in real life), sharing is often a compromise where the two partners sharing something are not sharing equally. This leads us to the next section.

3.3 Molecules with Polar Covalent Bonds and Lewis Structures

Obviously, the world around us is not made up of diatomic molecules, and it is necessary to extend the concept of Lewis structures to more complicated molecules, such as one of the simplest carbon compounds, methane (CH_4). Methane is one of the major constituents of natural gas and readily burns in air. Its formation from the elements can be summarized as

$$C(s) + 2H_2(g) \rightarrow CH_4(g). \tag{3.4}$$

Eq. 3.4 again is a balanced chemical equation in which the number of atoms on the left-hand side equals the number of atoms on the right-hand side. In Eq. 3.4, the symbols in parentheses, (s) and (g), denote the physical state of reactants or products, namely solid and gaseous, respectively.

The bonding in methane can be described as follows. Carbon has four valence electrons, two in the 2s and two in the 2p orbitals. The four hydrogen atoms also contribute four electrons, such that there are eight valence electrons, or four electron pairs, to form the four carbon–hydrogen bonds in methane. Thus, we can write a Lewis structure as follows:

$$\begin{array}{c} H \\ | \\ H-C-H \\ | \\ H \end{array} \tag{3.5}$$

where each line represents an electron pair. Based on similar consideration, we can also write a Lewis structure for ethane, C_2H_6, as follows:

$$\begin{array}{cc} H & H \\ | & | \\ H-C-C-H \\ | & | \\ H & H \end{array} \tag{3.6}$$

This molecule has 14 valence electrons, four each from the two carbon atoms and six from the hydrogen atoms. The molecule has one carbon–carbon and six carbon–hydrogen bonds. Notice that H can accommodate two electrons only; thus, structures with —H— groupings are strictly forbidden since this grouping implies four electrons around hydrogen. Ethane, by the way, is often found mixed in with methane in natural gas.

In the case of the C—H bonds discussed above, the electron pair is shared quite equally between the carbon and hydrogen atoms. However, some elements have the tendency of attracting electron pairs much more strongly than others. Let us consider the diatomic molecule hydrogen fluoride, HF, for example. Based on the discussion in the previous section, we can write its Lewis structure as H—F̄|, since there are eight valence electrons, one from the hydrogen and seven from the fluorine atom. Since H can only accommodate two electrons, or one bond, the remaining three pairs must be around the fluorine atom as nonbonding pairs.

However, fluorine is an element with many more positive protons in the nucleus than hydrogen. Thus, it will exert a much stronger pull on the shared electron pair, and we have here an example of a polar covalent bond, that is, a bond where the shared electron pair is more strongly attracted to the fluorine atom. In fact, most covalent bonds are at least partially polar, unless they are between atoms of the same element. So, even the bonds shown for CH_4 above are a bit polar, as we shall see later.

Another molecule with strongly polar covalent bonds is the water molecule. Oxygen has six valence electrons (two in the 2s and four in the 2p orbitals, see Figure 2.5). With the two electrons from the two hydrogen atoms, we again have eight valence electrons, or four electron pairs. Therefore, the Lewis structure of water can be written as H—$\bar{\text{O}}$—H. Notice that the Lewis structure, as written, does not imply any shape of the molecule: we will see later that the water molecule is nonlinear (bend), and that this bend structure actually is due to the four electron pairs around the oxygen atom, namely the two bonded and the two nonbonded pairs. However, when writing Lewis structures, this consideration is not yet relevant, but you must make sure that nonbonded electron pairs are included. Furthermore, we need to ascertain that elements in the first row of the periodic chart never can have more than two electrons, and elements in the second row can never have more than eight electrons. Please ascertain that the Lewis structures presented so far, H_2, F_2, CH_4, C_2H_6, HF, and H_2O, follow these rules.

There are three more aspects to polar covalent bonding we need to introduce before we can indulge in Lewis structure exercises. These are the concepts of electronegativity, multiple bonds, and resonance structures. Electronegativity is a numerical value assigned to each element such that the polarity of a covalent bond can be predicted. This idea goes back to Linus Pauling, one of the most eminent chemists of the first half of the twentieth century, who brought quantum mechanics into chemistry (prior to his efforts, quantum mechanics was mostly confined to physicists, and its applicability to chemistry had not yet occurred). Pauling devised the electronegativity scale based on the nuclear charge and the size of the valence electron shell. Fluorine, the most electronegative element, was assigned a value of 4, oxygen of 3.5, and most metals between 1.5 and 2. The alkali metals, which readily give up electrons, have electronegativity values between 0 and 1. Hydrogen and carbon have electronegativities of 2.1 and 2.5, respectively. The difference of the electronegativity values of two elements undergoing covalent bonding expresses the polar character of the bond. The C—H bonds discussed before in the example of methane and ethane are slightly polar with an electronegativity difference of 0.4, where the carbon atom is the more negative part. In the hydrogen–fluoride bond, this difference is 1.9, indicating a highly polar bond.

Whereas the previous topic, electronegativity, is an artificial (nonmeasurable) property, the next one, multiple bond formation, is a very real one. Experimental measurements show that a carbon–oxygen bond can have different values of bond length and bond strength, depending on the molecules in which they are found. The Lewis structure for methanol, CH_3OH, can readily be drawn from our knowledge acquired so far:

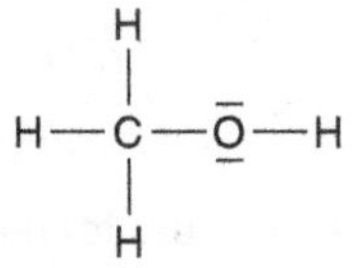

$$H-\underset{\underset{H}{|}}{\overset{\overset{H}{|}}{C}}-\bar{O}-H \tag{3.7}$$

As an organic alcohol, it combines elements of methane – the CH_3 or methyl group – with the OH group found in water, including the two free electron pairs. The C—O bond in methanol is typical for a carbon oxygen "single bond" and has a bond length of 143 [pm]. The term "single bond" implies a covalent bond formed by one electron pair.

Now let us look at another molecule containing the elements C, O, and H only, formaldehyde, H_2CO. Both hydrogen and the oxygen atom are bonded to the carbon atom. There is a total of 12 valence electrons, or six pairs. The two carbon–hydrogen bonds use up pairs, which leaves four pairs for the carbon and oxygen, and we arrive at something like

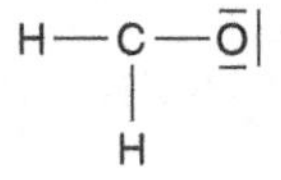

$$H-\underset{\underset{H}{|}}{C}-\bar{O}| \tag{3.8}$$

The problem with this structure is that the central C atom is surrounded by only three pairs, or six electrons, rather than the preferred eight. This problem is solved by sharing a second electron pair between carbon and oxygen according to

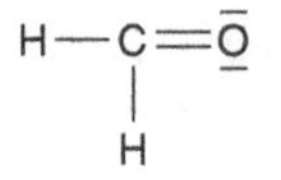

$$H-\underset{\underset{H}{|}}{C}=\bar{\bar{O}} \tag{3.9}$$

This drawing represents a valid Lewis structure for formaldehyde with a "double bond" (i.e. two shared electron pairs) between carbon and oxygen. This carbon–oxygen double bond, indeed, is much shorter (124 [pm]) and much stronger (stiffer) than a carbon–oxygen single bond. However, this drawing does not represent the true shape of the molecule that is revealed by spectroscopic methods, and a better approximation to the true shape of H_2CO is shown in Figure 3.2. We shall discuss how to model molecular shapes in Chapter 6 and the methods to deduce them experimentally in Chapter 15.

Molecules that incorporate the C=O double-bonded group are known as ketones. Acetone, used in finger nail polish remover, is a ketone in which the two hydrogen atoms in Figure 3.2 are replaced by two methyl groups. It is a molecule that will pop up here and there in the remainder of this text. Its (nontrivial) name would be dimethyl ketone or 2-propanone.

As mentioned before, the existence of the carbon–oxygen double bond is not a crazy way for chemists to justify an octet structure around both carbon and oxygen atoms, but it is a way to explain the experimental observation that the carbon–oxygen bond length is much smaller in H_2CO (124 [pm]) than it is in CH_3OH, and that the bond stiffness, as measured by the bond's resonance frequency (if one approximates the chemical bond by a spring obeying Hook's law), is nearly twice in H_2CO than it is in CH_3OH. These aspects will be discussed in more detail in Chapter 15. The statement in the last

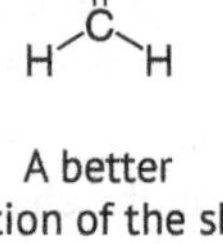

Figure 3.2 A better representation of the shape of the formaldehyde molecule.

sentence again emphasizes the interplay between theory and experiment: it is the experimental result that guides the development of a model!

The ability to form double bonds is restricted to the elements C, N, and O in the second row of the periodic chart, and to P and S in the third row. Double bonds are formed by and between atoms of these elements to achieve an octet configuration. Carbon, in fact, can form two double bonds at the same atom: in carbon dioxide, CO_2, the eight electron pairs are arranged as follows: $\overline{O}=C=\overline{O}$, giving each atom an octet structure. The three elements C, N, and O also are able to form triple bonds in a few selected molecules or molecular groups. Let us look at the nitrogen molecule, N_2, for an example. Nitrogen has five valence electrons, so the molecule N_2 has 10 electrons. The only way to distribute the electrons such that each nitrogen atom has an octet structure is

$$|N\equiv N| \tag{3.10}$$

Similarly, carbon monoxide, CO, has a triple-bonded structure:

$$|C\equiv O| \tag{3.11}$$

Its structure is "isoelectronic" to that of N_2, which implies it has the same arrangement of electrons. In the case of CO, four valence electrons are from the carbon atom and six from the oxygen, but the resulting five pairs are arranged in a triple bond and two nonbonding pairs just line in N_2. By the way, a correct Lewis structure of the diatomic molecule O_2 cannot be drawn, since the result, $\overline{O}=\overline{O}$, does not agree with experiment. More on this will be explored when we discuss molecular orbital theory (Chapter 6). This is not to be seen as a general failure of the Lewis structure approach, but rather as exceptional cases where a very simple model fails to predict experimental results.

A Lewis structure closely related to the one shown for methanol (Eq. 3.7) can be written for a compound known as methyl fluoride or fluoromethane:

$$\begin{array}{c} H \\ | \\ H-C-\overline{F}| \\ | \\ H \end{array} \tag{3.12}$$

This molecule contains a carbon–fluorine bond, which is one of the strongest chemical bonds that is only found in synthetic compounds. Since it does not occur naturally in chemicals, there are no bacteria in the environment that can break this bond, and molecules incorporating a C—F group are found in many places and are accumulating in the water supply since such compounds have been used in consumer products since the 1960s. These "forever chemicals" present an enormous environmental problem that has not been addressed.

When drawing the structure of molecular ions, the ionic charge represents the number of electrons that need to be added or subtracted from the number of valence electrons. Take the cyanide ion CN^- for example. The carbon atom provides four valence electrons, the nitrogen five, and the negative charge of the ion indicates that it has accepted an electron from a cation (Na^+, for example). Thus, the cyanide ion has five electron pairs (just as CO and N_2) and its Lewis structure indicates that it is isoelectronic with these two molecules $[|C\equiv N|]^-$. Notice the ionic charge is written here outside the brackets to indicate that its location is not specific to the nitrogen end of the ion. The compound, NaCN or sodium cyanide, is a (highly toxic) ionic compound that consists of Na^+ ions and CN^- cyanide molecular ions packed into an ionic lattice similar to the one shown in Figure 3.1.

The final of the three aspects mentioned above is the occurrence of resonance structures. Again, this is not a crazy scheme introduced by mad scientists, but an attempt to explain an experimental observation. We shall discuss resonance structure using ozone, O_3, as an example. Ozone is an allotrope of oxygen, with a different molecular structure. Many other elements exist in allotropic forms: white, red, and black phosphorous are allotropes of elemental phosphorous with different molecular structures, and different chemical properties, as are diamond and graphite. Ozone is highly important to life on earth, since the ozone layer in the high atmosphere absorbs ultraviolet light that would be enormously damaging to life.

The three oxygen atoms in O_3 contribute 18 valence electrons, or 9 pairs. One way to arrange them is shown here: $\overline{O}=\overline{O}-\overline{O}|$, which is a valid Lewis structure with all three oxygen atoms having an octet configuration. However, one could have just as easily written the structure as $|\overline{O}-\overline{O}=\overline{O}$. Both these structures imply that the molecule has one double and one single bond. However, experiment tells us that the two bonds have equal length and strength: an oxygen–oxygen single bond typically has a length of 148 [pm] and a double bond of 121 [pm]. The observed bond length in ozone is 128 [pm]. (By the way, we shall discuss how bond lengths can be determined experimentally in Chapter 15.) The truth here is that neither structure represents the real O_3 molecule, and the truth lies, as it does so often, in the middle. In chemistry, we write this situation as two "resonance structures" as follows:

$\overline{O}=\overline{O}-\overline{O}| \leftrightarrow |\overline{O}-\overline{O}=\overline{O}$, where the double-sided arrow indicates that the two structures shown are extreme forms of electron distribution and that the real situation is somewhere in between. This is often described as "delocalized electron pairs," indicated by the dashed line in Figure 3.3. These electrons are literally smeared over the entire three-atom framework, and the bond length observed corresponds to a bond strength between one and two.

To draw a resonance structure like the one shown in Figure 3.3, we take the electron pairs that are common to both structures (seven pairs) and add the remaining electrons to the pool of delocalized electrons. More on this will be explored when we consider valence theory and valence shell electron pair repulsion (VSEPR) in Chapter 6.

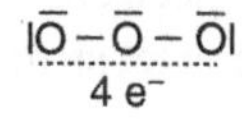

The discussion of resonance structures leads to the Lewis structures of polyatomic molecular ions that are very common in chemistry. As pointed out, anions such as carbonates, sulfates, silicate, etc. are abundant in the earth's crust. Many of them exhibit resonance structures such as the one worked out later for the carbonate anion, CO_3^{2-}.

First, we determine the number of valence electrons. Carbon contributes four, while each of the three oxygen atoms contributes six, for a total of 22 electrons. Add two electrons for the negative charges of the anion, to give a total of 24 electrons or 12 pairs. Most often, the center atom in a Lewis structure problem will be indicated; here, the carbon atom is the central ion (if it is not indicated, there is a rule involving "formal charges" that will be touched upon later). Thus, let us arrange the three oxygens around the carbon, and fill in 12 electron pairs. This will result in a structure shown in Figure 3.4a. However, in this structure, the central carbon does not have an octet structure, so any one of the free electron pairs on either oxygen atom can form a double bond, resulting in the three resonance structures shown in Figures 3.4b–d. Since it was found experimentally that all three carbon–oxygen bonds are equal, the true structure can be indicated by the drawing in Figure 3.5, where six electrons are delocalized. However, depending on your instructor in the course you are taking, it may be better to leave your answer in a test in the form of Figure 3.4b–d. The carbonate ion is isoelectronic with the nitrate ion, NO_3^-, since the nitrogen atom has one more valence electron than the carbon atom; thus, only one negative charge of the ion will suffice to result in the same total number of electrons for the ion. The resonance structures of the nitrate ion, consequently, will be the same as those of the carbonate ion.

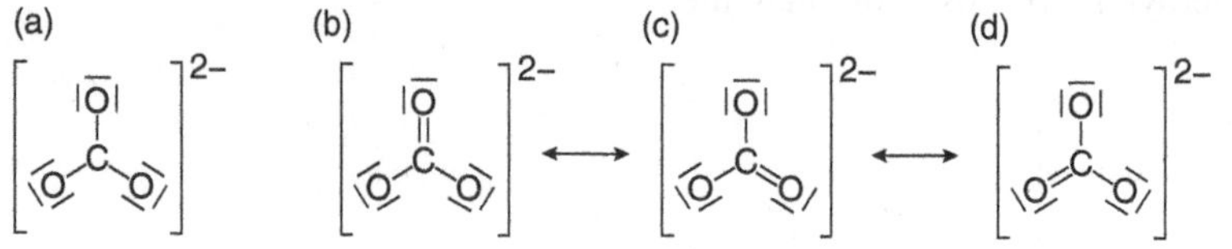

Figure 3.3 Resonance structure of the O_3 molecule showing that two electron pairs are delocalized.

Figure 3.4 Steps in constructing the Lewis resonance structures for the carbonate anion.

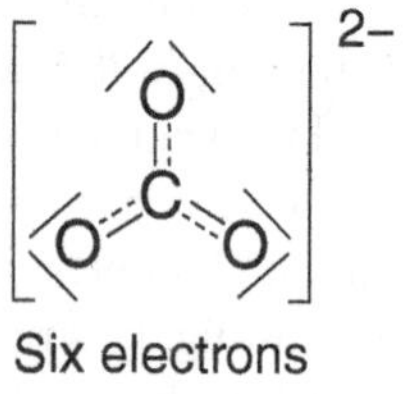

Figure 3.5 Lewis structure of the carbonate anion, indicating delocalized electrons.

There are a few molecules of second-row elements where an octet structure cannot be achieved because there are just not enough electrons around. An example for such a situation is borane, BH_3. Boron, element 5, contributes three valence electrons and the three hydrogen atoms, one each, for a total of six electrons or three pairs. Thus, the Lewis structure of BH_3 is

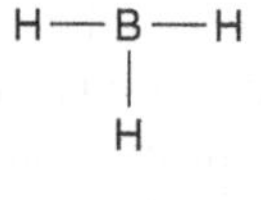

$$H-B-H$$
$$|$$
$$H$$

(3.13)

with the B atom, one electron pair short of an octet. Interestingly, this makes borane a "Lewis acid" (more on this in Chapter 10), a compound that is likely to accept an electron pair from a molecule that has an extra pair, such as ammonia, NH_3. Indeed, ammonia and borane form a compound for which a reaction scheme and Lewis structures can be written as

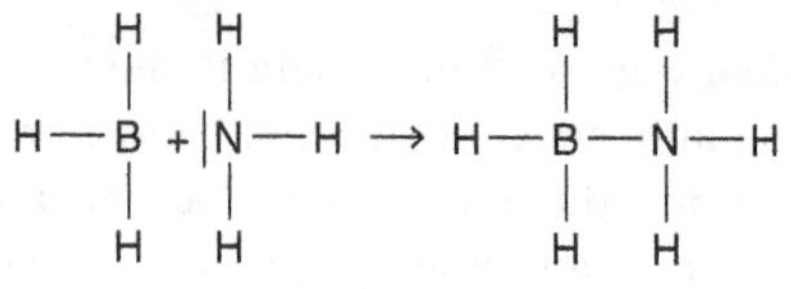

$$H_3B + |NH_3 \rightarrow H_3B-NH_3$$

(3.14)

Here, the ammonia molecule acts like a "Lewis base," a donor of an electron pair. The structure of the molecule written on the right-hand side of Eq. 3.13 is isoelectronic with a molecule introduced earlier in this chapter. Which molecule was that?

Another molecule that cannot achieve an octet configuration is boron trifluoride, BF_3, with a Lewis structure of

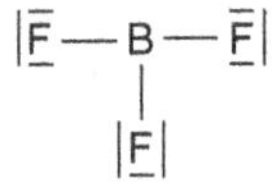

$$|\overline{\underline{F}}-B-\overline{\underline{F}}|$$
$$|\underline{\overline{F}}|$$

(3.15)

Do not attempt to draw a Lewis structure (and resonance structures) with a boron–fluorine double bond, since fluorine is not allowed to have multiple bonds.

So far, we have dealt with molecules or molecular ions that have an even number of valence electrons, and therefore, an integer number of electron pairs. What about the cases where this is not so, and we have an odd number of electrons? Let us take one of the oxides of nitrogen for an example, namely nitrogen dioxide, NO_2, a brown, toxic gas that contributes to smog in densely populated areas. Nitrogen contributes 5, and the two oxygen atoms together contribute 12 electrons, for a total of 17 valence electrons. We can write a Lewis structure for this using the rules we have discussed so far (and which are summarized later):

$$\overline{\underline{O}}=\dot{N}-\overline{\underline{O}}| \leftrightarrow |\overline{\underline{O}}-\dot{N}=\overline{\underline{O}}$$

(3.16)

where the dot over the nitrogen atom designates the single, unpaired electron. There a number of aspects that need to be discussed about the NO_2 molecule. First, a molecule with an unpaired electron is generally referred to as a free radical, since it is a reactive species. Second, we have assumed that nitrogen is the central atom when we drew the Lewis structure, and that the structure is not based on a O—O—N skeleton. More on this is explored later. Third, when drawing resonance structures of free radicals, the lone electron is generally placed on the less electronegative atoms, whereas the more electronegative atoms (here the oxygens) have the full octet configuration. Finally, the Lewis structure for NO_2 does not represent the true shape of the molecule, as we have indicated before. While the structure shown above suggests a linear molecule, this is not the case. Its true shape is a bend molecule, as shown in Figure 3.6a.

As indicated above, free radicals are normally quite reactive and try to acquire another electron to achieve paired electron configurations. What can a NO_2 molecule do to achieve a full octet around the nitrogen? Well, it could turn into an anion, $NO_2 + e^- \rightarrow NO_2^-$, the nitrite (an)ion you have probably encountered while learning the names of ions. The extra electron would be provided by a cation such as Na in an ionic compound sodium nitrite, $NaNO_2$. The Lewis structure of the nitrite ion then is

$$[\overline{\underline{O}}=N-\overline{\underline{O}}|]^- \leftrightarrow [|\overline{\underline{O}}-N=\overline{\underline{O}}]^-$$

(3.17)

Notice that we have indicated the negative charge of the ion outside the square brackets. Furthermore, the nitrite anion is isoelectronic with the ozone molecule discussed earlier.

Another way for the NO_2 molecule to reach a filled octet structure is by a process called "dimerization" or the formation of a dimer:

$2NO_2 \rightarrow N_2O_4$ (dinitrogen tetroxide), the structure of which is shown in Figure 3.6b.

Radical structures are quite common when we deal with molecules that contain nitrogen or chlorine atoms. For example, NO or nitrogen monoxide has a Lewis structure of $\overline{\underline{O}}=\dot{N}|$, where we, again, have placed the lone electron on the less electronegative atom.

Although we have not explicitly discussed the rules on how to draw Lewis structures, we will summarize next what we have done in the examples presented so far:

- H can never have more than one bond (or share one electron pair). So, H will not show up as a bridge between other atoms (such as —H—). Fluorine also forms single bonds only.
- Second row elements can accommodate up to eight electrons (the octet configuration given by two electrons in the 2s and six electrons in the 2p orbitals).
- Third row elements such as S and P can accommodate more than eight electrons, but prefer eight electrons. They can also form double bonds.

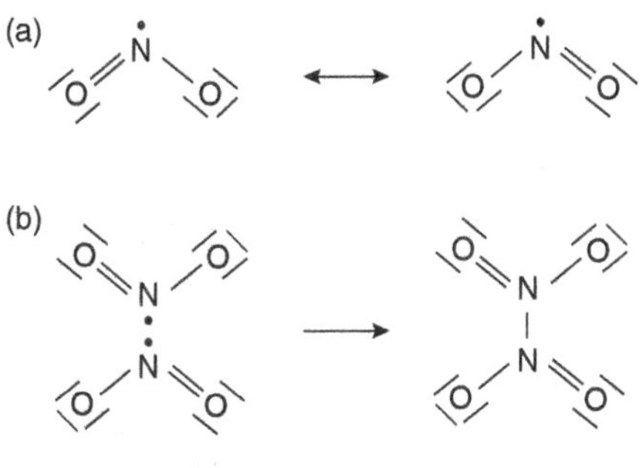

Figure 3.6 (a) Resonance Lewis structures for the free radical NO_2. (b) Dimerization of NO_2 to form N_2O_4.

- When drawing a Lewis structure, first determine the number of valence electrons from each atom's position in the periodic chart. Do not forget to add electrons for anionic species and subtract electrons for cationic species.
- Determine the center atom to which the other atoms are attached. The center atom may be given to you in the problem. Otherwise, you may have to resort to something called "formal charges." Since formal charges are an entirely artificial concept and have no theoretical foundations, they are not covered in this book. If your instructor insists on using them, please refer to your main text for details.
- Draw a skeletal structure, such as the one shown in Figure 3.4a that includes all electron pairs (or single electrons in the case of free radicals). Include nonbonding electron pairs. Remember that there cannot be more than one pair between H and any other atom. Draw multiple bonds when needed to complete octet structures, as shown earlier for the carbonate ion.

As stated above, S and P can have more than eight electrons, since they are third row elements that have orbitals with $n = 3$ and $\ell = 2$ (the 3d orbitals, see Figure 2.3 and Table 2.3) that are energetically close to the 3p orbitals with $n = 3$ and $\ell = 1$. Thus, sulfur can form a compound with fluorine with the molecular formula SF_6 (sulfur hexafluoride) that has all six fluorine atoms attached to the central sulfur atom *via* single bonds, and each fluoride has three nonbonded electron pairs.

The Lewis structure of the sulfate ion, SO_4^{2-}, could simply be drawn with four single bonds, as shown next:

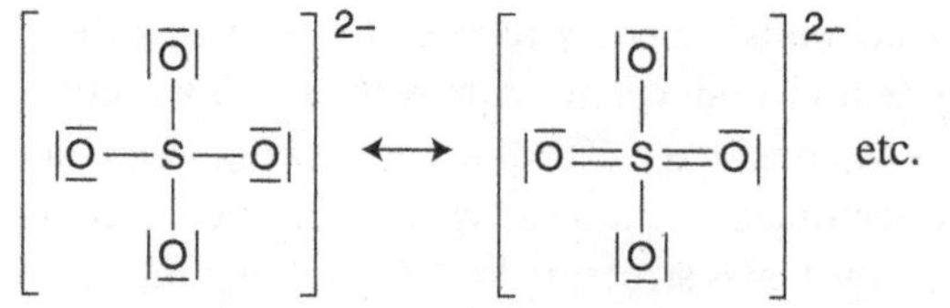

However, experiment tells us that the four sulfur–oxygen bonds are shorter and stronger than single bonds. Thus, we can draw resonance structures that include two double bonds and two single bonds between S and O atoms; in these structures, S is surrounded by 12 electrons.

One element, in particular, is known to readily form multiple (double and triple) bonds. This element is carbon, and an entire branch of chemistry – organic chemistry – is devoted to carbon-based chemistry. Thousands of organic compounds exist, many of which contain carbon-to-carbon double and triple bonds. The bonding in such molecules will be introduced in more detail in Chapter 6. Here, the Lewis structures of two carbon-based molecules with multiple bonds will be discussed. In Eq. 3.6, the Lewis structure of ethane, C_2H_6, was introduced. A molecule known as ethene (or ethylene) with the formula C_2H_4 has a Lewis structure that incorporates a carbon–carbon double bond, as shown in Figure 3.7a. Another molecule, ethine (or acetylene), with a formula C_2H_2 has a Lewis structure with a carbon–carbon triple bond, as shown in Figure 3.7b.

Although Lewis structures present a very simplistic view of chemical bonding, they are extremely useful in Chemistry. If you look at the structures of any molecules drawn in organic chemistry or biochemistry chapters, they most likely will be represented by Lewis structures like the one shown in Eqs. 3.5–3.8. In organic chemistry, in particular, chemical reactivity often can be predicted from Lewis structures, including the free electron pairs. This was shown before in Eq. 3.12 for an electron-rich compound, such as ammonia, reacting with an electron-deficient compound. In organic chemistry, similar interactions between "nucleophilic" and "electrophilic" compounds (i.e. electron-donating and electron-accepting groups, respectively) explain many reaction mechanisms.

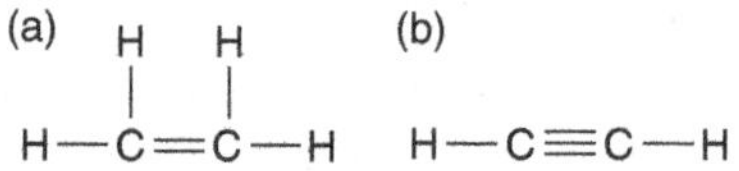

Figure 3.7 Lewis structures of (a) ethene (also known as ethylene) and (b) ethine (also known as acetylene). Notice that Panel (a) is a valid Lewis structure but does not adequately describe the molecular shape (see Chapter 6).

The naming of ionic compounds was discussed earlier in Section 2.2, and is straightforward since the ions have a well-defined charge, and the resulting compound must be neutral. Naming of small covalent compounds follows somewhat different rules, since the same elements can form several compounds. For example, sulfur forms two common oxides, SO_2 and SO_3 that are named sulfur dioxide and sulfur trioxide, respectively. In general, molecular compounds are named by using mono, di-, tri-, tetra-, penta-, and hexa- prefixes to indicate the number of atoms in the molecule. Thus, N_2O_5 is properly referred to as dinitrogen pentoxide. For oxides, there is also an alternative nomenclature, based on the oxidation state[2] of the element that binds to oxygen. This nomenclature uses the ending "ic" and "ous," such as in nitric acid (HNO_3) and nitrous acid (HNO_2). It gets worse in the case of the oxides of nitrogen, where dinitrogen monoxide (N_2O) is commonly called nitrous oxide or laughing gas, mono nitrogen monoxide or nitrogen monoxide (NO) is called nitric oxide, and NO_2 is properly called nitrogen dioxide. These inconsistences in nomenclature have historic origins but can make life miserable for students in early chemistry courses. Unfortunately, the common names for organic chemicals are even more confusing, even for professional chemists who are not organic chemists. Although memorization of facts is generally looked down upon by the author of this book, when it comes to nomenclature, there is no alternative to sitting down and memorizing names, alternate names, and formulae.

Example 3.2 Write Lewis structures, including resonance structures, when possible, for NO_2, NO, ClO_3, ClO_3^-, PCl_3, PCl_5, SF_4, SF_6, PO_4^{3-} (the first element listed is the center atom), and dinitrogen pentoxide N_2O_5 (with a O—N—O—N—O skeleton).

Answer: See answer booklet

3.4 Molecular Compounds and the (Gram) Molecular Mass

Just as we referred to an ensemble of like atoms as "element," we call an ensemble of like molecules a "compound." (Notice that the term "compound" was used in the context of "ionic compounds" such as NaCl earlier, although there are no distinct NaCl molecules in the crystal lattice of an ionic compound.) Thus, all the molecules mentioned so far in this chapter (methane, ethane, methanol, formaldehyde, carbon monoxide, etc.) also are referred to as "compounds" when we deal with an ensemble of like molecules.

We discussed before that each element is associated with an atomic mass number "$\mathcal{A}$," which is the mass that contains 1 mole of the atoms. In complete analogy, we define the molecular mass (formerly referred to as gram molecular mass, or GMM) as "$\mathcal{M}$," which is the sum of all atomic masses of the atoms in a compound.

Example 3.3 Determine the molecular mass "$\mathcal{M}$" for water, methanol, carbon dioxide, the cyanide ion, calcium phosphate, phosphorus pentachloride, glucose ($C_6H_{12}O_6$), and the sulfate ion.

Answers: Water: 18, methanol: 32, carbon dioxide: 44, cyanide: 26, calcium phosphate: 310, phosphorous pentachloride: 208.5, glucose: 180, sulfate: 96

The direct consequence of the definition of the (gram) molecular mass is that 44 g of carbon dioxide contains 1 mole of carbon dioxide molecules, or $6.022 \cdot 10^{23}$ carbon dioxide molecules, or $6.022 \cdot 10^{23}$ carbon atoms and $1.204 \cdot 10^{24}$ (or 2 moles) of oxygen atoms. Furthermore, the equation relating mass, number of moles, and molecular mass (Eq. 2.6) holds equally for molecules and compounds by substituting "$\mathcal{M}$" for "$\mathcal{A}$"

$$n = \frac{m}{\mathcal{M}} \left[\frac{g}{\frac{g}{mol}} = \frac{g}{g\ mol^{-1}} = mol \right]. \tag{3.18}$$

Example 3.4 (a) How many moles of iron (III) oxide are there in a sample of 1.20 g of the compound? (b) How many moles of iron atoms are there in this sample? (c) How many oxygen atoms are there in this sample?

Answer: (a) 7.51 [mmol], (b) 15.0 [mmol], and (c) $1.36 \cdot 10^{22}$ [atoms]

Example 3.5 The detection limit in some sensitive molecular assays is 1.0 [pmol]. How many molecules are being detected?

Answer: $6.0 \cdot 10^{11}$ [molecules]

---

2 Oxidation states will be discussed in detail in Chapter 12. For now, let us define the oxidation state as an indicator of the number of oxygen atoms in a compound.

3.5 Percent Composition and Empirical Formulae

When we calculated the molecular mass of glucose as 180 [g/mol] in Example 3.2, we concluded that 1 mole of glucose has a mass of 180 g, which is to say that 180 g of glucose constitute one mole of glucose. One mole of glucose, according to its molecular formula, contains 6 moles of carbon or 72 g of carbon, since the atomic mass of carbon is 12. Thus, there are 72 g of carbon in one mole of glucose. The percentage of carbon in one mole of glucose is then

$$(72/180) \cdot 100 = 40\%$$

By the same token, the percentage of oxygen is $(6 \cdot 16)/180$, or 53.33%, and the percentage of hydrogen is 12/180, or 6.67%. Adding up the percentages, we find that $40 + 53.33 + 6.67 = 100$; that is, we have accounted for all the percentages. In order to compute the "percent composition," we divided the masses of each element (that is, the stoichiometric coefficients multiplied by each element's atomic mass) by the molecular mass and multiplied by 100. So, let us do another example:

Example 3.6 Calculate the percent composition of all elements in (a) formaldehyde, CH_2O; (b) ammonium phosphate; and (c) sulfuric acid, H_2SO_4.

Answer: (a) 40%C, 6.67%H, 53.33%O; (b) 28.19%N, 8.05%H, 20.80%P, 42.95%O; (c) 2.04%H, 32.65%S, 65.31%O

The result of Example 3.6(a) demonstrated that glucose and formaldehyde have the same percent composition. A little thought shows why: in both, the ratio of carbon to hydrogen to oxygen is 1:2:1; so, they must have the same percent composition!

One of the earliest methods in chemistry to determine the molecular formulae of compounds was based on the experimental determination of the percent composition and proceed from there to calculate the molecular formula. Consider the case of a compound that consists of sodium, sulfur, and oxygen only (the word "only" in the last phrase indicates that the percentages of Na, S, and O must add up to 100). It was found that this compound consists of 32.4%Na, 22.5%S, and 45.1%O. (The percentages, indeed add up to 100.) Can we determine the molecular formula from this information? Well, 100 [g] of the sample would contain 32.4 [g] sodium, 22.5 [g] sulfur, and 45.1 [g] oxygen. Converting these masses to the number of moles gives us 1.4087 [mol Na] (worry about significant figures later!!), 0.7031 [mol S], and 2.8188 [mol O]. So, it appears that the molecular formula is $Na_{1.4087}S_{0.7031}O_{2.8188}$. This is, obviously, an undesirable molecular formula. However, if we divide by the smallest number of moles (0.7031 for S), we see that this molecular formula reduces to Na_2SO_4 where we rounded the coefficients to the nearest integer. Thus, we have found that the compound is sodium sulfate.

Now, consider the case of formaldehyde and glucose earlier. If we are given a percent composition of 40% C, 6.67% H, and 53.33% O for an unknown and apply the argument that 100 [g] of the compound contain 40 [g] = 40/12 = 3.33 [mol of C], 6.67 [g] = 6.67 [mol of H], and 53.33 [g] = 53.33/16 = 3.33 [mol of O], we arrive at a formula $C_{3.33}H_{6.67}O_{3.33}$. Dividing the coefficients by the smallest one gives CH_2O. This is of course the correct formula for formaldehyde, but it is the same for glucose or any other compound that has the formula $(CH_2O)_x$. For glucose, the factor x would be six. In other words, all compounds with the molecular formula $(CH_2O)_x$ have the same percent composition, and their exact chemical formula cannot be determined until their molecular mass is determined. We call the formula unit that can be determined from the percent composition the "empirical formula" unit. So, for glucose and formaldehyde, the empirical formula is CH_2O. If other experimental results (to be discussed later) reveal that the molecular mass of the unknown compound is 30 [g/mol], then the molecular formula would indeed be CH_2O. If the molecular mass is found to be 180, we divide the molecular mass by the molecular mass of the empirical formula unit (180/30 = 6) and find that the molecular formula is $(CH_2O)_6$ or $C_6H_{12}O_6$.

Example 3.7 (a) What is the empirical formula of a compound that consists of 30.43% nitrogen and 69.57% oxygen? (b) The molecular mass of this compound is 92 [g/mol]. What is its molecular formula?

Answer: (a) NO_2; (b) N_2O_4

Molecular masses classically can be determined by techniques such as boiling point elevation or freezing point depression of solutions, as compared to the pure solvent. This allows the concentration of the solution, and thereby, the molecular mass of the solvent to be determined. The boiling point elevation and freezing point depression both depend on the *molal* concentration of the solute in the solvent. This will be discussed in Chapters 4 and 7.

More modern techniques, now widely applied, include mass spectrometry, in which a vaporized molecule is ionized and passed through a magnetic field. The deflection the ions experience allows the determination of the mass of ion. More details about mass spectrometry will be discussed in Chapter 15.

Example 3.8 The percent composition of sulfur in sodium sulfate is approximately

(a) 31.1%
(b) 26.9%
(c) 22.5%
(d) None of the above

Answer: (c). Details in answer booklet.

Example 3.9 A compound consists of 5.88 mass percent of hydrogen and 94.12 mass percent of sulfur. Which of the following represents a possible empirical formula?

(a) HS
(b) HS_2
(c) HS_{16}
(d) None of the above

Answer: (d). Details in answer booklet.

Example 3.10 The number of iron atoms in 2.00 [g] $FeCl_3$ is approximately

(a) $1.2 \cdot 10^2$
(b) $7.4 \cdot 10^{21}$
(c) $9.4 \cdot 10^{21}$
(d) None of the above

Answer: (b). Details in answer booklet.

Further Reading

A summary of ionic bonding can be found at: https://en.wikipedia.org/wiki/Salt_(chemistry).
Lewis and resonance structures: https://en.wikipedia.org/wiki/Lewis_structure.
The concept of "the mole": https://en.wikipedia.org/wiki/Mole_(unit). This article should be used with caution, since it brings in much unnecessary information.

4

Chemical Reactions

4.1 Chemical Reaction and Stoichiometry

In the previous chapter, we have seen that molecules are formed when atoms undergo rearrangement of their valence electrons. In chemistry, we generally are interested in molecules with covalent bonds that are formed by what we call "chemical reactions." A typical chemical reaction might be the combustion of carbon with oxygen to form carbon dioxide, CO_2, according to the following equation:

$$C(s) + O_2(g) \rightarrow CO_2(g) \tag{4.1}$$

In Eq. 4.1, the symbols in parentheses indicate the physical state (solid, liquid, or gaseous) of the reactants and products as explained in Eq. 3.4. This chemical reaction has produced a molecule with covalent bonds, carbon dioxide, from the elements.

Eq. 4.1 implies that one atom of carbon will react with one molecule of oxygen to form one molecule of carbon dioxide. Furthermore, it implies that one dozen carbon atoms would react with one dozen oxygen molecules to form one dozen carbon dioxide molecules. It also implies that one mole of carbon atoms reacts with one mole of oxygen molecules to form one mole of carbon dioxide molecules, since the mole is nothing but a "chemists dozen" (albeit numerically quite different). If we write the last sentence in terms of masses involved, we obtain

$$12[g] \text{ of carbon react with } 32[g] \text{ of oxygen to yield } 44[g] \text{ of carbon dioxide} \tag{4.2a}$$

$$12[g] + 32[g] = 44[g] \tag{4.2b}$$

In this step, we have entered the atomic or molecular masses for each of the reactants and products; hereby, we have established the mass relationships between reactants and products. It is important to ascertain that the masses on the left side of Eq. 4.2a equal to the masses on the right-hand side, as shown by Eq. 42b. The reaction shown in Eq. 4.1, by the way, is the "combustion" of carbon to yield carbon dioxide. Combustions always are reaction with oxygen and are often referred to as "oxidation reactions." The reaction in Eq. 4.1 may also be called the formation reaction of carbon dioxide from the elements.

Thus, a balanced chemical equation such as Eq. 4.1 implies that

- one atom of carbon reacts with one molecule of oxygen to give one molecule of carbon dioxide;
- one mole of carbon atoms reacts with one mole of oxygen molecules to give one mole of carbon dioxide molecules;
- one mole of carbon atoms reacts with two moles of oxygen atoms to give one mole of carbon dioxide molecules; or
- 12 [g] of carbon atoms will react with 32 [g] of oxygen molecules to yield 44 [g] of carbon dioxide molecules.

Thus, we have found a way to take a balanced chemical (reaction) equation and translate it to the masses of reactants and products. This aspect of chemical reactions is called "stoichiometry" and will be discussed in more detail in the remainder of this chapter. The reason why this approach works is obvious if we think back at the definition of the mole (see Chapter 2) and extend it to include molecules: **one mole of an element or compound is the amount of the element or compound, in grams, that contains $6.022 \cdot 10^{23}$ individual atoms or molecules.** Therefore, if the amounts in Eq. 4.1 are expressed in grams, they all contain Avogadro's number of atoms or molecules.

Although this book, in general, does not delve into historical aspects of chemistry, a short paragraph here will summarize, in passing, the decades of work that went into establishing the aspects about moles, molar composition, and stoichiometry. It was found in the early nineteenth century that the proportion of elements in a well-defined chemical compound is always the same. This statement makes sense if you review in your mind the earlier discussions on percent

Understanding Essential Chemistry, First Edition. Max Diem.
© 2025 John Wiley & Sons, Inc. Published 2025 by John Wiley & Sons, Inc.

composition: when this percent composition is expressed in an empirical formula, we obtain constant ratios of the elements in a compound. The concepts involving the mole were established by determining the volumes of gases that undergo chemical reactions, and it was found that equal volumes of gases at the same conditions contain equal number of atoms (or molecules). This finding bears the name "Avogadro's law" and will be discussed in more detail in Chapter 8. Once the concepts of the mole and atomic and molecular masses were established, stoichiometry was a logical consequence.

Stoichiometric problems can be rather tricky and require careful evaluation of the coefficient in chemical equations. Let us consider the thermite reaction

$$Fe_2O_3(s) + 2\,Al(s) \rightarrow Al_2O_3(s) + 2\,Fe(l) \tag{4.3}$$

which describes a spectacular process that produces elemental iron when a mixture of iron (III) oxide and aluminum powder is ignited. The reaction produces molten iron that is so hot that the reaction is used to weld iron rails together and is therefore used to produce continuous railroad tracks that eliminate the clickety-clack of old railroad rides. If we translate Eq. 4.3 to mass units, we see that 160 [g] of Fe_2O_3 react with 54 [g] of Al to yield 102 [g] Al_2O_3 and 112 [g] of iron, since

$$Fe_2O_3(s) + 2\,Al(s) \rightarrow Al_2O_3(s) + 2\,Fe(l) \tag{4.3}$$
$$160\,[g] \quad 54\,[g] \rightarrow 102\,[g] \quad 112\,[g]$$
$$214\,[g] \rightarrow 214\,[g] \tag{4.4}$$

(To simplify matters, all atomic masses have been rounded to two significant figures.)

In Eq. 4.4, we ascertained that the mass of reactants equals the mass of products: we started with 214 [g] of reactants and obtained 214 [g] of products. This is an important concept in a chemical reaction: no mass is lost, and the overall mass balance must be zero. This step also affirmed that we calculated the molecular masses correctly.

A typical stoichiometry question may ask: How many grams of aluminum powder are required to produce 100.0 [g] of molten iron *via* the thermite reaction, assuming there is sufficient iron oxide? When confronted with such a question, one must decide whether the problem is going to be solved in terms of masses or moles.

Example 4.1 How many grams of aluminum powder are required to produce 100.0 [g] of molten iron *via* the thermite reaction, assuming there is a sufficient amount of iron oxide? Solve this problem (a) using a mole-based approach, and convert to mass units at the end, or (b) using mass-based approach.

Answer: (a) 1.786 [mol] of Al or 48.2 [g] of Al; (b) 48.2 [g] of Al (see answer booklet)

Example 4.2 How many grams of Fe_2O_3 are required to completely react with 25.0 [g] of Al *via* the thermite reaction. Solve this problem (a) using a mole-based approach, and convert to mass units at the end, or (b) using mass-based approach.

Answer: (a) 0.463 [mol] Fe2O3 or 74.1 [g] Fe2O3; (b) 74.1 [g] of Fe2O3 (see answer booklet)

There is near-infinite number of problems that can be constructed to test your knowledge of stoichiometry, and it is clear that you cannot memorize them all. Thus, it is important that you understand the methods we have elaborated to tackle such problems. The important aspect to remember is that the balanced chemical equation, i.e. an equation where all atoms of the reactants are accounted for in the products, holds for atoms and molecules, as shown in Eqs. 4.1 and 4.3. These equations also hold for moles of atoms and molecules, as pointed out by the bulleted items above. Thus, the balanced chemical equation can be translated to an equation that relates *masses* of reactants and products, as shown in Eqs. 4.3 and 4.4. Any problem can then be solved in terms of the masses, or moles, of reactants and products as shown by the separate "approaches" in Examples 4.1 and 4.2.

4.2 Limiting Reagents, Theoretical Yield, and Percent Yield

Things get a little more complicated when one of the reactants is in excess, and the other reagent limits the amount of product that can be obtained. Notice that the question in Example 4.1 was carefully worded to include the phrase "...assuming there is a sufficient amount of iron oxide?" This phrase implied that there was sufficient iron oxide, or even more iron oxide than required, in the reaction mixture. Thus, in this case, we assume that iron oxide was "in excess," and the amount of product was limited by the amount of Al, which would be the limiting reagent. Problems of this kind are often hard to grasp, so let us translate the concept back to our example of cartons of eggs in a grocery store.

Say you are the owner of a small cooperative farm and have enough hens running around in the yard to collect 480 eggs on a given day. These eggs need to be packed into 12-egg cartons to be sold in grocery stores. In terms of a "reaction equation," this could be written as

$$12 \text{ eggs} + 1 \text{ carton} \rightarrow 1 \text{ egg package} \tag{4.5}$$

On this given day, you only have 35 egg cartons left in your storage room, due to an unforeseen slowdown in egg carton delivery. How many egg cartons can you fill and deliver to the grocery store? There are basically two ways to attack this problem. In one approach, you base your calculations on the number of eggs and find out how many dozen eggs are there in 480 eggs. This gives

$$n = 480/12 = 40 \text{ dozen eggs}.$$

Since you have only 35 cartons, the number of filled cartons you can deliver is limited by the fact that you only have 35 cartons. Thus, the cartons are the "limiting reagent" and after filling these 35 cartons with $35 \cdot 12$ or 420 eggs, you are left with 5 dozen or 60 eggs that you cannot sell.

The other approach bases the calculations on the number of cartons you have. Since 35 cartons require $35 \cdot 12 = 420$ eggs, but you have 480 eggs, it is clear that the eggs are "in excess," and the cartons are the limiting reagent.

The next day, a shipment of 1000 empty egg cartons arrives, and the hens again produce 480 eggs. A little reflection shows that the 480 eggs, or 40 dozen eggs, now determine how many 12-egg cartons you can produce. In other words, now the eggs are the limiting reagent. In the next example, you should apply the same concept to a chemical reaction.

Example 4.3 What amount of phosphorus trichloride can be obtained from the (industrially important) reaction of 100 [g] of white phosphorus (P_4) with 200 [g] of chlorine gas?

Answer: Chlorine is the limiting agent. An amount of 258 [g] of PCl_3 can be obtained (see answer booklet).

To tackle a problem such as Example 4.3, start by writing a balanced chemical equation. The maximum amount of product that can be obtained in a chemical reaction, after considering the limiting reagent issue, is called the theoretical yield. It is important that the calculation of the theoretical yield is based on the limiting reagent, as shown above for the egg/carton case.

In a chemical preparation, the actual yield, that is, the actual amount of product you recover, is most likely less than the theoretical yield, since the transfer of product from the reaction flask into a container may waste a certain amount of product, or since some reactants may remain unreacted. We define the percent yield of a chemical reaction as the actual yield divided by the theoretical yield, multiplied by 100.

The conceptual approach delineated in Example 4.3 can easily be extended to reactions where more than two reactants produce more than one product. One more example to summarize the concepts of the last section is presented later (Example 4.9) after some introductory materials on solution stoichiometry.

4.3 Solutions: General Aspects

The reactions introduced in Eqs. 4.1, 4.3, and Example 4.3 are reactions that occur as written: solid carbon, such as a lump of coal, burns in gaseous oxygen to give gaseous carbon dioxide. Aluminum and iron (III) oxide react as solids to give products, and solid white phosphorus, P_4, reacts with chlorine gas to give liquid phosphorous trichloride. However, many reactions in chemistry and biochemistry occur in solution, either in water or in any other solvent. In fact, chemists prefer to carry out reactions in the solution phase, because reaction conditions can be adjusted, *via* varying the concentration of reactants, to control the reaction. The thermite reaction discussed earlier (Eq. 4.3) is a reaction that is not easily controlled since two solids react, and mixing the solid reactants to form a homogeneous mixture is required. This difficulty can be avoided when reactions are carried out in solutions.

In general, solutions are homogeneous mixtures of one or more compounds. Although we may think of solutions to be mostly in the liquid state, alloys can be considered solutions where one or more metals are dissolved in another metal.

Brass, for example, is a mixture of about 66% copper and 34% zinc and could be considered a (solid) solution of zinc in copper. Gases also dissolve in liquids to form solutions. We are all familiar with CO_2 being dissolved in water: when you open a can of seltzer water, some of the dissolved CO_2 leaves the solution to form small bubbles. Ocean water is a solution of solid sodium chloride (and a few other salts) in water.

The process of a solid or liquid dissolving in a liquid solvent is actually a quite complicated process. First, let us look at water as a solvent. Water is considered a polar (solvent) molecule, implying that the water molecule has an uneven distribution of electric charges: the oxygen atom, being very electronegative (see Chapter 3) has a higher electron density than the hydrogen atoms, and therefore represents the negative "end" of the water molecule. We shall discuss molecular polarity in much more detail in Chapter 6.

When an ionic compound dissolves in water, the compound completely dissociates into ions. It should be noted that ionic compounds may or may not dissolve in water, or dissolve only very sparingly. An example would be calcium carbonate, which is found in egg shells, lime stone, shellfish exoskeletons, and other everyday compounds. Clearly, it is not very soluble in water; otherwise, the Italian Dolomite mountains would have disappeared a long time ago. However, whatever amount of calcium carbonate dissolves in water will completely dissociate into calcium ions and carbonate ions. These ions are "solvated" by the water molecules, i.e. they are surrounded by the water molecules, thereby changing the structure of liquid water. This is shown schematically in Figure 4.1 for a solution of sodium chloride. We see that the positive sodium ions (small spheres marked with a "+" symbol) are surrounded by water molecules such that the negative ends, the oxygen atoms, point to the sodium ions, while the larger chloride ions (larger circles, marked with a "−" symbol) are surrounded by water molecules such that the more positive sides, the hydrogen atoms, point toward the chloride. Depending on the size of the ions and solvent molecules, the typical number of solvent molecules around an ion is about six.

Whether or not a salt, or any other compound, will dissolve in a given solvent depends on a number of factors: the attractive force between the ions in the lattice (lattice energy), the attractive forces between the solvent molecules and the ions, and the energy it takes to reorganize solvent structure. In general, an empirical rule holds that polar solvents dissolve polar solutes, and nonpolar solvents dissolve nonpolar solutes. Thus, water is an excellent solvent for ions, but it does not readily dissolve nonpolar compounds such as oil. You may have observed that in vinaigrette salad dressing, there are distinct aqueous and oily layers that do not mix well.

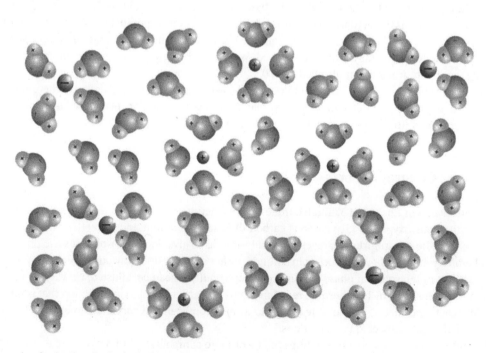

Figure 4.1 Example of solvation (hydration) of ions, for example, sodium (small positively charged spheres) and chloride ions (larger negatively charged spheres) in an aqueous solution of NaCl.

4.4 Solution Stoichiometry: Molarity, Molality, Mole Fraction, Dilutions

When chemistry is being carried out in solutions, there are several ways to specify the proportions of solvent and solute: the mass percentage of solute in a solvent, the molarity, the molality, and the mole fraction, to be discussed in detail later. The reasons we utilize different solution concentration measures is due to a rather unfriendly property of solutions: when solutions are formed, the volumes of solvent plus solute may not be additive, although the masses are. A typical example of this is the addition of methyl alcohol (a polar molecule, see Section 3.3) to water: the obtained volume after mixing equal volumes of the two components is about 4% smaller than the sum of the volumes we started with, although the masses of the two components are strictly additive. The reason for this "volume contraction" is that the interaction between the (polar) solvent and solute molecules perturbs the solution structures of each of the two liquids. Even if a solid is dissolved in a solute, there is no way to predict the volume of the solution. More on this will be explored later (Chapter 7) when intermolecular forces will be discussed. These forces exist between the solvent molecules themselves, the solute molecules, as well as between the solvent and solute molecules or ions.

The different descriptions of concentrations of solutions are based either on masses (which are additive) or volumes. The easiest description is just the mass percent composition of a solution: sea water has a concentration (of sodium chloride) of three mass percent, which implies that 100 [g] of sea water ($m_{solution}$) contains 3 [g] of solute, NaCl (m_{solute}). Here, we ignore other minor constituents. Thus, we write the definition of mass percentage C_{mp} of a solution as

$$C_{mp} = \left(\frac{m_{solute}}{m_{solution}}\right) \cdot 100 \tag{4.6}$$

or, in other words, the concentration is expressed in units of "parts per hundred" (= percentage).

Very low concentrations often are expressed in units of parts per million (ppm) or parts per billion (ppb). For example, the maximal concentration of the poisonous metal arsenic (As) in drinking water was set by the Environmental Protection Agency as 0.010 [mg/L] of water. At this low concentration, we may assume that 1 [L] of the As/water solution has a density of 1.00, and that the mass of the solution approximately equals the mass of solvent (in other words, the mass of As in the solution is so small that 1.00 [L] of solution contains 1.00 [kg] of water). Thus, this concentration could be written as

$$C = \left[\frac{1.0 \cdot 10^{-5}[g](As)}{10^3[g](solvent)}\right] \cdot 100 = 1.0 \cdot 10^{-8} = 10\,[ppb] = 1.0 \cdot 10^{-2}\,[ppm] \tag{4.7}$$

In analogy to mass percentage, the volume percentage C_{vp} is defined in terms of the volume of solute (v_{solute}) and volume of solution ($v_{solution}$) as

$$C_{vp} = \left(\frac{V_{solute}}{V_{solution}}\right) \cdot 100 \tag{4.8}$$

This is a seldomly used concentration unit in chemistry, because of the nonadditivity of volumes. However, the "proof" of liquors is just twice the volume percentage of alcohol in water.

There are two more mass-based concentration units that are used frequently in chemistry: the molality C_m and the mole fraction. The molality C_m is defined as the number of moles of solute divided by the mass of solvent in [kg]

$$C_m = \frac{n_{solute}}{m_{solvent}}\left[\frac{moles}{kg}\right] \tag{4.9}$$

Notice that this definition is based on the mass of pure solvent, not the solution. The molal concentration actually is a very useful unit of concentration, since it does not involve the contraction of volumes discussed above. There are several physical effects (for example, the freezing point depression or boiling point elevation mentioned in Chapter 3) that are proportional to the molal concentration of the solute. More on this will be explored in Chapter 9. The molal concentration is directly proportional to the concentration in mass percent units and can be interconverted easily, since both units depend on masses.

The mole fraction of solution is based on the number of moles of solvent and solvent according to

$$\chi_{solute} = \frac{n_{solute}}{n_{solvent} + n_{solute}} \tag{4.10}$$

and can be related directly to the molality. We shall encounter this concentration unit again in Chapter 7.

However, the concentration unit used most frequently in chemistry is the molar concentration C_M, defined as

$$C_M = \frac{n_{solute}}{volume_{solution}}\left[\frac{moles}{L_{solution}}\right] \tag{4.11}$$

Figure 4.2 Image of a volumetric flask with a volume fill mark.

This unit is useful since a certain volume of a solution can be prepared very easily by depositing the solute in a "volumetric flask" (see Figure 4.2) and adding sufficient solvent to fill the flask to the indicated fill mark, as shown for a typical 50 [mL] flask in Figure 4.2. At this point, we know that the volume of the solvent may have contracted as pointed out above in Section 4.3, but we do not care: by adding sufficient solvent such that a final volume of solution is established, we can eliminate the problem of volume changes in solute and solution.

Example 4.4 Assume that sea water is a solution that is 3.00% in NaCl (in reality, the concentration is approximately 3 %). Express this concentration in units of (a) molality and (b) molarity. Let the density d of sea water be $d = 1.03$ [kg/L].

Answer: (a) $C_m = 0.513$ [molal]; (b) $C_M = 0.528$ [M]

The advantage of using molar concentrations becomes obvious when we consider the process of dilution of a solution. In analytical chemistry, very dilute standard solutions are often required to calibrate instrumental methods of measurement. Consider the case of As in water that was discussed before. To make 1 L of a standard solution that has a concentration of 0.010 [mg/L] of As, one would have to weigh out 0.010 [mg] of As (or the equivalent amount of an As-containing compound), transfer it completely into a 1 L flask, and add the indicated amount of water. However, 0.010 [mg] of a compound is very hard to weigh out exactly, since it is a minute amount, and may be just a few crystals. A typical analytical balance in a chemistry lab may be able to offer an accuracy of 0.1 [mg]; thus, 0.010 [mg] is beyond the accuracy achievable using such an instrument. Thus, it is advantageous to prepare a much more concentrated solution, since it is much easier to weigh out amounts in the milligram or gram regime, and dilute the resulting "stock solution" by taking aliquots of the stock solutions and prepare the required standard solutions by dilution. It is much easier to measure out an exact volume aliquot, using standard, calibrated pipettes (see Figure 4.3), than weighing out very small amounts.

Figure 4.3 Image of a volumetric pipette.

In a dilution process, we start with an accurately known volume of a solution with accurately known concentration. From these known volume and concentration conditions, we can determine how many moles "n" of solute are contained in this aliquot, as we know by rewriting Eq. 4.10 that

$$n = C_M V \tag{4.12}$$

Next, this volume is transferred to another volumetric flask, and more solvent is added until the final volume of the diluted sample is reached. The volumes have changed between before (b) and after (a) the dilution, but what has not changed is the number of moles of solute before and after the dilution process:

$$n_b = n_a \tag{4.13}$$

Therefore, for dilution processes, we know that

$$(C_M \cdot V)_b = (C_M \cdot V)_a \text{ or } C_{Mb} \cdot V_b = C_{Ma} \cdot V_a \tag{4.14}$$

where C_{Mb} and C_{Ma} denote the molar concentrations before and after dilution, respectively. Eq. 4.14 is a convenient equation to calculate new concentration after a dilution process or the volume of solution required to establish a new concentration.

Therefore, let us compare how the dilution process compares to the method of using solid samples to prepare stock solutions. In Example 4.5, you will calculate how to prepare stock solutions from the solid sample.

Example 4.5 For the calibration of an "atomic absorption" (AA) instrument (see Chapter 15), solutions of copper ions may be required that are 0.0100, 0.100, and 1.00 [mM] in copper ions. Calculate the masses of solid copper (II) chloride, $CuCl_2$, that would be required to produce 20.0 [mL] each of (a) 1.00, (b) 0.100, and (c) 0.0100 [mM] solutions.

Answer: (a) $m = 2.69$ [mg]; (b) $m = 0.269$ [mg]; (c) $m = 0.0269$ [mg]

Next, let us discuss how you could prepare the same solutions by a dilution process. First, we have to prepare a stock solution of an accurately known concentration that is easy to measure out, that is, it requires an amount of the solid that can be accurately weighed out. We decide on 50.0 [mL] of a solution that is 10.0 [mM] in copper ions. This solution is prepared by weighing out 67.30 [mg] of copper (II) chloride and dissolving it in sufficient water in a volumetric flask to make 50.0 [mL] solution.

Example 4.6 Verify that the solution prepared in the last paragraph is, indeed, 10.00 [mM] in copper ions.

The next question is: what volume of this 10.0 [mM] stock solution is required to produce, by dilution, 20.0 [mL] of 1.00 [mM] solution of $CuCl_2$ (see the original task outlined in Example 4.5)? Using Eq. 4.11 and solving for V_b gives

$$V_b = \frac{C_{Ma} \cdot V_a}{C_{Mb}} = \frac{20.0 \,[\text{mL}] \cdot 1.00 \,[\text{mM}]}{10.0 \,[\text{mM}]} = 2.00 \,[\text{mL}] \quad (4.15)$$

Thus, we would use a calibrated 2.00 [mL] pipette, remove 2.00 [mL] of the stock solution, drain the pipette into a 20.0 [mL] volumetric flask, and fill up with pure water until the new volume is 20.0 [mL]. To check our work, let us ascertain that the amount of $CuCl_2$ is, indeed, the same before and after dilution. Before dilution, that is, in the 2.00 [mL] pipette, we have

$$n_b = C_{Mb} \cdot V_b = 10.0 \,[\text{mM}] \cdot 2.00 \,[\text{mL}] = 2.00 \cdot 10^{-5} \,[\text{mol}] \quad (4.16)$$

After dilution,

$$n_a = C_{Ma} \cdot V_a = 20.0 \,[\text{mL}] \cdot 1.00 \,[\text{mM}] = 2.00 \cdot 10^{-5} \,[\text{mol}] \quad (4.17)$$

Indeed, the amount of $CuCl_2$ is the same before and after we add the solvent to produce the new solution. Dilutions of stock solutions into less concentrated solutions are common in analytical laboratories. The examples shown here demonstrate how such standard solutions are prepared from stock solutions.

Example 4.7 Repeat the dilution calculations in Eqs. 4.15–4.16 for the preparation of the other two calibration solutions, namely 20 [mL] each of solutions that are 0.100 [mM] and 0.0100 [mM] in copper ions from the 10.0 [mM] stock solution.

Answers: 0.200 [mL] or 200 [μL] for the 0.100 [mM] calibration solution and
0.020 [mL] or 20 [μL] for the 0.0100 [mM] calibration solution.

Such volumes can be measured accurately with modern Eppendorf pipettes.

4.5 Precipitation Reactions

So far, we have dealt with aqueous solutions only, which are, at this level of introductory chemistry, by far the most common solutions. In addition, we have assumed that the solute, for example the sodium chloride or copper chloride, actually dissolves in water to form the ions, for which we have calculated solution concentrations. However, there are many ionic compounds that do not dissolve in water, such as the salts $CaCO_3$, $CaSiO_3$, and $CaSO_4$ that are part of everyday life (chalk in minerals, the silicates in rock, gypsum). If a salt is considered insoluble, it implies that the ions that make up this salt cannot coexist in solution. An example will illuminate this statement. Consider the following scenario: dissolving the soluble salt sodium sulfate (Na_2SO_4) in water will provide sodium and sulfate ions. Similarly, dissolving the soluble salt calcium chloride ($CaCl_2$) will create calcium and chloride ions. A precipitation of $CaSO_4$ will occur if the two solutions are mixed because the sulfate and calcium ion concentrations surpass a threshold value. Quantitative aspects about this threshold value will be discussed in Chapter 9.

The reactions discussed in the previous paragraph can be summarized as follows:

$$Na_2SO_4(s) \rightarrow 2\,Na^+(aq) + SO_4^{2-}(aq) \quad \text{(dissolving sodium sulfate in water)} \quad (4.18)$$

$$CaCl_2(s) \rightarrow Ca^{2+}(aq) + 2Cl^-(aq) \quad \text{(dissolving calcium chloride in water)} \quad (4.19)$$

$$2Na^+(aq) + SO_4^{2-}(aq) + Ca^{2+}(aq) + 2Cl^-(aq) \rightarrow CaSO_4(s)\downarrow + 2Na^+(aq) + 2Cl^-(aq) \quad (4.20)$$

In Eq. 4.20, the down arrow indicates that precipitation occurs. Since the sodium ions and the chloride ions in Eq. 4.20 do not participate in the actual formation of the precipitate, they are referred to as "spectator ions," and Eq. 4.20 is rewritten as a "precipitation reaction":

$$SO_4^{2-}(aq) + Ca^{2+}(aq) \rightarrow CaSO_4(s)\downarrow \quad (4.21)$$

In some text books, Eqs. 4.18 through 4.21 are represented as

$$Na_2SO_4 + CaCl_2 \rightarrow CaSO_4 + 2NaCl \tag{4.22}$$

and are referred to as "replacement reactions." However, this description does not represent what is happening, because it omits the solvent. When the two solids, Na_2SO_4 and $CaCl_2$, are mixed, nothing happens. Only in the presence of the solvent, water, will the reaction occur, and is best represented by Eq. 4.20 or 4.21.

The discussion so far should lead to the next questions: which salts are soluble in water, and why are some salts soluble while others are not? Neither of these questions has a simple answer. Let us start with the second question. Whether or not a salt dissolves depends on a number of energetic considerations: how much energy is needed to overcome the ionic interactions in the crystal, and how much energy is gained by solvating the ions formed. The ionic interactions in the crystal depend on the ionic charges and the distance between ions (i.e. the crystal structure). The solvation energy gained depends on the ionic size and charge. Neither of these effects can be easily quantified but will be tackled in Section 11.6.1 (see Figure 11.6).

Therefore, the answer to the first question, which salts are soluble in water, is very unsatisfactory: there is really no easy way to predict solubility, and you need to do a bit of (despised) memorization, such as:

- Most salts of the alkali metals (Group 1A) are **soluble** (NaCl, KBr, etc.).
- Most salts of the alkaline earth elements (Group 2A) are **insoluble**, except most of their halides and nitrates. Thus, CaO, CaS, $CaCO_3$, $BaSO_4$, etc. are **insoluble**, whereas $CaCl_2$ and $Ba(NO_3)_2$ are soluble.
- Most sulfates are **insoluble**, except those of the alkali metals (Na_2SO_4 is soluble).
- Most oxides and sulfides are **insoluble**, except those of the alkali metals.
- Most metal halides (Group 7A) are soluble, except those of silver (AgCl, AgBr, and AgI are insoluble; AgBr was used as a major component of photographic film).
- Most nitrates are soluble.

Example 4.8 Write a net ionic equation for the process that occurs when 0.1 [M] aqueous solutions of potassium bromide and silver nitrate are mixed. Identify spectator ions.

Answer: $Ag^+(aq) + Br^-(aq) \rightarrow AgBr(s)$. The potassium and nitrate ions are the spectator ions.

Historically, introductory chemistry laboratory courses included many precipitation reactions, for example, for a quantitative determination of solution concentrations by precipitation of a given ion. This is demonstrated in Example 4.9, which was previously promised and combines aspects of stoichiometry, precipitation reactions, actual yield, and solution concentrations:

Example 4.9 Consider the reaction that occurs upon mixing 100.0 [mL] of an aqueous solution of 0.0100 [M] iron (III) chloride with 50.0 [mL] of an aqueous solution of 0.200 [M] sodium sulfide to give iron (III) sulfide and sodium chloride. Iron (III) chloride, sodium sulfide, and sodium chloride are soluble in water, whereas iron (III) sulfide is insoluble, and can be recovered as a solid and dried. If the amount of iron (III) sulfide recovered was 83.0 [mg], what was the percent yield of the preparation?

Answer: $m(Fe_2S_3) = 104$ [mg]; 79.8% yield.

Example 4.10 A solution of magnesium chloride is prepared by dissolving 3.70 [g] of the solid in sufficient water to make 500 [mL] of solution. The molar chloride ion concentration in the solution is:

(a) $7.78 \cdot 10^{-2}$ [M]
(b) $1.24 \cdot 10^{-1}$ [M]
(c) $1.57 \cdot 10^{-1}$ [M]
(d) None of the above

Answer: (c) Details in answer booklet

Example 4.11 What is the theoretical yield (in grams) of nitrogen dioxide if 34 [g] of ammonia and 68 [g] of oxygen are reacted to produce nitrogen dioxide and water?

Answer: An amount of 55.8 [g] of nitrogen dioxide. Oxygen is the limiting reagent.

Further Reading

For a summary of chemical reactions and solution reactions, see https://chem.libretexts.org/Bookshelves/Inorganic_Chemistry/Supplemental_Modules_and_Websites_(Inorganic_Chemistry)/Chemical_Reactions/Chemical_Reactions_Examples/Chemical_Reactions_Overview.

This website contains material that is related to classification of reactions, which is of lesser importance than the actual understanding of the stoichiometry.

For rules and examples for precipitations, check out https://chem.libretexts.org/Bookshelves/Inorganic_Chemistry/Supplemental_Modules_and_Websites_(Inorganic_Chemistry)/Descriptive_Chemistry/Main_Group_Reactions/Reactions_in_Aqueous_Solutions/Precipitation_Reactions.

5

Electronic Structure of Atoms

In Chapter 2, we briefly introduced the quantum numbers that define atomic orbitals in order to discuss the "Aufbau principle" that governs the periodic chart of elements. We did not, however, delve into the origin of these quantum numbers, the nature of the orbitals, and other aspects that are required to get a better idea of chemical bonding. These topics, and the resulting shape of molecules, will be the subject of this and the next chapter. Some of the mathematical principles required to understand this discussion are quite involved and will be elaborated in more detail in Sections 15.2 and 15.3. But even in these sections, there are certain mathematical aspects that are beyond the level of introductory chemistry and only will be mentioned in passing.

The basic principles of atomic structure, chemical bonding, and molecular structure as discussed in Chapters 2 and 3 were originally obtained by observing the interaction of light (within and outside the visible region) with matter. This is a subject treated in lectures in physics, and you may ask yourself why this topic is brought up in an introductory chemistry book. The answer here is that the basic structure of atoms and molecules – which is responsible for Chemistry as we know it – is based on physical and mathematical principles; thus, it is necessary to describe the structure of atoms in terms of physical models.

When the aforementioned interactions of light with atoms were first observed by the end of the nineteenth and the beginning of the twentieth centuries, there were no adequate models in place to explain the perplexing experimental results. To interpret the results of these experiments, entirely new branches of science needed to be developed, quantum mechanics and spectroscopy, which in turn revealed more and more details on atomic and molecular structure. We shall return to a discussion of spectroscopy in Chapter 15. It is also necessary for us to understand some basic properties of light before we can discuss the interaction between matter and light.

5.1 Description of Light as an Electromagnetic Wave

The description of light as a form of electromagnetic radiation was published in the early 1860s in terms of Maxwell's equations. The solution of these complicated equations reveals that light is a transverse wave (see below) of electric and magnetic fields that oscillate at a given frequency ν and a wavelength λ, as shown in Figure 5.1. The term "transverse wave" indicates that the changes in the electric field $\boldsymbol{E}$ occurs in a direction perpendicular to the propagation direction, for example, along the y-axis.[1] The oscillating electric field $\boldsymbol{E}$ of such a wave, propagating along the positive z direction can be described by the following equation:

$$\boldsymbol{E} = E_0 \sin(2\pi z/\lambda - 2\pi\nu t) \quad (5.1)$$

with an equivalent equation for the propagation of the magnetic field:

$$\boldsymbol{B} = B_0 \sin(2\pi z/\lambda - 2\pi\nu t) \quad (5.2)$$

which oscillates in the x-direction. Here, E_0 and B_0 are the amplitudes of the electric and magnetic fields $\boldsymbol{E}$ and $\boldsymbol{B}$ which are perpendicular to each other, as shown in Figure 5.1. They oscillate in phase at the frequency ν, measured

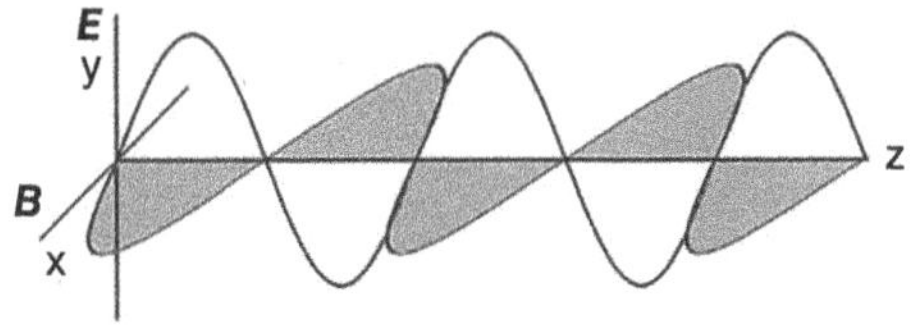

Figure 5.1 Description of the propagation of an electromagnetic wave as the oscillation of electric (E) and magnetic (B) fields.

---

[1] Sound waves are longitudinal waves where the pressure change that makes up the sound wave oscillates along the propagation direction.

Table 5.1 Wavelength ranges and energies of electromagnetic radiation.[a]

	ν	λ	E_{photon} [J]	E_{photon} [kJ/mol]
Radio waves	300 MHz	1 m	$2.0 \cdot 10^{-25}$	$1.2 \cdot 10^{-4}$
Microwave	10 GHz	3 cm	$6.6 \cdot 10^{-24}$	$4.0 \cdot 10^{-3}$
Infrared	$3 \cdot 10^{13}$ Hz	10 µm	$2.0 \cdot 10^{-20}$	12
Visible	$6 \cdot 10^{14}$ Hz	500 nm	$4.0 \cdot 10^{-19}$	240
Ultraviolet	$1 \cdot 10^{15}$ Hz	300 nm	$6.6 \cdot 10^{-19}$	400
X-ray	$1 \cdot 10^{18}$ Hz	0.3 nm	$6.6 \cdot 10^{-16}$	$4 \cdot 10^{5}$

a. Medium wavelengths and frequencies for each spectral range. Infrared radiation, for example, extends from about 1–50 µm.

in units of $[s^{-1}]$ = [Hz]. λ is the wavelength of the wave, measured in units of length. λ is defined by the distance between two consecutive peaks (or troughs) of the electric or magnetic fields.

The electromagnetic wave can travel through vacuum, as well as through media. This aspect originally presented serious difficulties in the understanding of light, since it was thought that any wave requires a medium to propagate, such as sound waves through air or liquid or solid media, or water waves through water. However, light creates its own "medium" in the form of the oscillating electric and magnetic fields, and thus, can travel through a vacuum (think about sunlight reaching the earth!). In a medium, light travels at a lower velocity, dependent on the refractive index of the medium.

In general, any wave motion can be characterized by its wavelength λ, its frequency ν, and its propagation velocity c, which is defined by

$$c = \lambda \nu \tag{5.3}$$

For light, this propagation velocity in vacuum is $c = 2.998 \cdot 10^8$ [m/s]. For all practical purposes, we will use $c = 3.0 \cdot 10^8$ [m/s]. Light visible to the human eye has wavelengths between about 430 (violet) and 690 (red) [nm]. From Eq. 5.3, we can easily calculate the frequency of (mid-range) visible light with a wavelength of 500 [nm]:

Example 5.1 What is the frequency (in Hz) of visible light with a wavelength $\lambda = 500$ [nm]?
Answer: $6.0 \cdot 10^{14}$ $\left[\frac{1}{s} = Hz\right]$

We see that visible light has very short wavelengths and very high frequencies. Other wavelength ranges of electromagnetic radiation are listed in Table 5.1. Notice that radio waves, the longest wavelength radiation listed below, have wavelengths in the range of fractions of meters to many meters; therefore, antennas to absorb these waves have lengths in the meter range. We will talk about the last two columns in Table 5.1 shortly.

When light encounters a narrow slit or a pinhole, it shows diffraction patterns similar to those of a water wave that impinges on a barrier with a narrow aperture. Such diffraction patterns are shown in Figure 5.2. These patterns are due to constructive and destructive interference of the amplitudes of individual wavelets, as was shown by C. Huygens in the seventeenth century. This observation confirms the wave properties of light, as predicted by Maxwell's equation.

5.2 Particle Properties of Light and Wave-particle Duality

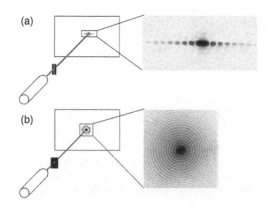

Figure 5.2 Diffraction of light from a single slit (a) or a pinhole (b) creates diffraction patterns akin to those of water waves.

However, the view of light as an electromagnetic wave needed to be augmented by an alternate description after A. Einstein reported the photoelectric effect in 1905. As pointed out earlier, this was one of the experiments that required new theories to be developed because it defied any previously established models.

In this experiment, light of variable color (frequency) illuminated a photocathode contained in an evacuated tube, as shown in Figure 5.3. An anode in the same tube was connected externally to the cathode through a current meter and a source of electric potential (such as a battery). Since the cathode and anode were separated by vacuum, no current was observed. However, if light with a frequency above a threshold frequency was illuminating the photocathode, a current was observed. Einstein concluded that light with a frequency above this threshold value had sufficient kinetic energy to knock out electrons from the metal atoms of the photocathode, that is, they acted like a "particle" of a certain energy. The ejected electrons subsequently travel to the anode, thereby creating a photocurrent. Furthermore, Einstein reported that the photocurrent produced by the irradiation of the photocathode was proportional to the intensity of light, but that increasing the intensity of light with a frequency below the threshold did not produce any photocurrent. Einstein concluded that the (kinetic) energy of a "light particle" or "photon" is given by

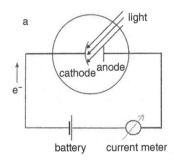

Figure 5.3 Schematic of the photoelectric experiment.

$$E = h\nu \tag{5.4}$$

or, using Eq. 5.3

$$E = h\,c/\lambda \tag{5.5}$$

where ν is the frequency of light, and h is Planck's constant, $h = 6.62 \cdot 10^{-34}$ [Js].

Eq. 5.4 defines light as a form of energy. Furthermore, it postulates that light, on a microscopic level, exists in finite "quanta" or photons, just as matter exists in finite amounts that we call atoms or molecules. As we shall see later (Eq. 5.14), there is a relationship that describes the interaction of light and matter at the atom–photon level.

According to Eqs. 5.4 and 5.5, the energy of a photon is proportional to the frequency – the higher the frequency, the more energy a photon has – and indirectly proportional to the wavelength; i.e. photons with longer wavelengths have lower energy. The quantization of light, i.e. the existence of "light particles" had been postulated by M. Planck (at about the same time) when he was formulating the theory of "black body radiation" (see below); therefore, "h" bears his name. Since h is a very small number, the energy of a photon is very small, even though the frequency of photons is a large number (see Table 5.1). In Table 5.1, the photon energy, $h\nu$, is listed in the fourth column and is seen to increase with increasing frequency or decreasing wavelength. The fifth column lists the energy of light expressed in [kJ per mol of photons], that is, the entries of the fourth column multiplied by Avogadro's number. This column shows that UV-visible photons have energies of the same order of magnitude as the energy of a chemical bond, which typically is between 100 and 400 [kJ/mol] for (single) bonds. Consequently, these photons can break chemical bonds which leads to two important biochemical processes. One of these is the capture of a photon in the retina of the eye that involves the breakage of a bond in the compound rhodopsin and is responsible for vision. The other is the photodamage of the human epidermis and dermis causing tanning, sunburn, and serious skin diseases.

Einstein's experiment demonstrated that light, in addition to its wave properties, has particle character with the kinetic energy of the photons given by Eq. 5.4. This observation led to the concept of wave–particle duality of light. This duality implies that light can exhibit both wave and particle properties, depending on the experiments being performed. Macroscopically, light exhibits wave-like properties that account for the effects encountered in classical optics, such as diffraction, refraction, etc. Microscopically, we need to think of light as a stream of light particles or photons that can interact with atomic and molecular species giving rise to a multitude of spectroscopic phenomena.

At this point, we need to discuss a few more aspects of the quantization of light, such as the momentum of a photon and photon mass, and the wave–particle duality observed for moving masses, in general. L. de Broglie showed that photons have momenta as well as well as kinetic energy: the momentum p of a photon is given by

$$p = h/\lambda = h\,\nu/c \tag{5.6}$$

Eq. 5.6 is known as the de Broglie equation. Thus, a light particle has kinetic energy and momentum given by Eqs. 5.5 and 5.6, respectively. If a photon has momentum, it follows that it must have a nonzero mass m, since the momentum is defined as

$$p = m\,v \tag{5.7}$$

Here, v is the velocity of the moving particle. Since a photon can only move at the velocity of light (see below), its momentum is given by

$$p = mc \qquad (5.8)$$

and its mass is

$$m_{photon} = p/c = h/(c\lambda) \qquad (5.9)$$

Example 5.2 Calculate the energy in (a) [J/photon] and (b) [J/mol of photons] of the photons in Example 5.1.

Answer: (a) $E = 4.0 \cdot 10^{-19}$ [J]; (b) $E = 240$ [kJ/mol]

Example 5.3 Calculate the (a) momentum and (b) the mass of the photon of Example 5.1.
(c) Compare the momentum of the photon in part (a) to that of a 10 [g] bullet traveling at 1000 [m/s]

Answer: (a) $p = 1.32 \cdot 10^{-27} \left[\frac{kg\,m}{s}\right]$; (b) $m = 4.4 \cdot 10^{-36}$ [kg]; (c) $p = 10 \left[\frac{kg\,m}{s}\right]$

The wave–particle duality was confirmed experimentally in 1927 by C. Davisson and L. Germer for moving electrons. In the Davisson and Germer experiment, a beam of electrons was scattered by a metal lattice, and a diffraction pattern was observed. This confirmed the de Broglie hypothesis that any moving object has a wavelength associated with it that is also given by Eq. 5.6.

$$\lambda = h/p \qquad (5.6)$$

It is hard to visualize that a baseball, traveling at 100 [mph], has a wavelength associated with it. But as we shall see in the examples below, for macroscopic items moving at velocities of everyday life, this wavelength is imperceptibly small. For an electron traveling at near the velocity of light, however, these wave properties become relevant as shown by the electron diffraction experiment by Davisson and Germer. The wave properties of fast-moving electrons are exploited in electron microscopy, a tool that allows visualization of items much smaller than those detected in visible light microscopy by using a beam of electrons at high speed and very short wavelength.

When we talk about the masses of electrons and photons, we need to consider one further complication. A particle, such as an electron, has a certain mass when it is not moving. We call this mass the "rest mass," m_0 (or "mass at rest") which is, for the electron, $9.1094 \cdot 10^{-31}$ [kg]. However, as a result of relativistic effects, the mass of any moving body increases with the velocity. This increase is given by Eq. 5.10:

$$m_v = \frac{m_0}{\sqrt{1 - \left(\frac{v^2}{c^2}\right)}} \qquad (5.10)$$

In Eq. 5.10, m_v is the mass of any particle with rest mass m_0 traveling at a velocity v. Thus, as the velocity v of a particle approaches c, the denominator of Eq. 5.10 becomes very small, and m_v gets very large. In Example 5.4, you will show that the mass of an electron increases enormously when it reaches 99% of the velocity of light. This effect, that mass is a function of velocity, was first postulated by Einstein in the specific theory of relativity and has been experimentally verified. Eq. 5.10 also shows why it is darn difficult for us to travel at the velocity of light since our masses would approach infinity when we travel at the velocity of light. Since the energy required to accelerate a mass to a certain velocity depends on the square of the velocity, we see that the energy required to accelerate us to the velocity of light becomes infinite as well (tell this to Han Solo and other Sci-Fi space travelers). A photon, on the other hand has a rest mass of zero and can only move at the velocity of light. Therefore, the mass of a photon can only be defined at the velocity c.

We now turn to a number of example calculations to get a feel for the wavelength, frequency, momentum, and mass of photons, as well as the wavelength, momentum, and mass of a moving electron.

Example 5.4

(a) Calculate the mass of an electron moving at 99.0% of the velocity of light (such velocities can easily be reached in a synchrotron).

(b) Calculate m_v for a 10 [g] bullet travelling at 1000 [m/s] (see Example 5.3).
(c) Calculate the de Broglie wavelength of the bullet in part (b).

Answer:

(a) $m_v = 6.457 \cdot 10^{-30}$ [kg], or about seven times its rest mass
(b) $m_v = 1.0000000000055 \, m_0$ (ignore significant figure aspects)
(c) $\lambda = 6.6 \cdot 10^{-35}$ [m]

Example 5.5 In electron microscopy, "electron waves," rather than light waves, are used to visualize microscopic objects. (a) Using the de Broglie relation for matter waves, calculate the velocity to which an electron needs to be accelerated such that its wavelength is 10 [nm]. (b) What is the momentum of such an electron? (c) What percentage of the velocity of light is the velocity in (a)?

Answer:

(a) $v = 0.725 \cdot 10^5 = 7.3 \cdot 10^4$ [m/s]
(b) $p = 6.6 \cdot 10^{-26}$ [kg m/s]
(c) 0.024%

Example 5.6

(a) What is the mass of a photon with a wavelength of 30 [nm]?
(b) What is the momentum of the photon in (a)?
(c) Comment on the masses and momenta of the moving particles in problems 5.5 and 5.6.

Answer:

(a) $m = 2.2 \cdot 10^{-34}$ [kg]
(b) $p = 6.6 \cdot 10^{-26}$ [kg m/s]
(c) The momentum of an electron with a wavelength of 30 [nm] is the same as that of a photon with a wavelength of [30] nm.

5.3 The Hydrogen Atom Emission Spectrum: Stationary Atomic States

Now that we have discussed some of the properties of light, we can begin to understand how the information we get from the interaction of light and matter was used to deduce information on the structure of matter. At the beginning of the twentieth century, experimental evidence was amassed that pointed to the necessity to redefine some aspects of classical physics, which led to a field known as quantum mechanics. One of the experiments that led to the formulation of quantum mechanics was the observation of "spectral lines" in the emission spectra of the hydrogen atom (see Figure 5.4b). Let us recapitulate some principles of classical optics and the concept of spectral lines.

When any material is heated, it begins to glow red at about 800 [K], then turning orange, yellow and white when the temperature is increased. This is a manifestation of the so-called black-body radiation law which describes the radiative energy emitted by an (idealized) material, and which is summarized in Figure 5.4a. We see in this Figure that the peak emission shifts toward lower wavelength as the temperature of the black body increases (hence its color turns from red to white), and that the total energy radiated increases as well. In fact, the total energy radiated increases with the temperature to the fourth power. The formulation of the theory of the black-body radiation led M. Planck to the conclusion that light must exist in the form of light particles, or photons, with the energy given by Eq. 5.4. By the way, the sun, whose surface temperature is ca. 5700 K appears white to the human eye because it is a superposition of all visible Wavelengths.

However, when light emitted from atomic hydrogen atoms at high temperature was analyzed by researchers of the late nineteenth century, a totally different emission profile was observed. By the way, atomic hydrogen can be produced when an electric discharge passes through hydrogen gas. This breaks the H—H bond and leaves the hydrogen atoms "thermally excited," or in a high energy state. Instead of a broad emission spectrum observed in the black-body experiment, sharp emission lines at discrete wavelengths were observed, as shown in Figure 5.4b which were referred to earlier as "spectral

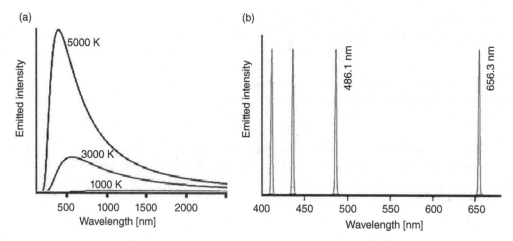

Figure 5.4 (a) Black body emission spectra at 1000, 3000, and 5000 K. (b) Schematic of the hydrogen atom emission spectrum in the visible spectral range.

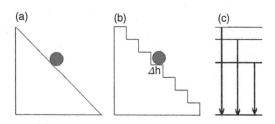

Figure 5.5 (a) Continuously variable potential energy of a mass on an inclined plane. (b) "Quantized" potential energy of a mass on steps. (c) Quantized energy differences ("transitions") for the mass in (b).

lines." This led the researchers in the early twentieth century to the following conclusion: the hydrogen atoms can only exist in certain "quantized energy states." When light is emitted from these states of hydrogen atoms, it transitions from one into another quantized energy state and sheds the energy difference as a photon of light. These last two sentences need some serious explanation.

First, what is a "quantized energy state"? The best classical analog would be the potential energy "quantization" of a mass "m" on steps, as opposed to a mass on an inclined plane, as shown in Figures 5.5a and 5.5b. On the inclined plane, the potential energy, given by $E_{pot} = m\,g\,h$, can vary in infinitesimally small increments, since the height can vary in infinitesimally small increments. Here, the symbol "h" refers to the height of the ball, not to Planck's constant, and g is the gravitational acceleration.

On the other hand, on the steps, the potential energy of the mass is given by

$$E_{pot} = m\,g\,(n\,\Delta h) \tag{5.11}$$

where "n" denotes the step on which the ball rests, and Δh is the step height. Therefore, the potential energy can vary only in finite increments. In this case, "n" could be seen as a "quantum number," that is, an integer number that identifies the energy state. The difference in potential energy between two states on the steps would be

$$\Delta E_{pot} = (n_f - n_i)\,m\,g\,\Delta h, \quad \text{(the subscripts } f \text{ and } i \text{ denote final and initial states)} \tag{5.12}$$
$$= \Delta n\,m\,g\,\Delta h \tag{5.13}$$

So, for the mass-on-the-staircase model, the change in potential energy between several upper states and the "ground state" can only occur in multiples of $m \cdot g \cdot \Delta h$. These energy differences are shown by the down arrows in Figure 5.5c.

Secondly, if the hydrogen atom only can exist in similar discrete energy states, the energy it emits when it drops from a higher to a lower state can only have certain discrete, quantized values, just as the energy difference between the mass resting at different steps can only have discrete values. The energy difference between the two states, $\Delta E_{atom} = (E_f - E_i)$, is carried away by a photon that is created in this process:

$$\Delta E_{atom} = (E_f - E_i)_{atom} = E_{photon} = h\,\nu = h\,c/\lambda \tag{5.14}$$

Therefore, an emission spectrum of sharp lines is observed, since only discrete energy packages are emitted by the hydrogen atoms. Consequently, ΔE_{atom} in Eq. 5.14 somehow depends on the change in energy level (but we do not know yet how the energy depends on "n" in the hydrogen atom), and there can only be discrete energy jumps. Thus, the photons created when a hydrogen atom undergoes the energy "jumps" (known as "transitions" in the jargon of spectroscopists) can have only certain discrete energies. Since energy and wavelengths of photons are (inversely) related to each other, only discrete colors are observed in the hydrogen emission spectrum, a small part of which is shown in Figure 5.4b.

It was found out by empirical data analysis that in the hydrogen atom, the energy levels are not proportional to "n" as in the case of the mass-on-the-staircase model, but are proportional to $1/n^2$. "n" is the integer main quantum number discussed before in Chapter 2 (see Figure 2.3). Thus, we may write

$$E_{\text{H-atom}} \propto 1/n^2 \quad (n \neq 0) \tag{5.15}$$

or, by writing a proportionality as an equation with a proper proportionality constant

$$E_{\text{H-atom}} = R_y/n^2 \quad (n \neq 0) \tag{5.16}$$

where R_y is the Rydberg constant, $2.179 \cdot 10^{-18}$ [J] (take Eq. 5.16 with a grain of salt for the moment). As we discussed earlier, n, the main quantum number, also referred to earlier as the "electron shell," determines the energy level, as well as the distance r between the electron shell and the nucleus. So, when "n" gets large, the distance r between the electron and the nucleus gets large as well. The (attractive) energy U between nucleus and electron is given by Coulomb's law as

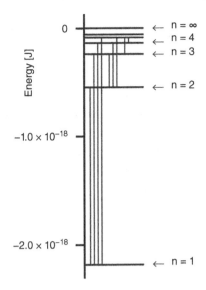

Figure 5.6 Hydrogen atom energy levels as determined by the Rydberg equation. The ground state of hydrogen with $n = 1$ is defined as a negative energy as compared to the ionized atom with $n = \infty$.

$$U = k \, e_e \, e_N / r \tag{5.17}$$

where k is Coulomb's constant. In Eq. 5.17, e_e and e_N are the electronic and nuclear charges, which are equal and opposite in value. Thus, we may write Eq. 5.17 as

$$U = -e^2 \, k/r \tag{5.18}$$

where "e" is the elementary charge. We may argue that when n approaches infinity, "r" becomes very large as well, and the potential energy becomes zero. For $n = 1$ (n cannot have the value of zero), the attraction is strongest, and the distance between the nucleus and the electron is smallest; therefore, the potential energy should have the largest negative value (because of the negative sign in Eq. 5.18). Thus, we rewrite Eq. 5.16 as

$$E_{\text{H-atom}} = -R_y/n^2 \tag{5.19}$$

With this sign convention, a plot of the hydrogen atom energy levels is shown in Figure 5.6. The horizontal lines in Figure 5.6 represent the energy levels obtained when values 1, 2, 3,... are substituted into Eq. 5.19 for the quantum numbers n. The vertical lines indicate the transitions between these energy levels. If a transition occurs between higher to lower values of n, a photon is emitted. If a transition occurs between lower to higher values of n, a photon with the exact energy difference (cf. Eq. 5.14) is absorbed and annihilated.

Combining Eqs. 5.14 and 5.19, we obtain an equation for the wavelength of a photon emitted when a hydrogen atom undergoes a transition from an energetically higher state n_i to an energetically lower state n_f as

$$\begin{aligned} \Delta E_{\text{atom}} &= (E_f - E_i)_{\text{atom}} = -R_y \left(\frac{1}{n_i^2} \right) - \left(-R_y \left(\frac{1}{n_f^2} \right) \right) = E_{\text{photon}} = h \, \nu = h \, c/\lambda \\ &= R_y \left[\left(\frac{1}{n_f^2} \right) - \left(\frac{1}{n_i^2} \right) \right] = E_{\text{photon}} = h \, \nu = h \, c/\lambda \end{aligned} \tag{5.20}$$

Here, we see again how empirical observations, namely the hydrogen atom emission spectrum, paved the way for the development of models to explain the observation. These models subsequently were refined and allowed the determination of the numeric value of the Rydberg constant that confirmed the experimentally obtained value.

Example 5.7 Use the wavelengths of the transition indicated in Figure 5.5b, 656.3 [nm], to calculate the initial state of the hydrogen atom responsible for this transition. The final state is $n_f = 2$.

Answer: $n_i = 3$

5.4 Hydrogen Atom Orbitals

The take-home message from the hydrogen atom emission experiment is the existence of quantized energy states of atoms (and molecules, as we shall see later), and that the interaction with light can either promote an atom from a lower to a higher energy state in a process where a photon, or packet of energy is destroyed, or from a higher to a lower energy state in a process where a photon is created. This was formulated previously in Eq. 5.14

$$\Delta E_{atom} = \left(E_f - E_i\right)_{atom} = E_{photon} = h\nu = h\,c/\lambda \tag{5.14}$$

which is a fundamental equation governing the interaction of light and atoms or molecules. This equation allows us to determine the energy levels from the frequencies or wavelengths of the photons absorbed or emitted.

Scientists of the early twentieth century were confronted with the facts that we have discussed so far: first, light exists as individual light particles, or photons, that have mass, momentum, and energy. Second, atoms exist in discrete energy states and can interact with light. However, a theoretical framework explaining the existence of these discrete energy states was not established until quantum mechanics was formulated independently by two scientists: E. Schrödinger and W. Heisenberg for the hydrogen atom, and later for more complicated systems.

Quantum mechanics was developed originally for the simplest atomic system, the hydrogen atom, and had to explain the experimental data discussed in the previous section. In the quantum mechanical formalism, the lone electron of the hydrogen atom is described as a standing wave (see Chapter 15), in accordance with the concept of matter waves introduced above (see Eq. 5.6). The square of the amplitude of this matter wave (the "wavefunction") is the actual probability of finding the electron at a given time and in a spatial volume element. Schrödinger's reasoning was, in part, guided by the fact that W. Heisenberg had shown earlier that on the microscopic level, the position and momentum of a particle cannot be determined simultaneously and therefore, the properties of a particle must be described by a "probability function" or "wave function" that incorporated the uncertainty in position and momentum. This step, along with the quantization of light and the observation of the hydrogen emission spectrum were the three major experimental results that required a new branch of physics to be formulated.

This line of reasoning led to the so-called Schrödinger equation, a complicated differential equation, the solutions of which define energy levels the electron can occupy. Associated with these energy levels are shapes in space that describe where the electron is likely to be found. These regions are the square of the wavefunction and are referred to as orbitals. An orbital is a mathematical construct based on the solutions of the Schrödinger equation that describes the amplitude of the wavefunction in space. The calculations are generally performed in terms of the spherical polar coordinates r, θ, and φ, rather than the Cartesian coordinates x, y, and z (for the definition of spherical polar coordinates, see Chapter 1). The solutions of the Schrödinger equation are polynomials in these variables r, θ, and φ. A polynomial is a function that contains the independent variable at different powers, e.g.

$$f(x) = a_0 + a_1 x + a_2 x^2 + a_3 x^3 + \cdots \tag{5.21}$$

is a polynomial in the single variable x. For the 1s, 2s, and 3s orbitals, for example, these polynomials are 1, $\left(1 - \frac{r}{2a}\right)$ and $\left(1 - \frac{2r}{3a} + \frac{2r^2}{27a^2}\right)$, respectively, where "a" is a constant. These functions will be discussed in more detail in Chapter 15. Similarly, the angle dependent parts are polynomials in the variables θ and φ. These angle-dependent solutions of the Schrödinger equation are functions that had been described by two French mathematicians (P. Laplace and A. Legendre) nearly 150 years before quantum mechanics and define, for example, the waves of a spherical planet covered by an ocean, or the electric charge oscillations on a charged metal sphere. These functions are known as the spherical harmonic functions, $Y_l^m(\theta, \varphi)$ that will be discussed next, since their shapes and relationships profoundly influence chemistry. You may want to consult Chapter 15 for more details.

The spherical harmonic functions $Y_l^m(\theta, \varphi)$ are polynomials in the polar coordinates θ and φ, and the super- and subscripts l and m determine the highest power of the variables θ and φ in the polynomials, respectively. There exist rules that follow from the mathematical solutions of the Laplace and Legendre differential equation that determine the values and relationship between l and m: l (the power of the θ-dependent term) is an integer that can assume all (positive) values from zero to infinity. It is known as the orbital angular momentum quantum number. However, m may only assume certain values that depend on the values of l:

If $l = 0$, m must also be zero, and the only allowed spherical harmonic function is Y_0^0.

If $l = 1$, m may assume values of $-1, 0, +1$. Therefore, three spherical harmonic functions are allowed, namely Y_1^{-1}, Y_1^0, and Y_1^1

If $l = 2$, m may assume values of $-2, -1, 0, +1, +2$, and five spherical harmonic functions are allowed: Y_2^{-2}, Y_2^{-1}, Y_2^0, Y_2^1, and Y_2^2. For higher values of l, analogous relationships hold.

It is the relationship between the spherical harmonic functions that determines the allowed electronic configuration of the elements, and thereby, the shape of the periodic chart: we see that the scheme of sublevels in the energy level diagram Figure 2.3 follows exactly the naming of the spherical harmonic functions. These rules are not imposed arbitrarily by mathematicians but follow from the fact that the θ- and φ-dependent differential equations need to be solved simultaneously, and are related by what is known in mathematics as a "recursion formula" that determines subsequent polynomials (that is, with higher powers of the variables) from previous ones.

If you go back to Figure 2.3 in Chapter 2, you will see immediately that the number of sublevels for each of the $l = 0$, $l = 1$, $l = 2$, and $l = 3$ levels, and even their nomenclature ($-1, 0, 1$, or $-2, -1, 0, 1, 2$, etc.) exactly follow the nomenclature of the spherical harmonic functions. Obviously, the spherical harmonic functions profoundly influence the energy states of atomic systems.

This brings us to the next aspect of the spherical harmonic functions: what do they look like? Figure 5.7 shows the shapes of these functions in space. Y_0^0 can be depicted as a spherical balloon. The next spherical harmonic functions, shown in the second row of Figure 5.7 show a perturbation of the spherical balloon, which may be viewed as an elongation of the balloon into a shape containing a nodal plane (a region in space where the function is zero and changes sign) and two "bulges," as shown in the second row of Figure 5.7 and marked Y_1^0. The third row in Figure 5.7 depicts the five spherical harmonic functions with l–values of 2, namely Y_2^{-2}, Y_2^{-1}, Y_2^0, Y_2^1, and Y_2^2. The spherical harmonic functions with higher l–values are more complicated and will not be discussed here any further. Notice that these spherical harmonic functions are just the solutions of the angle dependent part of the Schrödinger equation, but we can see how the relationship between the "quantum numbers" or index numbers l and m determines the shape of the periodic chart.

The radial part of the Schrödinger equation (the part of it that depends on the coordinate r) is a polynomial in r as well, as indicated before, and is denoted by $R_{nl}(r)$. Here, the main quantum number n denotes the highest power of r in the polynomial. The final solutions of the Schrödinger equation are the products of the angle-dependent (or spherical harmonic)

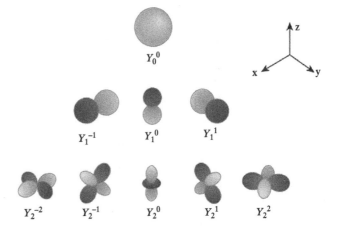

Figure 5.7 First nine spherical harmonic functions. The light gray colors denote regions where the sign of the spherical harmonic function is positive, and the darker gray regions where the sign is negative.

functions and the radial part and are usually denoted by the symbol of wavefunctions, ψ. The wavefunctions are defined by three subscript quantum numbers n, l, and m, written as ψ_{nlm}, and also are called orbital wavefunctions of the hydrogen atom.

$$\psi_{nlm} = R_{nl}(r)\, Y_l^m(\theta, \varphi) \tag{5.22}$$

The indices m and l in Eq. 5.22 refer to the quantum numbers discussed before for the spherical harmonic functions, and the integer n denotes the dependence of the functions $R_{nl}(r)$ on r and is the main quantum number defined in the Rydberg equation (Eq. 5.19). This quantum number may assume integer values from 1 to ∞. The occurrence of the index l in the expression $R_{nl}(r)$ again indicates that the equations that determine the radial and angle-dependent parts of the Schrödinger equation need to be solved simultaneously, and therefore, a rule exists that links n and l. This rule implies that

n may take integer values between 1 and infinity; i.e. $n = 1, 2, 3, 4 \ldots \infty$ (5.23)

l may assume any integer value from 0 to $(n-1)$: $l = 0, 1, 2, 3 \ldots (n-1)$ (5.24)

m may assume any integer value from $-l$ to $+l$: $m = 0, \pm 1, \pm 2, \ldots \pm l$ (5.25)

The last rule was encountered before in the definition of the spherical harmonic functions. The orbital shapes defined by the wavefunctions in Eq. 5.22 differ from the spherical harmonic functions in that the p- and d-orbitals have somewhat more elongated shapes, as compared to the shapes of the spherical harmonic functions. This is due to the multiplication with the radial part, $R_{nl}(r)$, (see Eq. 5.22) of the wavefunction, and is shown for a $2p_y$ orbital (see Table 5.2) in Figure 5.8 (see also Figure 6.2).

Historically, orbitals have been given names, as follows: in the name of an orbital, the main quantum number is listed first, followed by a code for the quantum number l. This code is summarized below:

l:	0	1	2	3
	s	p	d	f

The orbital names and associated wavefunctions, are given in Table 5.2, which is an extension of Table 2.3 introduced in Chapter 2.

Table 5.2 Quantum numbers (QNs), atomic orbital names, and associated wavefunctions.

Main QN	Orbital QN	Magnetic QN	Orbital	Wavefunction	Alternate wavefunctions[a]
$n = 1$	$\ell = 0$	$m = 0$	1s	ψ_{1s}	
$n = 2$	$\ell = 0$	$m = 0$	2s	ψ_{2s}	
$n = 2$	$\ell = 1$	$m = -1, 0, 1$	2p	$\psi_{2p_{-1,0,1}}$	$\psi_{2p_x}\, \psi_{2p_z}\, \psi_{2p_y}$
$n = 3$	$\ell = 0$	$m = 0$	3s	ψ_{3s}	
$n = 3$	$\ell = 1$	$m = -1, 0, 1$	3p	$\psi_{3p_{-1,0,1}}$	$\psi_{3p_x}\, \psi_{3p_z}\, \psi_{3p_y}$
$n = 3$	$\ell = 2$	$m = -2, -1, 0, 1, 2$	3d	$\psi_{3d_{-2,-1,0,1,2}}$	[b]
$n = 4$	$\ell = 0$	$m = 0$	4s	ψ_{4s}	
$n = 4$	$\ell = 1$	$m = -1, 0, 1$	4p	$\psi_{4p_{-1,0,1}}$	$\psi_{4p_x}\, \psi_{4p_z}\, \psi_{4p_y}$
$n = 4$	$\ell = 2$	$m = -2, -1, 0, 1, 2$	4d	$\psi_{4d_{-2,-1,0,1,2}}$	[b]
$n = 4$	$\ell = 3$	$m = -3, -2, -1, 0, 1, 2, 3$	4f	$\psi_{4f_{-3,-2,-1,0,1,2,3}}$	[c]

a. Some of these functions ($\psi_{2p_{-1}}$ and $\psi_{2p_{+1}}$) are complex and cannot be plotted in real space. However, an algebraic manipulation can be performed (see Chapter 1 and Problem 15.6) that transforms these functions into real functions along the x- and y-axes that here are denoted as "alternate wavefunctions" ψ_{2p_x} and ψ_{2p_y}. The same holds for the 3p and 4p orbitals.
b. The 3d and 4d wavefunctions are written as ψ_{3d_a} or ψ_{4d_a} with the subscript a being xy, xz, yz, $x^2 - y^2$, or z^2.
c. The f orbitals are even more difficult to depict and are omitted here.

We have encountered in the discussion of the hydrogen atom, a number of rules that appear arbitrary at this point. However, these rules are implicit in the mathematics underlying the solutions of the differential equations involved. We need not concern ourselves with these rules, but Chapter 15 discusses more details. However, even the advanced discussion omits many mathematical details that are quite complicated, to say the least.

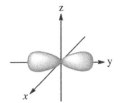

Figure 5.8 Shape of a $2p_y$ orbital.

The three-dimensional plots of the hydrogen atom wavefunctions resemble, in overall appearance, those of the spherical harmonic functions and are referred to as the electron "orbitals" of the hydrogen atom. Notice that the "nodal planes" found in all but the first spherical harmonic functions denote regions where the wavefunction changes sign, and in which the electron cannot be found, since the square of the wavefunction is zero.

There are a few more points to be discussed about the hydrogen atomic orbitals listed in Table 5.2. First and foremost, wavefunctions associated with the orbitals are mathematical solutions of the Schrödinger equation, and as presented here, only hold for the hydrogen atom (and a few other 1-electron systems, such as He^+ or Li^{2+}). As soon as there are two or more electrons in an atom, the Schrödinger equation cannot be solved explicitly (but approximated methods to do so exist). Second, an orbital *per se* is a mathematical construct and cannot be observed. Therefore, the announcement "Orbitals Observed" which was a *large* headline and appeared in the journal *Nature* was misleading. What was observed in a rather revolutionary scientific report was an experimental three-dimensional electron density map[2] that closely resemble the shapes of the squared wavefunctions that, in fact, indicate the probability of finding an electron. Thus, experimental evidence was presented in this report that supports Schrödinger's approach to quantum mechanics. In addition, the validity of the quantum mechanical model introduced so far is confirmed by the overall structure of the periodic chart, discussed in more detail in Section 5.5.

The rules relating to the allowed quantum numbers for the hydrogen atom (Eqs. 5.23–5.25) can be summarized in a graphical scheme shown in Figure 5.9. In this "orbital energy diagram" for the hydrogen atom, the energies of the orbitals depend on n only, and are identical to those shown in Figure 5.7 (the Rydberg energy scheme). What is different between Figures 5.6 and 5.9 is the splitting of the atomic energy levels into sublevels with different l and m values. In the hydrogen atom, all orbitals within the same main quantum number are degenerate, that is, they have the same energy. Although these sublevels result from the solutions of the Schrödinger equation (and, as we saw, from the form of the spherical harmonic functions), their existence can be demonstrated experimentally by carrying out the hydrogen emission experiment in the presence of a magnetic field: in this case, the p and d orbital energy levels split into 3 and 5 energy levels, respectively, for the different quantum numbers m, and the emission lines split into multiplets. Hence, the quantum number m is often referred to as the magnetic quantum number because it affects the orbital energies in a magnetic field. Here again, it is the experimental evidence that confirms the validity of the theoretical conclusions.

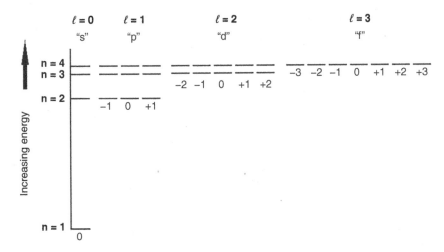

Figure 5.9 Partial energy level diagram of the hydrogen atom orbitals. Notice that all orbitals within a "shell" with a given main quantum number are degenerate, as compared to the situation for many electron atoms shown in Figure 2.3.

---

2 Zuo, J.M., Kim, M., O'Keefe, M., and Spence, J.C.H. (1999), *Nature* **401**, 49–52.

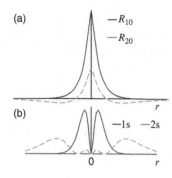

Figure 5.10 (a) Plot of the wavefunctions R_{10} and R_{20} vs. the distance r from the nucleus. (b) Plot of the radial distribution function vs. the distance r from the nucleus.

The orbital energy scheme for the hydrogen atom shown in Figure 5.9 differs from the scheme shown in Figure 2.3 in that, for atoms with more than one electron, the energies of the sublevels also depend on the value of the orbital quantum number l; that is, the 2p orbitals differ energetically from the 2s orbitals. This gives rise to some important aspects that determine the shape of the periodic chart, to be revisited in Section 5.5.

Plots of the radial part of the wavefunctions, R_{10} and R_{20}, are shown in Figure 5.10a. These plots indicate that the wavefunction has a maximum at $r = 0$, that is, at the cusp shown in Figure 5.10a, which is at the nucleus. Since the square of the wavefunction is an indication of the most probable occurrence of the electron, as pointed out before during the introduction of the concepts of wavefunctions, the plots of the R_{10} and R_{20} functions seem to suggest that the electron's most probable location is in the nucleus. This result contradicts the experimental findings discussed in Chapter 2 that indicated that the electron(s) surround the very small nucleus at a distance much larger than the size of the nucleus. This apparent discrepancy is addressed as follows. The probability P of finding an electron (in the 1s orbital, for example) in a thin spherical shell of thickness dr at a distance r from the nucleus (and independent of θ and φ), is obtained by multiplying the square of the radial part $(R_{10})^2$ by the volume of the spherical shell, which is $r^2 dr$.

$$P = (R_{10})^2 r^2 dr \tag{5.26}$$

The expression $r^2 (R_{10})^2$ is called the radial distribution function and presents the probability of finding the electron in a shell with thickness $r + dr$. At $r = 0$, the volume of this shell is zero; thus, the probability of finding the electron at the nucleus is zero. The argument presented here holds for all orbitals, not just the "s" orbitals. A plot of the radial distribution function as a function of r is given in Figure 5.10b, which shows that the 1s orbital has its maximum at the Bohr radius. Again, the radial distribution function shown in Figure 5.10b has the same appearance in all directions from the nucleus. Thus, the electron is most likely found in a spherical shell around the nucleus at a distance a.

Both the wavefunctions themselves and the radial distribution functions are quite important for the description of various effects. We will use the wavefunctions themselves, and the overlap between wavefunctions on different atoms, to describe chemical bonding. Thus, the plots shown in Figure 5.10a will reappear in the discussions in Chapter 6. The radial distribution functions are important for the interpretation of X-ray and electron diffraction results. In these experiments, solid or gaseous molecules are exposed to high-energy electromagnetic radiation (X-rays) or beams of electrons that are scattered off the electrons in an atom or molecule. These scattering phenomena actually allow the experimental verification of the radial distribution functions of electrons in compounds, and thereby, present a view of the three-dimensional electron distribution and the shape of molecules. The electron density maps as previously mentioned (see footnote 2 in this Chapter) are a typical result of such scattering measurements that can reveal detailed electron distributions around atoms.

Example 5.8 What is the name of a hydrogen atomic orbital with $n = 3, \ell = 2$?

Answer: Any one of the five 3d orbitals.

Example 5.9 Is there a hydrogen-like orbital with $n = 3$ and $\ell = 3$? Explain.

Answer: No. ℓ cannot exceed $n - 1$

5.5 Atoms with Multiple Electrons: The Aufbau Principle Revisited

When describing atoms with more than one electron, a few additional comments and rules need to be introduced. First, as mentioned above, the Schrödinger equation cannot be solved explicitly for a three-particle system (a nucleus and two electrons) and becomes even more difficult as the number of electrons increases. Therefore, approximations have to be made to describe atoms from helium. However, these approximate methods have been refined such that computational methods available today can predict electronic energies that agree with experimental energies to a very high degree. These approximations include the use of hydrogen-like orbitals, electron correlation (basically, an approach that stipulates that the negatively charged electrons try to stay out of each other's way), effective nuclear charges (electrons in lower shells shielding the nuclear charge experienced by outer electrons), and a few others. But, most importantly, is the requirement of an additional quantum number, the electronic "spin" quantum number, which is needed to describe electrons in a multi-electron atom. This "spin" quantum number is not required to describe the electron in the hydrogen atom using Schrödinger's approach (since there is only one electron), but is included in Heisenberg's and Pauli's alternate description of the hydrogen atom (both approaches were later seen as completely equivalent, so there is no controversy of which one is better). We introduced the concept of the spin quantum number in Chapter 2 when the Aufbau principle was first introduced.

There probably are a few concepts in physics that routinely are described as incorrectly as the "spin". What is this magical property, and how does it influence chemistry and physics? Again, let us look at the experimental evidence of this property that was discovered about the same time as the original formulation of quantum mechanics in the mid- to late 1920s. There are two major pieces of experimental evidence that electrons (and certain other elementary particles) possess an inherent property that can interact with an external magnetic field. One of these was carried out by passing a beam of electrically neutral Ag atoms through an inhomogeneous magnetic field (the so-called Stern–Gerlach experiment). Ag atoms have a lone electron in the 5s orbital, and the same results could have been obtained using a beam of electrons. In the Stern–Gerlach experiment, the beam of silver atoms was split into two beams when passed through the magnetic field, deflected either toward, or away from the magnetic field gradient. This result implies that the single electron in Ag atoms has a magnetic moment and that this magnetic moment is analogous to the angular momentum of a spinning object. Such a magnetic moment would result if the electron was "spinning" around an axis, either clockwise or counterclockwise because a spinning (electric) charge would produce a magnetic moment. Thus, the term "spin" was born. However, the electron's description of a wave does not go well with the concept of a spinning particle; thus, one should refer to the spin of an electron as a spatial quantization of the angular momentum into two states that interact oppositely with a magnetic field.

Spin properties are not restricted to electrons. As we shall see later in Chapter 15, protons (but not neutrons) have spins as well. The proton spin gives rise to a spectroscopic technique called "nuclear magnetic resonance" spectroscopy, one of the most widely used structural tools in chemistry and biochemistry. It is also applied quite a bit in modern medicine: the "magnetic resonance imaging" (MRI). This technology is based on the fact that proton spins (mostly those in water molecules) interact with a magnetic field. To obtain an MRI image, a patient is subject to a strong magnetic field, and the transitions from one to the other spin states of the hydrogen atoms in water are observed.

The other experiment that suggested a quantized angular momentum of the electron was the aforementioned hydrogen atom emission experiment carried out in a magnetic field. In addition to the splitting of the energy levels with different values of the quantum number m, another even smaller splitting of the hydrogen emission lines suggested that the electron existed in two quantized states that can be simply described by a quantum number s that can have the values $s = +\frac{1}{2}$ or $s = -\frac{1}{2}$ (in units of angular momentum). This spin quantum number s is also referred to as m_s in some books. Thus, an electron in the ground state of a hydrogen atom can be described by either set of four quantum numbers as given in Table 5.3:

Table 5.3 Possible quantum numbers for the single electron in a hydrogen atom ground state.

n	l	m	s		n	l	m	s
1	0	0	$+\frac{1}{2}$	or	1	0	0	$-\frac{1}{2}$

However, if there is more than one electron in an atom, a further rule comes into play that is known as "Pauli's exclusion principle" which states that no electron in an atom can have an identical set of all four quantum numbers. If two electrons occupy the same orbital (i.e. if the three quantum numbers n, l, and m are the same) then the spin quantum number s must be different for the two electrons. This rule is based on mathematical considerations as well that follow from quantum mechanics.

As pointed out in the last paragraph of Section 5.4, in a multi-electron system, the orbital energies do not only depend on n but also on l, as shown in Figures 2.3 and 5.11. The consequences of loss of degeneracy between the 2s and 2p (as well as the 3s, 3p, and 3d) orbitals are as follows: when advancing in the second row of the periodic chart, from Li to Ne, each successive element has one more proton and one more electron. In Li, the third electron is found in the 2s orbital, with the spin function either plus or minus ½. The next element, beryllium (Be), will also fill the 2s orbital since it is energetically

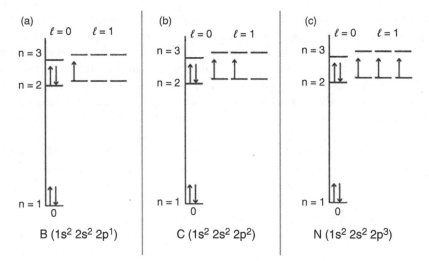

Figure 5.11 Schematic atomic orbital energy diagrams, orbital occupancy, and electron configurations for (a) boron, (b) carbon, and (c) nitrogen.

the lowest-lying orbital (see Figure 2.3); however, the two electrons in the 2s orbital must have opposite spins to fulfill the Pauli exclusion principle. In the next element, boron (B), the additional electron can fill any of the three 2p orbitals; traditionally, one fills the electron into the left-most of the three 2p orbitals, but this is totally arbitrary (see Figure 5.11a). The next element, C, fills another 2p orbital with an electron with the same spin quantum number as the previous electron. This is shown in Figure 5.11b. Whether the two spins in the 2p orbitals are drawn with "spin up" or "spin down" symbols is immaterial; what is important is that the two electron spins in the two 2p orbitals are the same. This is the consequence of another rule, known as "Hund's" rule, that implies that degenerate orbitals are filled to maximize unpaired spins (maximal spin "multiplicity"). By the same token, nitrogen has the 2p orbitals filled as shown in Figure 5.11c, with three unpaired spins. This is due that pairing spins to allow electrons to occupy the same orbital requires a certain amount of energy, known as "spin-paring energy." One could paraphrase Hund's rule to imply degenerate orbitals are filled with electrons with parallel spins to avoid the spin-pairing energy. Again, this is not a crazy rule to make your understanding of the subject more complicated, but is based on the measurement of magnetic spin momenta: if the spins of the two electrons in the 2p orbitals in the C atom were opposite (that is, if the two electrons were in the same 2p orbital with opposite spin quantum numbers), a naked C atom would have zero magnetic moment, but in reality, it has a magnetic moment in line with two unpaired electrons. We will talk more on the experimental determination of magnetic moments at the end of this section.

Next, a shortcut will be introduced to designate the occupancy of atomic orbitals. This so-called electron configuration is a sequence of the atomic orbitals (1s, 2s, 2p, 3s, 3p, etc.) with the occupancy of the orbital given as a superscript, as shown in Figure 5.11 in the parentheses next to the element symbol. The element iron, Fe, for example, would have an electron configuration of $1s^2 2s^2 2p^6 3s^2 3p^6 4s^2 3d^6$ which often is abbreviated as [Ar] $4s^2 3d^6$, where [Ar] denotes completely filled orbitals for the first 18 electrons: [Ar] $\equiv 1s^2 2s^2 2p^6 3s^2 3p^6$.

The order in which the orbitals are filled is given by the scheme in Figure 5.12 where the arrows indicate the sequence of orbitals to be filled as one goes from one element to the next, namely 1s, 2s, 2p, 3s, 3p, 4s, etc. There are a few exceptions in the order of filling atomic orbitals. One of the examples is the element chromium (Cr) which, according to the previous discussion, should have an electronic configuration of [Ar] $4s^2 3d^4$, but has an electronic configuration of [Ar] $4s^1 3d^5$. This is due to the fact that half-filled shells offer the advantage that no spin-pairing energy is required; in the case of the Cr atom, the energy difference between the 4s and 3d orbitals is sufficiently small that an arrangement with six parallel spins is energetically more suitable than the expected [Ar] $4s^2 3d^4$ configuration. Another example of the electron configuration not following the scheme in Figure 5.13 is the Cu atom, for which the expected electron configuration is [Ar] $4s^2 3d^9$, but is found to be [Ar] $4s^1 3d^{10}$. Also, a few exceptions occur with the rare earth (lanthanide) group.

Some of the atomic properties expected from the Aufbau principle can actually be determined experimentally. As stated before, the nicest theory is worthless if it does not agree with the experiment.

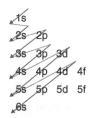

Figure 5.12 The order in which atomic orbitals are filled is given by following the arrows and lines connecting them.

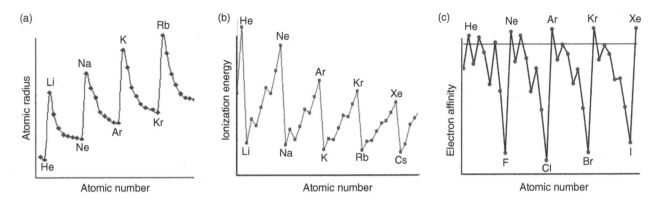

Figure 5.13 Atomic radii (a), first ionization energies (b), and electron affinities (c) for the main group elements.

Here, three important properties will be discussed that confirm many of the aspects we discussed in this section so far. These properties are the atomic radii, ionization energies, and electron affinities, shown in Figure 5.13.

The atomic radii of the main group elements (see Chapter 2) are shown in Figure 5.13a. Whenever the main quantum number of a new electron added in the Aufbau scheme increases (e.g. between Ne and Na, or Ar and K), the atomic radius increases drastically. This is because the new electron in an "s" orbital perceives a nuclear charge that is shielded by the electrons in the previous shell, and therefore, experiences a much smaller attractive force to the nucleus; hence, the larger radius. In this graph, the transition metals are omitted, since their atomic radii do not follow an easily perceived trend. This is, of course, due to the fact that successive electrons are filled into the 3d orbitals that lie well to the inside of the 4s orbitals. The same holds for the second and third-row transition metals. Another trend that can readily be observed in Figure 5.13a is the overall increase of atomic radii with atomic number Z. Although the positive charge of the nucleus increases steadily with Z, the attraction between the nucleus and outer electrons gets smaller and smaller due to the shielding effect of the inner electrons. Furthermore, the negatively charged electrons in the orbitals will avoid each other. Both these effects contribute to the larger size of atoms with increasing Z.

Figure 5.13b shows another periodic trend for elements, namely the first ionization energy (IE_1). This energy, also referred to as the ionization potential, describes the energy required for an ionization process of a gaseous atom M according to

$$M(g) \rightarrow M^+(g) + e^- \quad IE_1 \tag{5.27}$$

In accordance with the general sign convention (processes that require energy have positive energy values), all the ionization potentials shown in Figure 5.12b are positive. They range in value for the ionization of Li according to

$$Li(g) \rightarrow Li^+(g) + e^- \quad IE_1 = 520 \text{ [kJ/mol]} \tag{5.28}$$

to the ionization of He according to

$$He(g) \rightarrow He^+(g) + e^- \quad IE_1 = 2372 \text{ [kJ/mol]} \tag{5.29}$$

These numbers indicate how tightly the electrons are bound in the He atom, where the electrons are in close proximity to the nucleus and unshielded by inner electrons. In contrast, the 2s electron in Li is located further out from the nucleus, and is shielded by the two 1s electrons, and therefore, has a much lower IE_1. The ionization energy increases for the Be atom to 900 [kJ/mol] due to the increased nuclear charge, but drops to 801 [kJ/mol] for boron, due to the fact that the first 2p electron perceives a reduced nuclear charge due to the filled spherical 2s orbital. The next two elements, C and N, have increasing values of IE_1 before there is a slight dip when the O atom is reached. This is because the fourth electron that is accommodated in the 2p orbitals needs to pair its spin with either of the electrons in the half-filled 2p. Thus, this dip, which is very apparent in Figure 5.12b, is a direct manifestation of the spin-pairing energy. Between O, F, and Ne, there is a steady increase in IE_1, and Ne, after He, has the second highest IE_1. A little reflection will allow you to make predictions for the second ionization potential IE_2, defined by

$$M^+(g) \rightarrow M^{2+}(g) + e^- \quad IE_2 \tag{5.30}$$

When transition metals are ionized, the order in which electrons are lost does not conform with the Aufbau principle. Whereas the 3d orbitals are filled before the 4s orbitals, as discussed above, the outer electrons (the 4s-electrons) are lost in an ionization process before 3d-electrons are lost. Thus, when a transition metal, such as Ti ([Ar]$4s^2 3d^2$) forms a cation, the Ti$^+$ ion has an electronic configuration of [Ar]$4s^1 3d^2$ and the Ti^{2+} ion has an electronic configuration of [Ar]$4s^0 3d^2$. Only after the 4s electrons are lost in an ionization will the 3d electrons be removed. The ionization process and the Aufbau principle seem to contradict, but this is not the case. Rather, they represent different scenarios: when proceeding from one atom to the next, following the Aufbau principle, a proton and an electron are added to an atom, as compared to the previous atom in the periodic chart. In an ionization process, the number of protons in the nucleus is maintained, but electrons are removed. Thus, this process follows a rule that is different from the building-up approach discussed for successive elements in the periodic chart.

Example 5.10 Write electron configurations for the following species:
F, O^{2-}, Ti, Fe^{2+}, Cu

Answer: See answer booklet

Example 5.11 In your own words, explain why the 1s electron in a hydrogen atom is not found at the nucleus, although the 1s wavefunction has a maximum at the nucleus.

Answer: See answer booklet

Example 5.12 Which of the following processes has the highest ionization energy?
(a) Ca → Ca$^+$ + e$^-$; (b) Ca$^+$ → Ca^{2+} + e$^-$; (c) Ca^{2+} → Ca^{3+} + e$^-$

Answer: (c)

Another property shown in Figure 5.13c is the electron affinities, E_a. Electron affinities are energies for the process in which an atom A accepts an electron to form an ion, A$^-$ according to

$$A(g) + e^- \rightarrow A^-(g) \quad E_a \tag{5.31}$$

First of all, do not confuse the electron affinity, which is a measurable quantity, with the electronegativity of atoms discussed in Chapter 3. The electronegativity is an assigned quantity to assess whether or not a covalent bond in a molecule will have polarity, whereas the electron affinity is a physical quantity relating to an atom. Second, for many elements, the electron affinity is a negative number, indicating that the reaction described by Eq. 5.31 releases energy. This is particularly for the halogen atoms that have large negative electron affinities: F: −328 Cl: −349, Br: −324 [kJ/mol], indicating that the reaction releases energy. This accounts for the fact that the halogens frequently occur as anions since the anions formed are energetically more favorable than the neutral atoms. Many of the cations in ionic compounds, such as the alkali metals, have low ionization energies and readily give up electrons. The total energy balance of forming an ionic compound often is dominated by the large negative electron affinity.

The noble gases such as He, Ne, or Ar have positive electron affinities; thus, the process

$$Ne(g) + e^- \rightarrow Ne^-(g) \tag{5.32}$$

is energetically unfavorable.

Another experimental verification of the Aufbau principle and Hund's rule is provided by the observation of an atom's magnetic moment. A magnetic moment is the result of one or more unpaired electron spins on an atom and causes the atom to be attracted toward increasing magnetic field strength. Such an atom is called paramagnetic. If there are no unpaired electron spins, an atom is called diamagnetic, and such atoms are repelled by a magnetic field. Certain transition metals, such as iron with an electronic configuration of [Ar] $4s^2$ $3d^6$, exhibit a particularly strong magnetic response referred to as ferromagnetism. Here, the magnetic moments of small domains of iron atoms in pure iron (and certain iron oxides) can be aligned such that neutral iron may exhibit permanent magnetism. A similar situation arises for the element neodymium which has four unpaired spins and exhibits an extremely strong magnetic dipole moment. Neodymium magnets, about the size of a dime, exhibit very strong magnetic forces and have found numerous applications in such fields as computer hard disk drives or separation technologies to recover certain metals from plastic waste.

So far, we have discussed the shape of the periodic chart strictly from the viewpoint of the electronic structure of the elements. There is, however, another important consideration of the shape of the periodic chart. If, indeed, the rules discussed above determine the allowed electronic configurations around an atomic nucleus, what about the nuclei themselves? We know from previous discussions that the nucleus determines an element by the number of protons, Z, in the nucleus. As the number of protons increases with increasing Z, the number of neutrons also increases, even faster than the number of protons. There are rules that determine the stability and energetics of the nuclei. Interestingly, the neutrons and protons in the nucleus follow similar rules with "nuclear quantum numbers" that define a "nuclear shell model." We shall revisit this aspect briefly in Chapter 14, Nuclear Reactions. It is noteworthy, again, to point out how the mathematical constraints on electronic and nuclear quantum numbers profoundly influence chemistry and nuclear chemistry, and thereby, the structure and properties of matter.

Example 5.13 Which of the following sets of spin quantum numbers are illegal for a three-electron atom?

(a)

n	l	m	s
1	0	0	$+\frac{1}{2}$
1	0	0	$-\frac{1}{2}$
1	1	0	$+\frac{1}{2}$

(b)

n	l	m	s
1	0	0	$+\frac{1}{2}$
1	0	0	$-\frac{1}{2}$
2	1	0	$+\frac{1}{2}$

(c)

n	l	m	s
1	0	0	$+\frac{1}{2}$
1	0	0	$-\frac{1}{2}$
2	0	0	$+\frac{1}{2}$

(d)

n	l	m	s
1	0	0	$+\frac{1}{2}$
1	0	0	$+\frac{1}{2}$
2	0	0	$+\frac{1}{2}$

Answer: (a) and (d) are illegal. Details in answer booklet

Example 5.14 Which of the following is the proper electron configuration of Ca^{2+}?

(a) $1s^2 2s^2 2p^6$
(b) $1s^2 2s^2 2p^6 3s^2 3p^6$
(c) $1s^2 2s^2 2p^6 3s^2 3p^6 3d^{10}$
(d) $1s^2 2s^2 2p^6 3s^2 3p^6 4s^2$

Answer: (b) Details in answer booklet

Example 5.15 The photodissociation of chlorine according to the equation $Cl_2 \xrightarrow{light} 2Cl$ requires one photon of wavelength 498 nm or shorter. What is the bond energy (i.e. the energy required to break the bond) of Cl_2 in [kJ/mol]?

Answer: 240 [kJ/mol]

Further Reading

For a short history of our understanding of light, and present theories and application of optical technology, see https://en.wikipedia.org/wiki/Light.

Atomic structure and periodic properties are resented in a very comprehensive review at: https://en.wikipedia.org/wiki/Periodic_table.

6

Chemical Bonding: Covalent Bonding, Molecular Geometries, and Polarity

6.1 General Aspects of Covalent Bonding

In Chapter 5, the structure of atoms was introduced, starting with the simplest system, the hydrogen atom, for which the Schrödinger equation can be solved explicitly. The solutions of the Schrödinger equation predict experimentally verifiable energy levels associated with orbitals that are regions in space in which the electron is most likely found. Subsequently, more complicated atoms were introduced, for which no explicit solutions of the Schrödinger equation exist; however, approximate methods have been developed that predict energy levels to any degree of agreement with experimental results. Furthermore, these methods predict accurately the Aufbau principle, that is, the recipe that predicts the form of the periodic table and therewith, the periodic chemical properties of elements.

In our quest to further understand chemistry and the properties of molecules, rather than atoms, we encounter the same problem as before: the Schrödinger equation for any molecular system cannot be solved explicitly. The simplest imaginable molecule, the H_2^+ molecular ion, is the starting point of a myriad of books and articles covering chemical bonding and consists of three interacting charged particles, just like the simplest atom with more than one electron, the He atom. In both cases, it is the interplay between positions and attractions/repulsions that render a "three-body problem" unsolvable. Yet, enormous progress has been made in the development of approximate methods to obtain numerical answers for the energies and orbital occupation in molecules that form the chemical bonds. To this end, as so often in science, models were developed that contain simplifications and approximations. In the case of chemical bonding, there are two major models, the valence bond (VB) theory and the molecular orbital (MO) theory that are used to describe covalent bonds. Both models have their pros and cons: the VB model is more intuitive from a chemist's viewpoint and follows the Lewis concept of shared electron pairs, see Chapter 3, whereas MO theory is more suitable for quantitative prediction of molecular energies. The VB model is extremely useful for the qualitative prediction of molecular shapes, including bond angles and molecular polarity. MO theory is particularly useful in computational chemistry and predicts molecular energies such as the standard enthalpy of formation, ΔH_f^0 (see Chapter 11), molecular interactions, and reaction pathways (see Chapter 13), and the conformation and solvation of many macromolecules, such as enzymes.

In this chapter, we shall briefly review the Lewis model of chemical bonding and refine this model using concepts of VB theory. A qualitative extension of VB theory is the so-called Valence Shell Electron Pair Repulsion (VSEPR) model, which is useful for predicting molecular shapes, and from the shapes, molecular polarity. As we shall see in Chapter 7 in the discussion of intermolecular forces, molecular dipole moments are extremely important in the discussion of properties and reactivities of molecules.

6.2 Lewis and VB Theory

VB theory is rooted in the view introduced in Chapter 3 that chemical bonds are formed when electrons are shared between two atoms – the Lewis model of chemical bonding. In Chapter 3, we saw that the orbitals from different atoms can physically overlap in space, creating MOs that can accommodate two electrons (with opposite spins). For the simplest (neutral) molecule, H_2, this involves the overlap of the two spherical 1s orbitals of the individual hydrogen atoms to form a region in space, located between the two nuclei, in which the electron density increases, or in which the electrons can be found, as shown in Figure 6.1. This region may be visualized in the shape of a blimp and is symmetric about the axis between the two

Figure 6.1 Representation of the overlap of two 1s orbitals to form a region in space associated with a bonding electron pair.

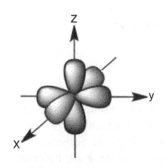

Figure 6.2 Representation of the three 2p orbitals. Notice that the shapes of these orbitals are somewhat different from those of the spherical harmonic functions shown in Figure 5.7.

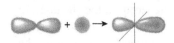

Figure 6.3 Representation of the overlap of a 1s and a 2p orbital in HF to form a region in space associated with a bonding electron pair.

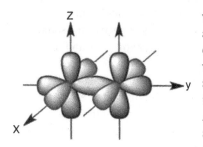

Figure 6.4 Representation of the overlap of a two 2p orbitals in F_2 to form a region in space associated with a bonding electron pair.

nuclei. This region is what we previously called a chemical bond or a bonding electron pair. In Section 6.5, we shall re-investigate the shape of the bonding orbital and its energetics. The scheme shown in Figure 6.1 holds for the overlap of all spherical symmetric orbitals (i.e., s orbitals).

Next, let us consider another diatomic molecule, HF, for which the Lewis structure was written previously as H—$\overline{\underline{F}}$| (see Section 3.3). Before we go any further in the discussion of the VB model of HF, we need to clarify one important point. In the following figures, pictures of orbitals are drawn for elements other than the hydrogen atom, and we assume that the shapes of the orbitals, for the fluorine atom for example, look like those of the hydrogen atom. This is quite a leap of faith, for it was pointed out before that the Schrödinger equation cannot be solved explicitly for any atom other than hydrogen. However, approximate methods, using generally accepted simplifications, indicate that the shapes of orbitals predicted for the hydrogen atom hold for other atoms as well. Thus, we use an image as shown in Figure 6.2 to indicate the 2p orbitals in an atom such as fluorine. Also note that the orbital shapes shown here are somewhat different from the shapes of the spherical harmonic functions, as indicated in the discussion of Figure 5.8.

Fluorine has an electron configuration of $1s^2 2s^2 2p^5$, that is, it has five electrons in the three 2p orbitals, which are the outermost and highest energy orbitals (see Figure 2.3). One of the three 2p orbitals is occupied by one electron only. This orbital can overlap with the (spherical) 1s orbital of the hydrogen atom, as shown in Figure 6.3. The other two 2p orbitals of the F atom will be unchanged in this model and point along the axis as shown in Figure 6.2. The single bond formed *via* the overlap of two s orbitals, or the overlap of an s- and a p orbital, are referred to as σ-bonds (sigma bonds), as opposed to π bonds (see later).

In a molecule such as F_2, for which the Lewis structure was given before (see Section 3.3) as |$\overline{\underline{F}}$—$\overline{\underline{F}}$|, the overlap of two 2p orbitals (here shown as the $2p_y$ orbitals) also leads to the formation of a σ bond, as shown in Figure 6.4. The pairs of valence electrons in the $2p_x$ and $2p_z$ orbitals on each fluorine atom are assumed to be relatively unaffected by the formation of the bond and represent the free electron pairs indicated in the Lewis structure.

The model introduced so far represents the bonds formed in H_2, HF, and F_2 quite well, and it is based on the assumption that the electrons involved in the bonds are associated with the orbitals of the individual atoms participating in a bond. This is one of the criteria of the VB model. However, this model requires some modifications when it comes to other, more complicated molecules. Consider the water molecule for an example. The central oxygen atom has an electron configuration of $1s^2 2s^2 2p^4$ with two unpaired electrons in the 2p orbitals, according to Hund's rule (see Section 5.5). A simple extension of the arguments presented so far suggests that the two hydrogen atoms in water would latch on the two partially filled 2p orbitals and form a water molecule with the two O—H bonds at 90° with respect to each other. However, the observed bond angle in a water molecule is much larger at 104.5°. Thus, the bonding does not appear to be a simple overlap of the oxygen 2p orbitals with the 1s orbitals of hydrogen, and the concept of hybridization needs to be introduced.

6.3 Hybridization and Multiple Bonding

The concept of hybridization resulted from experimental observations of molecular geometries. As it is so often, experimental results guide theoretical developments and models to the point when the models properly predict experimental results. One of these models in VB theory is the concept of hybridization, which can nicely be demonstrated using the methane molecule, CH_4, for an example. We have written the Lewis structure of methane earlier (see Eq. 3.5) and have concluded that the molecule has four equivalent C—H bonds. However, the Lewis structure does not give any indication of the molecular shape. Molecules with the general formula of XY$_4$ can have different shapes, some of which are shown in Figure 6.5. These shapes are referred to as square planar (Figure 6.5a), trigonal pyramid (Figure 6.5b), and tetrahedral (Figure 6.5c). More details on these shapes will be presented in Section 6.4; at this point, it is again the experiment that reveals that the structure methane is given by the tetrahedral shape shown in Figure 6.5c.

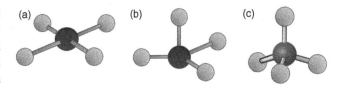

Figure 6.5 Possible molecular shapes of molecules with the general formula XY$_4$: (a) square planar, (b) trigonal pyramid, and (c) tetrahedron.

A tetrahedron is one of the classical Platonic solids, shown in Figure 6.6a. It has four faces, each of which is an equilateral triangle, six edges of same length, and four corners. The four lines pointing from the center of the tetrahedron to its corners form six internal angles, all of which are 109.5°. The shape shown in Figure 6.6a is exactly the same regardless of which face of the tetrahedron forms the base.

In the structure for methane shown in Figure 6.5c, the carbon atom is located at the center of the tetrahedron, and the four hydrogen atoms are at the four corners. All H—C—H bond angles are 109.5°. This brings up a question: how can we get the four bonds around the carbon atom to point toward the corners of a tetrahedron, given the shapes of the 2p orbitals

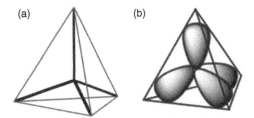

Figure 6.6 (a) Edges and faces of a tetrahedron. The four heavy black lines are the connectors from the center of the tetrahedron to its corners. (b) Shape of the sp^3 hybrid orbitals.

that are at 90° with respect to each other? In the VB model, this question is answered by mixing the 2s and the three 2p orbitals to form four "hybrid," or mixed orbitals known as the four "sp^3 hybrid orbitals" shown in Figure 6.6b. At this point, we need not concern ourselves with the mathematics of mixing of the 2s and 2p orbitals, but it is important to point out that four energetically equivalent sp^3 orbitals are formed, each accommodating one electron, as shown in Figure 6.7a. The energy of the four 2p^3 hybrid orbitals, shown in gray, is between that of the 2s and 2p orbitals, as shown in Figure 6.7a. Each of the four valence electrons of the carbon atom are in one of the four sp^3 orbitals and overlap with the 1s orbitals of the four hydrogen atoms to form four (covalent) σ bonds.

We assume that exactly the same bonding situation arises in the ammonium ion, [NH$_4$]$^+$, which is isoelectronic with methane. In the neutral ammonia molecule, NH$_3$, we still have eight valence electrons, with one nonbonding electron pair at the nitrogen atom (for the Lewis structure of ammonia, see Eq. 3.13), and the nitrogen atom is sp^3 hybridized as well. Thus, we assume that the four electron pairs around the nitrogen atom – the three that form the covalent bonds plus the nonbonding pair – point toward the four corners of a tetrahedron as well, and the shape of this molecule is referred to as a trigonal pyramid (more on this, and molecular shapes in general, in Section 6.4). A nonbonding electron pair occupies more space than a bonding electron pair (see Section 6.4); thus, the H—N—H bond angle at 107° in ammonia is somewhat smaller than that in methane.

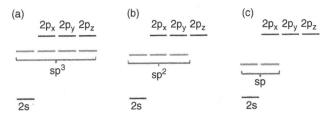

Figure 6.7 Energies of the (a) sp^3, (b) sp^2, and (c) sp hybrid orbitals formed from the 2s and 2p orbitals.

In the water molecule, we assume that the oxygen's four electron pairs again point toward the four corners of a tetrahedron, and that the two hydrogen atom 1s orbitals overlap with two hybrid orbitals that contain a single electron, whereas the two nonbonding orbitals have two electrons each. Since the two nonbonding electron pairs occupy even more space than the one nonbonding electron pair in ammonia, the H—O—H bond angle in the water molecule is even further reduced to 104.5°. This is shown in Figure 6.8. Notice that the nitride ion, NH$_2^-$, is isoelectronic with water, and that the H$_3$O$^+$ ion is isoelectronic with ammonia. The molecular shapes of the isoelectronic species are very similar. We conclude with the observation that whenever an atom is surrounded by four electron pairs that point toward the corners of a tetrahedron, the central atom is sp^3 hybridized.

Ethane, for which the Lewis structure was introduced in Eq. 3.6, may be thought of being formed by the head-on overlap of two sp^3-hybridized carbon atoms, which results in a σ bond

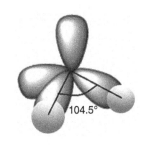

Figure 6.8 Schematic of the bonding in water.

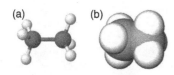

Figure 6.9 (a) Ball-and-stick model and (b) space-filling model of ethane, showing the staggered arrangement due to the spatial requirement of the H atoms.

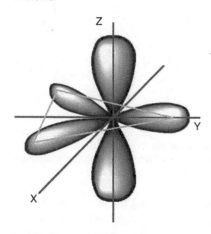

Figure 6.10 Shape of the sp^2 hybrid orbitals in the X–Y plane. The $2p_z$ orbital is not involved in the hybridization (see also Figure 6.7b).

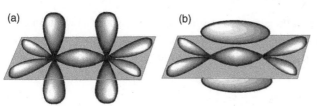

Figure 6.11 Valence orbital picture for ethene (ethylene). (a) Two sp^2 hybridized carbon atoms forming σ-bond between them. (b) Lateral overlap of the $2p_z$ orbitals to form π-bond.

between the two carbon atoms, each of which has three hydrogen atoms attached to the other sp^3-hybrid orbitals. This results in a structure of ethane shown as a "ball-and-stick" model in Figure 6.9a. Notice that the two sp^3 hybrids of the two C atoms are rotated by 60° with respect to each other, resulting in a staggered conformation, as shown in Figure 6.9b. This representation is called a space-filling model. Figure 6.9b indicates the size of electron clouds around each of the atoms and it is evident that the orbitals due to the C—H bonds would interfere with each other if the molecule is in a nonstaggered shape.

Next, the hybridization in molecules with multiple bonds will be discussed, using ethene and $H_2C=CH_2$ (ethylene) as an example. The Lewis structure of this compound was introduced earlier (Figure 3.7). There are, indeed, four electron pairs around each of the carbon atoms, but these electron pairs do not point toward the corners of a tetrahedron, like in methane, but rather assume a structure known as a sp^2 hybrid at each carbon atom. In sp^2 hybridization, only two of the 2p orbitals – the $2p_x$ and the $2p_y$ orbitals – mix with the 2s orbital, as shown in Figure 6.7b, and form a hybrid in which the three sp^2 orbitals point toward the corners of an equilateral triangle, as shown in Figure 6.10 by the light gray lines. The angle between the sp^2 orbitals is 120°. The remaining $2p_z$ orbital lies along the z axis, which is perpendicular to the plane of the sp^2 hybrid. (The assignment of the x, y, and z direction here is arbitrary.)

In ethene, two sp^2-hybridized carbon atoms form a σ bond from the head-on overlap of one of the sp^2 orbitals on each carbon atom. The two remaining sp^2 orbitals on each carbon atom form single bonds with the hydrogen atoms. Consequently, the H—C—H bond angle is close to 120°. The two $2p_z$ orbitals at each carbon atom overlap laterally to form a blimp-shaped π orbital, indicated by the two ellipses, above and below the plane of the molecule, as indicated in Figure 6.11. The two electrons in the π orbital form the second bond between the carbon atoms.

In many small molecules and molecular ions, the central atom is sp^2-hybridized as well. In the formaldehyde molecule, introduced in Chapter 3 (see Figure 3.2), the carbon atom is sp^2-hybridized. Therefore, the H—C—H bond angle as well as the H—C—O bond angles are expected to be close to 120° (observed angles are 116 and 122 degrees, respectively). The π orbital is located above and below the plane of the molecule. Similarly, the central atoms in ozone (O_3, see Figure 3.3), SO_2, CO_3^{2-}, NO_2^-, and many others are sp^2-hybridized and, consequently, have bond angles around 120°. Interestingly, the nitrogen atom in the nitrogen dioxide radical molecule $\dot{N}O_2$ (see Figure 3.6) is also $2p^2$ hybridized, with the lone electron in one of the sp^2 orbitals on the nitrogen atom. The lone electron appears to occupy significantly less space than an electron pair, since the observed O—N—O bond angle is 134°. Here, we see again how experimental data guide the development of models, in this case the spatial requirements of certain orbitals. We shall pick up on structural parameters in the next section on VSEPR.

For molecules incorporating triple bonds, we find another hybridization scheme, in which only one of the 2p orbitals (the $2p_x$ orbital, for example) mixes with the 2s orbital (see Figure 6.7c). The resulting sp hybrid is linear and assumed to lie along the y axis in Figure 6.12a. In acetylene (ethine) with the Lewis structure of H—C ≡ C—H (see Figure 3.7), two sp-hybridized carbon atoms form a σ bond by a head-on overlap of the sp hybrid, with the H atoms bonding to the other lobe of the sp hybrid. The $2p_z$ orbitals from both C atoms overlap laterally and form a π orbital similar to the situation shown in Figure 6.11. In addition, the $2p_x$ orbitals from both C atoms also overlap laterally and form a second π orbital as shown in Figure 6.12b. A totally analogous situation is found in the N_2 and CO molecules and the cyanide molecular ion, CN^- (all of which are isoelectronic with acetylene).

It is interesting to compare certain properties, such as bond lengths and bond strengths of single, double, and triple bonds, using ethane, ethylene, and acetylene for examples. The distance between the two carbon atoms in these three molecules is 154, 135, and 120 pm, respectively, indicating that the extra electron pairs in the double and triple bond actually do pull the carbon atoms closer together. Another measure of the bond strength is the energy required to break a bond: the C—C single bond requires about 345 [kJ/mol] to break the bond, the double bond requires about 615 [kJ/mol], and the triple bond requires about 840 [kJ/mol]. This indicates that each of the π bonds contributes nearly as much bond energy as the σ bond.

Finally, a quantity easily measured in a chemistry laboratory (see Chapter 15 and Example 15.12) is a "spring constant" of a chemical bond. A chemical bond can be approximated by a physical model of a spring connecting two masses. These masses are in constant (vibrational) motion at the resonance frequency of the mass/spring system. Elongation of a spring requires a force, measured in Newton (which is the unit of force) per unit of elongation (in units of meter). For a C—C single bond in ethane, the resonance frequency is about $2.98 \cdot 10^{13}$ [Hz], corresponding to a spring constant of about 440 [N/m]. For a double bond in ethylene, the resonance frequency is about $4.87 \cdot 10^{13}$ [Hz] and a force constant of about 1090 [N/m]. Finally, for a triple bond in acetylene, these values are $5.9 \cdot 10^{13}$ [Hz] and about 1500 [N/m]. Again, this comparison shows that the bond stiffness increases roughly linearly with the number of chemical bonds. More on this topic will be presented in Section 15.4.2 and Problem 15.12. This little excursion into thermodynamic and spectroscopic results shows that the VB model developed so far very well agrees with measurable (observable) laboratory results.

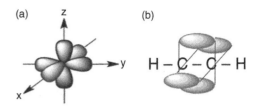

Figure 6.12 (a) sp hybrid (along the y axis) with the unhybridized $2p_x$ and $2p_z$ orbitals shown as well. (b) Formation of two π orbitals from the $2p_x$ and $2p_z$ orbitals on each carbon atom.

6.4 VSEPR Model

The VSPER model closely follows the discussion of the previous section on hybridization. This model is highly successful in predicting molecular shapes, following a few very simple rules. VSEPR states that the valence electron pairs around a central atom – whether bonding or nonbonding – arrange in such a way that they are as far away from each other as possible. Thus, if you have two electron pairs on a central atom, such as in beryllium hydride, BeH_2, they will be at 180° with respect to each other, and the BeH_2 molecule consequently will be linear, with Be at the center. If there are three electron pairs such as in boron hydride, BH_3, the electron pairs will point at the corners of an equilateral triangle, with the boron atom at the center of the triangle, and the three hydrogen atoms at the corners of the triangle. This molecule is planar, that is, all four atoms are in the same plane. The shapes of the electron pairs for 2, 3, 4, 5, and 6 valence electron pairs are depicted in Figure 6.13.

Figure 6.13 Electron-pair and possible molecular geometries predicted from the VSEPR model. Bonding electron pairs are shown in black; nonbonding pairs in gray. The electron pair geometries are referred to as linear, triangular, tetrahedral, trigonal pyramidal, and octahedral for two, three, four, five, and six electron pairs, respectively, surrounding the central atom.

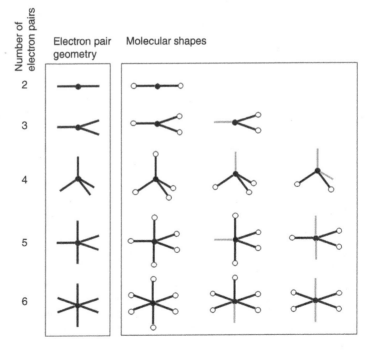

An important second rules for the prediction of molecular shapes using VSEPR is that in molecules containing multiple bonds, only the σ bonds are considered when determining the shape around an atom. Thus, in carbon dioxide, CO_2, with a Lewis structure of $\overline{O}=C=\overline{O}$ (see Example 3.2), one only counts the two σ-bonding electron pairs and concludes that the molecule is linear, with the π-electrons not contributing to the VSEPR count. A similar argument is used in a molecule such as ozone, O_3, for which one resonance structure was given earlier (Section 3.3) as $|\overline{O}-\overline{O}=\overline{O}$. Here, one argues that the central atom has a VSEPR electron pair count of three (not four), and these three pairs point toward the corners of an equilateral triangle. Thus, the ozone molecule has a bent shape with a bond angle of about 120° (observed value: 117°) (see Figure 6.13, second row). The reason for not counting multiple bonds in the VSEPR electron pair count can be understood from the discussion in the previous section: the π orbitals are located above and below the plane of the molecule, and therefore, do not contribute to the congestion of electron density around the central atoms.

When the central atom in a molecule is sp^3-hybridized, the four electron pairs point toward the corner of a tetrahedron. If there are three atoms attached, like in NH_3, a triagonal pyramidal shape is obtained, as shown by the central structure in row 3 of Figure 6.13. If there are only two atoms attached to a sp^3-hybrid, such as in water, the molecular shape obtained is also referred to as "bend," but with a bond angle less than 109° (see later).

Observed molecular parameters, such as the bond angles mentioned above, lead to the conclusion that nonbonding electron pairs on a molecule (such as the lone pair on the nitrogen atom in ammonia) occupy a larger volume than each of the three bonding pairs. The consequences of this statement can be demonstrated using the case of four electron pair systems around a central atom. In methane, the four (bonding) electrons point toward the corners of a tetrahedron, with all six H—C—H bond angles exactly of 109.5°. In ammonia, the three bonding pairs and the lone pair on the N atom point toward the corners of a slightly distorted tetrahedron, with all three H—N—H bond angles slightly less than 109.5°, namely 107°. The nonbonding pair on the nitrogen atom in ammonia occupies a larger volume than the bonding electron pairs and "pushes" the bonding electron pairs further away. This can be understood from the viewpoint that for a bonding pair, there are two nuclei exerting attractive forces on the bonding electrons, whereas the nonbonding pairs experience the attractive force of one nucleus only, and consequently occupy a larger volume.

In the water molecule, the H—O—H bond angle is even lower at 104.5°, due to the two nonbonding electron pairs located at the oxygen atom. The geometries of the third row of molecular shapes in Figure 6.13 are referred to as tetrahedral, trigonal pyramidal, and bend, with the bond angles decreasing from 109.5° to about 104°.

The fourth row of molecular shapes shown in Figure 6.13 includes species that have five (bonding or nonbonding) electron pairs around a central atom. In the phosphorous pentachloride (PCl_5) molecule, there are 40 valence electrons or 20 pairs. The Lewis structure is shown in Figure 6.14, top row. The molecular shape is described as a trigonal bipyramid. In this structure, the P and three Cl atoms are in one plane with the three bonding orbitals pointing toward the corners of an equilateral triangle. The two remaining Cl atoms are located below and above the plane of the equilateral triangle. Thus, there are three Cl—P—Cl bond angles of 120° and six Cl—P—Cl bond angles of 90°. The bonding scheme allowing more than four electron pairs around a central atom involves d orbitals; the hybridization of the P atom is generally referred to as dsp^3. Although the ground state of the P atom does not fill any d orbitals (see the discussion of the Aufbau principle in the previous chapter), d orbitals are available, albeit unoccupied, to third row elements.

Another molecule that has five electron pairs around the central atom is sulfur tetrafluoride, SF_4. There are 34 valence electrons or 17 pairs. The Lewis structure of SF_4 is shown in the second row of Figure 6.14. Spectroscopic experiments (see Chapter 15) show that the right-most structure (the "seesaw" structure) is the correct one. In terms of the VSEPR model, this can be explained by the fact that in the seesaw structure, there are two 90° interactions of the free electron pair with S—F bonds, whereas in the trigonal pyramidal structure, there would be three 90° interactions. Thus, the seesaw structure is energetically more favorable. Experiment also shows that the free electron pair slightly deforms the two axial S—F bonds such that their bond angle is slightly less than 180°.

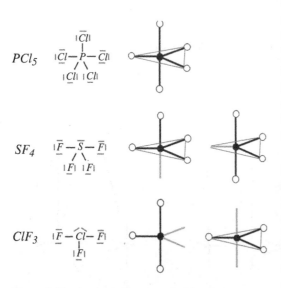

Figure 6.14 Lewis structures and molecular geometries of three species that have five electron pairs around the central atom. Nonbonding electron pairs are shown in gray.

Finally, a molecule will be introduced that has five electron pairs around the central atom, but forms a planar, triangular molecular shape. This molecule, chlorine trifluoride, ClF_3, has 28 valence electrons or 14 pairs. Its Lewis structure is shown in the third row of Figure 6.14. Three electron pairs are directly involved in the chemical bonds that point toward the corners of an equilateral triangle, forming the planar structure. The other two electron pairs are directed toward points above and below the central chlorine atom.

If there are six electron pairs and six ligands, as in sulfur hexafluoride, SF_6, the electron pairs point toward the corners of an octahedron, as shown in Figure 6.13, bottom row. If there are six electrons around the central atom, but only five or four ligands, a square pyramidal or a square planar molecular shape is achieved as shown in Figure 6.13, bottom row.

Example 6.1 Draw VSEPR-based structures, name the species, and describe the molecular shapes of NO_2, NO_2^-, NO_3^-, BF_3, PF_3, PF_5, and H_3O^+. What is the hybridization of the central atom in these species?

Answer: See answer booklet

6.5 Molecular Polarity

We discussed earlier (Chapter 4) that some elements, such as fluorine and oxygen, exert a stronger attractive force on electrons than other elements in covalent bonds, due to their high electronegativity. This results in polar covalent bonds. In diatomic molecules incorporating a polar covalent bond, such as in H—F, the unequal distribution of electrons leads to a molecular dipole moment, which is a physically measurable quantity and determines how such a molecule will react in a static electric field. Molecules that have a dipole moment, or are "polar," align in an electric field such that the negative end of a molecule is attracted to the positively charged plates of a capacitor and *vice versa*. Polar molecules also tend to line up in the liquid and solid phases as shown in Figure 7.8 to cause what we shall refer to later as "intermolecular forces."

Dipole moments are vectorial quantities, implying that they have a magnitude and direction. This aspect is important when considering a polyatomic molecule, since the individual bond dipole moments need to be added up vectorially. In order to do so, the geometry of the molecule needs to be established first, since the geometry strongly influences whether or not a molecule is polar. Let us look at carbon dioxide for an example. The C—O bond is polar, since oxygen has a larger electronegativity value than carbon. However, the CO_2 molecule is nonpolar, since the two C—O bond dipole moments cancel out, as shown in Figure 6.15a.

However, other triatomic molecules may be polar, for example, sulfur dioxide. In SO_2, the bond angle is approximately 120°, since the central S atom is sp^2-hybridized. Thus, when the two S—O bond dipole moments are added vectorially, the resulting dipole moment lies in the direction indicated by the fat black arrow in Figure 6.15b. In the water molecule, we also need to add the two H—O bond dipole moments vectorially, but the two bond dipole moments point toward the oxygen. Hence, the molecular dipole moment is in the same orientation as shown in Panel (b), but in opposite direction (the negative end is toward the oxygen atom).

A molecule such as sulfur trioxide, SO_3, has 24 valence electrons, and the central sulfur atom is sp^2 hybridized. Consequently, the three σ-bonding electron pairs point toward the corners of an equilateral triangle, and the molecule is planar. The π orbital is below and above the plane of the molecule. Each S—O bond is polar, but the molecule is nonpolar, as shown in Figure 6.15c. This happens because the vectorial addition of any two bond dipole moments exactly equals in magnitude to that of the remaining bond dipole moment, but with an opposite sign.

A similar situation arises in tetrahedral molecules, such as carbon tetrachloride, CCl_4. Although the C—Cl bond is polar, CCl_4 is a nonpolar molecule since the pairwise vectorial sum of any two C—Cl bond dipole moments is exactly equal in magnitude, but opposite in sign, to the sum of the other two bond dipole moments, as shown in Figure 6.15d. If one chlorine atom in CCl_4 is substituted by another atom, such as hydrogen, a new molecule is obtained, in this case trichloromethane, also known as chloroform. The bonding electron pairs

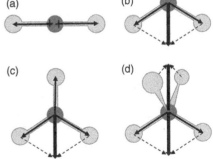

Figure 6.15 Vectorial addition of bond dipole moments for (a) linear, (b) bend (angular), (c) triangular, and (d) tetrahedral molecular geometry. In all drawings, the arrowhead is the negative end of the bond dipole moment.

still point toward the corners of a tetrahedron, but the bond dipole moments no longer cancel out, and chloroform is a polar molecule with the direction of the dipole moment along the C—H bond.

The previous discussion strongly emphasized the concept of molecular polarity, since this quantity allows us to predict many physical properties and chemical interactions. In polar molecules, the interactions between the molecular dipole moments of individual molecules lead to strong interactions between these molecules. These "intermolecular" forces affect physical properties such as boiling point and vapor pressure that will be discussed in detail in Chapter 7.

Example 6.2 Which of the species in Example 6.1 are polar?

Answer: NO_2, NO_2^-, PF_3, and H_3O^+

Example 6.3

(a) Which elements discussed so far can form triple bonds?
(b) Name molecules or ions that incorporate triple bonds for each of these elements.
(c) What is the hybridization of the atoms involved in triple bonds?

Answer: See answer booklet

6.6 MO Theory

VB theory, as discussed in Sections 6.2 through 6.4, allows us to predict molecular structures and properties and is extremely useful in obtaining an intuitive grasp of covalent chemical bonding. However, as presented, it actually is somewhat vague in exactly how a chemical bond forms: a bond was just described as a spatial overlap of atomic valence orbitals. Furthermore, the energetics of bond formation has not been discussed so far, nor has been another property of chemical bonds: the existence of "antibonding" orbitals. This latter aspect again is important from the viewpoint of spectroscopy: we shall see that electrons from a bonding orbital can be promoted by light into antibonding orbitals, just like electrons can be promoted into higher atomic orbitals, as discussed for the hydrogen atom. These spectroscopic results reveal the energy difference between bonding and antibonding orbitals, and consequently a spectroscopic tool to measure the energy advantage of bond formation. The good news is that MO and VB theory are not contradictory, but rather complementary, with MO theory more likely to give quantitative and computational results, and VB theory more likely to give intuitively understandable structural results.

MO theory is based on a description of MOs formed by mathematically overlapping atomic orbitals. MOs are delocalized over the entire molecule, in contrast to the localized description of chemical bonds in the VB model. However, the MOs often resemble the localized VB orbitals, as we shall see later. MO theory tends to become mathematically quite complicated, so we restrict ourselves to a qualitative discussion of its principles. Furthermore, MO calculations for larger molecules used to be impractical due to the computational complexities. However, the enormous computational power provided by modern computer cores and the development of advanced computational algorithms have brought MO calculations to the realm of home computers, and many advanced Chemistry laboratory experiments actually involve MO calculations implemented in software such as HyperChem or Gaussian.

We start the discussion of MO methodologies with the smallest molecule possible, the H_2^+ molecular ion. By the way, the H_2^+ species does exist and can be observed, for example, in interstellar space where it is produced from H_2 by cosmic ray bombardment, which knocks one electron out of the covalent bond. In MO theory, one starts with a hydrogen atom and adds a second proton to describe this species. Thus, we have two nuclei, which we label nucleus "a" and "b," and one electron.

Next, to form a bond between the two nuclei, we assume that the electron must be delocalized between the two nuclei. Thus, its wave function must somehow reflect this delocalization. This can be achieved by writing the wave function of the single electron as a sum of two hydrogen 1s wave functions:

$$\psi = c_a \phi_{1s_a} + c_b \phi_{1s_b} \tag{6.1}$$

For the distinction of atomic and molecular wave functions, we denote the atomic wave functions by the symbol ϕ and the molecular wave function by the symbol ψ. Furthermore, we add the subscripts $1s_a$ and $1s_b$ to indicate that the atomic orbitals are 1s orbitals located at nucleus "a" or "b." We call the approach shown in Eq. 6.1 the linear combination of atomic

Figure 6.16 Bonding (a) and antibonding (b) molecular orbitals formed from the linear combination of hydrogen 1s atomic orbitals at nuclei a and b. (c) Energy of these molecular orbitals, in relation to the two atomic orbitals ϕ_{1s_a} and ϕ_{1s_b}.

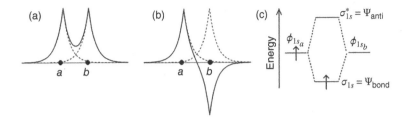

orbitals (LCAO) method, and the coefficients c_a and c_b indicate how much each of the two atomic wave functions contribute to the molecular wave function. Since the electron is equally shared by the two nuclei, it follows that c_a and c_b are numerically equal, but may have opposite signs:

$$c_a = \pm c_b \tag{6.2}$$

This leads to the formation of two MOs, a "bonding" orbital ψ_{bond} and an "antibonding" orbital ψ_{anti}

$$\psi_{bond} = c_b(\phi_{1s_a} + \phi_{1s_b}) \tag{6.3}$$
$$\psi_{anti} = c_a(\phi_{1s_a} - \phi_{1s_b}) \tag{6.4}$$

where the values of the expansion coefficients c_b and c_a ascertain that both wave functions represent a probability of 1 of finding the electron (this is referred to as a "normalization" of wave functions). A plot of these two wave functions is presented in Figure 6.16, with each of the two 1s wave functions identical to the curves labeled R_{10} in Figure 5.10.

Figure 6.16a shows an area of increased overlap of the atomic orbitals between the nuclei "a" and "b," which represents the chemical bond. In the antibonding orbital (Figure 6.16b), there is a nodal plane or a region with a zero value of the wave function, between the nuclei, which is typical for antibonding orbitals. Panel (c) shows a typical energy level diagram depicting the quantitative nature of MO theory. It shows, in an energy graph, the energies of the two atomic orbitals, ϕ_{1s_a} and ϕ_{1s_b}, located at nuclei "a" and "b," before bond formation, and the energies of the MOs with respect to these atomic orbitals. In accordance with the nomenclature of bonds introduced in the discussion of VB theory, we call the bonding orbital a σ bond. Since it arises from a 1s orbital, we designate it a σ_{1s} orbital. The antibonding orbital is designated a σ_{is}^* orbital. The arrows in Panel (c), as usual, denote a single electron. In Figure 6.16c, we have one electron on one of the hydrogen nuclei, and it will be found in the σ_{1s} orbital once the bond has formed. The MO theory actually reveals the energy difference between the σ_{1s} and the ϕ_{1s_a} orbitals; thus, it is a quantitative description of the chemical bond energy.

The treatment of the hydrogen molecule, H_2, will be similar, and a second electron is added to the system discussed so far. Without going into theoretical details, the result can be summarized by Figure 6.17a. In this diagram, both hydrogen

Figure 6.17 Molecular orbital diagrams for (a) H_2, (b) H_2^-, and (c) Li_2. Notice that the energy in Panel (c) is not drawn to scale: the energy splitting between the 1s and the 2s orbitals is larger than indicated.

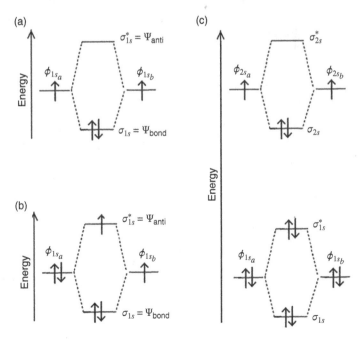

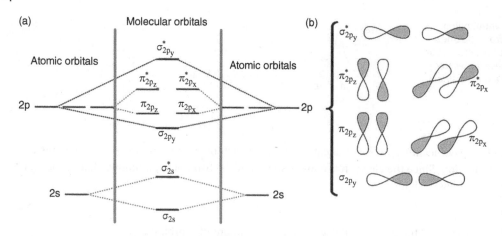

Figure 6.18 (a) Energy level diagram of the MO's formed from the overlap of 2s and 2p orbitals in diatomic homonuclear molecules such as N_2, O_2, and F_2, (b) visualization of the "head-on" overlap of the $2p_y$ orbitals to form the σ_{2p_y} and the lateral overlap of the $2p_x$ and $2p_z$ orbitals to form the two π_{2p} orbitals. The direction of the Cartesian axes is the same as in Figure 6.10. Shaded areas of the orbitals correspond to the sign of the wavefunction.

1s orbitals now are occupied by one electron, and the σ_{1s} bond formed has the two electrons in it, with opposite spin directions. Obviously, MOs follow the same rule as atomic orbitals that they can accommodate maximally two electrons, but only if their spins are opposite.

We define the bond order BO at this point by Eq. 6.5:

$$BO = \tfrac{1}{2}\{n_{\text{bonding orbitals}} - n_{\text{antibonding orbitals}}\} \tag{6.5}$$

where $n_{\text{bonding orbitals}}$ denotes the number of electrons in bonding orbitals and $n_{\text{antibonding orbitals}}$ the number of electrons in antibonding orbitals. Thus, we see that the H_2^+ molecular ion has a bond order of ½, and the hydrogen molecule has a bond order of 1, corresponding to the single bond in the Lewis structure, and the VB model of a simple overlap of the two spherical 1s orbitals. Another species, the short-lived H_2^- molecular ion has an MO diagram shown in Figure 6.17b and has a bond order of ½.

For elements of the second row in the periodic chart, both the inner and valence electrons need to be considered. This is shown for the gaseous molecule Li_2 (dilithium) in Figure 6.17c. Again, this is not a fictitious molecule, but exists at about 1% abundance in lithium vapor. Here, we see that the σ_{1s} and the σ_{1s}^* orbitals are completely filled and do not contribute to the covalent bond. The bond is formed from the overlap of the ϕ_{2s} orbitals on the lithium atoms, each of which has one electron in the 2s orbital. The resulting σ_{2s} orbital is filled with the two 2s electrons of each Li atom, but the σ_{2s}^* orbital is unoccupied. Thus, dilithium has a bond order of 1.

Things get more complicated for the diatomic molecules of nitrogen, oxygen, and fluorine since these elements fill 2p orbitals, and the overlap of these orbitals needs to be taken into account. An MO schematic of these molecules is shown in Figure 6.18a. In this figure, the left and right parts denote the atomic orbitals that are involved in forming the MOs, and the center part shows the MOs formed. The MOs formed from the 1s orbitals are omitted in this figure, since the bonding and antibonding orbitals are completely filled and do not contribute to the bond. The $2p_y$ orbitals on each atom form two MOs, denoted as σ_{2p_y} and $\sigma_{2p_y}^*$ as shown in Figure 6.18a, by head-on overlap, as discussed in the VB model. (See Figures 6.10 and 6.11. The orientation of the orbitals is the same in Figures 6.10 and 6.18.)

The $2p_x$ orbitals on both atoms overlap laterally and form the π_{2p_x} and $\pi_{2p_x}^*$ MOs. Similarly, the $2p_z$ orbitals overlap laterally and form the π_{2p_z} and $\pi_{2p_z}^*$ MOs. The π and π^* orbitals lie energetically between the σ_{2p_y} and $\sigma_{2p_y}^*$ orbitals, and the π_{2p_x} and the π_{2p_z} orbitals are degenerate, as are the $\pi_{2p_x}^*$ and the $\pi_{2p_z}^*$ orbitals. Figure 6.18b shows the shapes and signs of the atomic orbitals involved in the formation of these MOs.

Next, the occupied MOs in the nitrogen, oxygen, and fluorine molecules will be discussed. The MO scheme for N_2 is shown in Figure 6.19a. Nitrogen has five valence electrons, two in the 2s and three (unpaired) electrons in the 2p orbitals, as shown in Figure 6.19a. In the N_2 molecule, these six electrons occupy the three MOs that were labeled σ_{2p_y}, π_{2p_x}, and π_{2p_z} in Figure 6.18. The bond order, derived from this diagram, is 3, in agreement with the triple-bonded Lewis structure discussed earlier. In analogy to the atomic electron configuration introduced in Section 5.5, one would write the electronic configuration of the nitrogen molecule as

$$(\sigma_{1s})^2, (\sigma_{1s}^*)^2, (\sigma_{2s})^2, (\sigma_{2s}^*)^2, (\sigma_{2p_y})^2, (\pi_{2p_x})^2, (\pi_{2p_z})^2 \tag{6.6}$$

Notice that the orbitals corresponding to the first two terms in the expression 6.6 are not depicted in Figure 6.19a.

The MO diagram for the oxygen molecule, O_2, is given in Figure 6.19b. The oxygen atom has six valence electrons, two in the 2s and four in the 2p orbitals. In the O_2 molecule, the three MOs σ_{2p_y}, π_{2p_x}, and π_{2p_z} are filled with electron pairs, just as in the nitrogen molecule. The remaining two electrons occupy the $\pi^*_{2p_x}$ and $\pi^*_{2p_z}$ MOs that are degenerated. Thus, the two electrons are unpaired in the two orbitals, with parallel spins. The energy level diagram for the oxygen molecule shown in Figure 6.19b explains very impressively two facts about this species: First, the molecule has a bond order of two, and second, it is paramagnetic. The bond order of 2 is expected from a naive attempt to write a Lewis structure for O_2 (see Chapter 3). However, such a Lewis structure would have no unpaired electrons, which contradicts the fact that molecular oxygen is paramagnetic, and therefore, must possess unpaired electrons.

The F_2 MO scheme is presented in Figure 6.19c. Here, we see that four electrons occupy the $\pi^*_{2p_x}$ and $\pi^*_{2p_z}$ MOs. Consequently, the bond order in F_2 is 1, as predicted from the Lewis structure.

Example 6.4 Consider the peroxide ion, O_2^{2-}. Based on the MO energy level diagram shown in Figure 6.18a, (a) draw an MO scheme for this ion, and answer the following questions. (b) Is the species paramagnetic or diamagnetic? (c) What is the bond order in O_2^{2-}? (d) With what molecule is the peroxide ion isoelectronic? (e) Write the electronic configuration of O_2^{2-}.

Answer: (a) For an MO diagram, see answer booklet; (b) diamagnetic; (c) 1; (d) F_2; (e) $(\sigma_{1s})^2$, $(\sigma^*_{1s})^2$, $(\sigma_{2s})^2$, $(\sigma^*_{2s})^2$, $(\sigma_{2p_y})^2$, $(\pi_{2p_x})^2$, $(\pi_{2p_z})^2$, $(\pi^*_{2p_x})^2$, $(\pi^*_{2p_z})^2$.

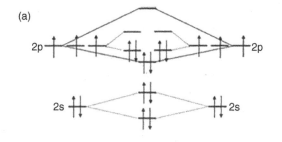
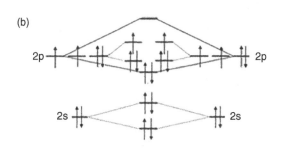
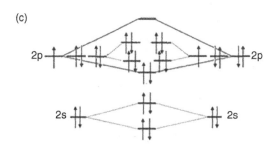

Figure 6.19 MO schemes for (a) N_2, (b) O_2, and (c) F_2.

MO theory, as mentioned earlier, has been implemented into several computer programs that predict the energies of MOs, as well as their shapes and other properties. To conclude, this section of Chapter 6, some of the results taken from a website http://ursula.chem.yale.edu/~chem220/chem220js/STUDYAIDS/MO-HTMLs/formaldehyde12mo-2-log.html will be reviewed. This website depicts the bonding and antibonding orbitals of a small molecule, formaldehyde, H_2CO, which was introduced in Chapter 3 (see Figure 3.2). In Figure 6.20, four of the computed MOs are shown to illustrate the shapes of these MOs (the computations also reveal the energies of these MOs). Panel (a) of Figure 6.20 shows a molecular orbital very close to the C atom, which – although it represents an MO – is composed mostly of the carbon 1s orbital. Panel (b) depicts an MO that has axial symmetry about the C—O bond and represents the σ bond. Panel (c) shows an MO that encompasses the two C—H bonds. Panel (d) depicts an MO that lies above and below the plane of the molecule and resembles the π-electron clouds drawn in the VB approach. This last panel illustrates a comment in the first paragraph of Section 6.4 that VB and MO theory are complementary and are two different models with similar outcomes.

Since these calculations also reveal the energies of the orbitals, they also predict the spectroscopic properties of molecules. Species containing the carbonyl functionality are known as ketones, and most ketones exhibit a

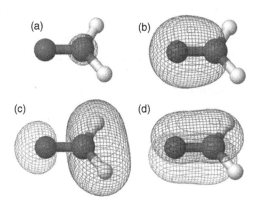

Figure 6.20 Shapes of four of the MOs from computational results. *Source*: From http://ursula.chem.yale.edu/~chem220/chem220js/STUDYAIDS/MO-HTMLs/formaldehyde12mo-2-log.html.

spectral feature in the ultraviolet absorption spectrum (see Chapter 15) where an electron is promoted from the π to the π^* orbital. This transition occurs at about 200 [nm] wavelength. The photon energy of this transition can be determined from the observed wavelength and is reproduced closely by the computational methods discussed above. An important aspect of MO-based molecular structures is that the geometry of molecules is obtained from a minimization of the total molecular energy: MO theory does not assume the shape of molecules based on hybridization, or empirical arguments, but obtains the minimum energy geometry by variation of nuclear coordinates.

Example 6.5 Referring to the discussion of hybridization and multiple bond description in VB theory (Section 6.3), and the discussion of MO theory and Figure 6.20, compare and contrast the two bonding theories when applied to formaldehyde, CH_2O.

Answer: See answer booklet

To close this chapter on bonding theory, a molecule will be introduced that incorporates many concepts discussed so far. This molecule, benzene, C_6H_6, is the prototypical species for a property known as aromaticity and will be discussed in much more detail in organic chemistry. The structure of this molecule eluded chemists for quite some time, and its structure was written as a resonance situation as shown in Figure 6.21a. However, this representation is inaccurate in the sense that it implies that there are single and double bonds between the carbon atoms. In reality, all six carbon–carbon bonds are the same. In the modern structural explanation, all six carbon atoms in benzene are sp^2-hybridized and form a planar hexagon structure, with the p_z orbitals (using the coordinate convention shown in Figure 6.10) to form ring-shaped (toroidal) MOs above and below the hexagon. This is indicated by the circle drawn inside the hexagon shown in Figure 6.21b. This representation is commonly used as a shortcut to indicate a benzene structure. It emphasizes that all C—C bonds have the same length, whereas the structure in Figure 6.21a suggests single and double bonds. The MO scheme of benzene describes in detail the bonding and antibonding energy levels of the conjugated π-electron system. Benzene is nonpolar due to its high symmetry, but polar aromatic molecules occur frequently in organic chemistry, such as pyridine, in which one of the CH groups of benzene is substituted by a nitrogen atom. This molecule, C_5H_5N, has a near-identical hexagonal structure as benzene, but it is slightly polar due to the lower symmetry and the presence of a heteroatom (the nitrogen atom). Many other cyclic molecules with nitrogen atoms in the ring structure are aromatic as well and play important roles in biochemistry as well: the four bases in DNA (see Chapter 7) contain N atoms in the ring systems, and the π-electron systems in these bases contribute to the base-stacked structure of DNA.

Graphite, to be discussed in more detail in Chapter 7, is a black allotrope (see Section 3.3) of the element carbon that consists of layers of fused, aromatic six-membered carbon rings as shown in Figure 6.22. Each of these layers is referred to as a "graphene" sheet that is a covalent network solid, to be discussed in more detail in the next chapter. In these graphene sheets, the delocalized π electrons shown in Figure 6.21b extend over the entire graphene sheet, which make graphite a good conductor of electricity with some metallic properties. In graphite, these layers are loosely stacked and can be easily cleaved from each other. The "lead" in a pencil, which is a mixture of graphite and a binder such as clay (and contains no lead) leaves a black line on paper that consists of graphite bits rubbed off by the roughness of the paper.

However, each graphene sheet has surprisingly strong mechanical properties. The carbon-fiber material, a high-tech composite of carbon fibers embedded in an epoxy-based matrix, owes its strength to the carbon fibers that consist of long stretches of structures shown in Figure 6.22. This material is less dense than many metals, yet exhibits tensile strength comparable to that of metals. Consequently, parts of airplanes, or entire assemblies like bicycle frames, are manufactured from carbon fiber composite materials.

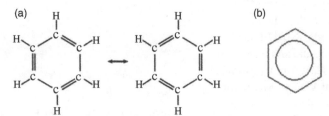

Figure 6.21 (a) Resonance structures for benzene and (b) shorthand benzene structure, indicating delocalized electrons in an aromatic system.

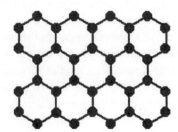

Figure 6.22 Structure of a graphene sheet.

Example 6.6

(a) What is the hybridization of the carbon atoms in benzene and graphene?
(b) How does this hybridization determine the properties of graphene sheets?
(c) Read up on carbon nanotubes on the web. How are carbon nanotubes related to graphene?

Answer: (a) sp^2; (b) see answer booklet; (c) see answer booklet

Further Reading

https://en.wikipedia.org/wiki/Chemical_bond

7

Solids and Liquids: Bonding and Characteristics

In this chapter, we shall build on our knowledge of chemical bonding to discuss the substances that surround us in everyday life. A quick look around you will convince you that, aside from the air that you breathe, most of the matter around you is in the solid state at ambient temperature. But we know from everyday experience that at increasing temperature, many solids – for example, all metals – melt and even vaporize at sufficiently high temperatures. Many other solids, such as salts or plastics, melt as well or decompose at increasing temperature. If the temperature is lowered, everyday liquids, such as water or diesel fuel, will solidify. Even gases, such as the components of air, will liquefy at low temperatures and even solidify when the temperature is lowered far enough. The transitions of one and the same compound from solid to liquid to gaseous states are known as "phase transitions" since we refer to the solid, liquid, and gaseous states as "phases." The three phases mentioned above – solid, liquid, and gaseous – are associated with temperatures typically encountered in chemistry; at very high temperatures or other extreme conditions, another phase, known as the "plasma" phase is encountered that consists of ionized molecules or atoms that have quite different chemical properties and will only be mentioned in passing throughout these chapters.

The phase transitions mentioned above have specific terms associated with them: the transition from solid to liquid state is generally referred to as melting, and from liquid to gaseous state as vaporization or evaporation. The reverse processes are known as condensation and solidification. Condensation also refers to the direct transition from gaseous to the solid phase, whereas the transition from solid to gaseous state is referred to as sublimation. These processes will be discussed in more detail in Chapter 9 in the context of vapor pressure and Chapter 11 in the context of the energetics of phase transitions.

The discussion of the states of matter will begin with the solid state, which is, as mentioned, the most common phase of matter at typical conditions encountered in everyday life. We classify pure solid substances as ionic solids, metals, covalent solids, and molecular solids, including macromolecular solids.

7.1 Metals and Semiconductors

We start the discussion of solids with the properties and structure of metals. At ambient temperature, all elemental metals are solids, except mercury, which is a liquid at room temperature. Gallium is a solid, but melts at about 29 [°C]. Metals are historically the most versatile materials for making implements such as tools, and they can be shaped in the solid state relatively easily since they are malleable. The structure of metals, and metallic bonding, is very different from other solids, such as the ionic solids discussed previously, and covalent solids to be introduced in Section 7.3.

There are a few properties of metals that distinguish them from other solids and make them useful for everyday life. All metals are good conductors of electricity, with Al, Cu, and a few others being used extensively for this purpose. You may be familiar with copper wire being used in household wiring because it has particularly high conductivity. Metals generally are malleable and are good conductors of heat. Pure metals not covered with an oxide layer exhibit metallic luster, hence they are used in jewelry and mirrors. These properties, including luster, malleability, and electric and heat conductivity, are all consequences of the particular bonding forces that hold metal atoms in their lattice. These bonding forces are generally referred to as metallic bonding. We shall discuss metallic bonding along with the bonding that is found in semiconductors, since metals and semiconductors share common bonding principles.

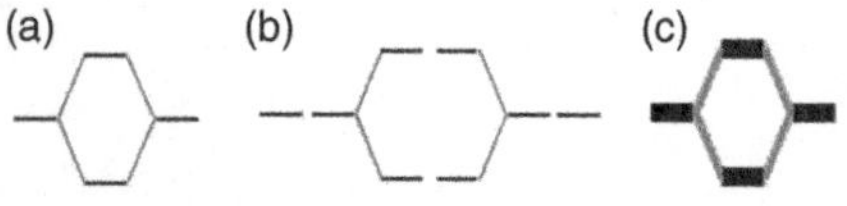

Figure 7.1 Bonding and antibonding molecular orbitals from (a) 2, (b) 4, and (c) multiple interacting atomic orbitals.

Along the zigzag line in the periodic chart (Figure 2.1) are a few elements that are considered semiconductors, such as Si and Ge. A piece of pure silicon actually appears nearly metallic in appearance, since it is a shiny material, but it is much more brittle and not malleable like a metal. Many mixtures of metals and nonmetals are semiconductors as well. As the name of this class of elements implies, they do not conduct electricity as efficiently as metals, but have some properties that make semiconductors some of the most important materials in modern technology, since every computer chip contains millions of tiny semiconductor "gates," or switches.

To understand bonding in metals and semiconductors, we go back to the discussion of how bonding and antibonding orbitals are formed from atomic orbitals. Figure 7.1a shows the energy levels of bonding and antibonding orbitals in a situation such as H_2. From the two atomic orbitals, we obtained a lower energy bonding and a higher energy antibonding orbital. In the case of more than one orbital on each atom, as shown in Figure 7.1b, we obtain (generally degenerate) pairs of bonding and antibonding orbitals. If we take this approach a step further and have a large number of degenerate or near-degenerate atomic orbitals interacting, they form a large number of degenerate or near-degenerate bonding or antibonding molecular orbitals, as shown as the black "bands" in Figure 7.1c.

Solid state physicists call the lower of these "bands" the "valence band," and the higher one the "conductivity band." The energy difference between these bands is referred to as the "band gap." In the valence band, the electrons are relatively confined to their respective atoms, whereas in the conductivity band, the electrons are delocalized and relatively free to move around over the dimensions of the crystal. In the context of this description, there are three classes of compounds to consider: insulators, semiconductors, and metals. In insulators, the band gap is so large that under normal circumstances, electrons cannot be promoted into the conductivity band; therefore, insulators are – as the name implies – not conducting electricity under normal conditions. However, if the energies are very high, even air, which is generally a good insulator, becomes conductive, for example, in a lightning strike.

In semiconductors, the band gap may be between 10 and 100 [kJ/mol]. This is an energy range that is accessible under everyday conditions. This makes semiconductors important materials in modern electronics. When photons with energies within the band gap fall onto a semiconductor crystal, they can promote electrons from the valence band into the conductivity band, thereby changing the resistance of the crystal. This principle is used in photodiodes that can serve as detectors of light. Conversely, if electrons are promoted into the conductivity band, for example by an electric potential, they can emit a photon of energy corresponding to the band gap. This principle is used in light emitting diode (LED) displays of cell phone or TV screens. In LEDs, the band gap of a semiconductor is adjusted by suitable doping of the semiconductor material to emit light of different colors. "Doping" here implies that small "impurities" of metals are introduced into the semiconductor crystal to create positive or negative sites in the crystal.

In metals, the valence and conductivity bands basically overlap to allow electrons to move freely over the entire crystal lattice of a piece of metal. Thus, we may envision a metal to consist of an arrangement of (spherical) atom "cores," which are to atoms devoid of the valence electrons. The valence electrons from each atom are delocalized to form a "sea of electrons" that permeate the entire piece of metal and provide sufficient energy to hold all the cores in place. These delocalized electrons provide the conductivity of electric current, since a small potential applied over the metal will move the negative electrons toward the positive potential.

The "sea of electrons" also explains the good thermal conductivity of metals. The electrons can easily transmit thermal energy by moving around freely. The malleability of metals results from the fact that, when the local arrangement of the metal cores is perturbed, the nondirectional "sea of electrons" provides the force to hold the cores in place, even if the lattice is perturbed. This is not the case in an ionic lattice, where the forces between ions is strictly determined by local charge distribution, that is, neighboring effects of positive and negative charges. The luster of metals is also determined by the broad energy level distribution.

There is an interesting difference in the electric conductivity of metals and semiconductors. In metals, the electric resistance increases with increasing temperature. Within the "sea of electrons" model, we can interpret this effect by the more disordered state, and larger amplitude of motion, of the metal core arrangement with increasing temperature. We shall see later in the next chapter that temperature is a measure of the motion of atoms and molecules in all phases. Thus, increasing the temperature of a metal causes the metal lattice to become less ordered, thus making it harder for the electrons to move through the metal lattice. In semiconductors, on the other band, the increasing disorder of the lattice is of secondary importance compared to the fact that more electrons are promoted into the conductivity band. Therefore, semiconductor exhibit lower resistance, or higher conductivity, at increasing temperature.

7.2 Ionic Solids

Ionic solids were first introduced in Chapter 3 when we discussed chemical compounds. Ionic compounds consist of positively charged cations and negatively charged anions held together by electrostatic forces between the ions in a crystal lattice (see Figure 3.1). There are several physical properties of ionic compounds that directly result from the ionic structure. Ionic compounds generally are more brittle than metals and tend to cleave along crystal faces. This is in contrast to metals that deform, rather than cleave, since the "sea of electrons" in metals counteract the breaking of metals into fragments. Ionic crystals are poor conductors of electricity. As we have seen above, electric conductivity results from the relatively unrestricted motion of charges such as electrons; however, in an ionic crystal, the electrons are tightly held by the atoms. However, when an ionic crystal is dissolved (in water, for example), the resulting solution is a good conductor of electricity, as the (solvated) ions in solution are free to move to the cathode and anode, thus allowing charge to be transported. However, the solubility of ionic compounds varies enormously between different compounds. Whereas NaCl readily dissolves in water to quite high concentrations (over 300 [g] of NaCl can be dissolved in 1 [kg] of water at room temperature), AgCl is sparingly soluble, and only a little over 1 [mg] of it will dissolve in 1 [kg] of water. Both NaCl and AgCl have the same crystal structure, and the differences in solubility is attributable to different ionic forces in the solid and the different solvation energy of the ions by the solvent molecules. However, it is important to emphasize that the dissolved material will completely dissociate in aqueous solution. Solubility and hardness of ionic compounds are strongly influenced by the "lattice energy." This is the energy change associated with the process of forming the lattice, i.e. the ionic arrangements in the solid from gaseous ions. For solid NaCl, the lattice energy would be the energy change for the process

$$Na^+(g) + Cl^-(g) \rightarrow NaCl(s) \tag{7.1}$$

The sign of this energy change (−786 [kJ/mol]) indicates that the process is exothermic (i.e. releases heat, see Chapter 11) and includes the attractive and repulsive forces, according to Coulomb's law, between oppositely and equally charged ions, respectively, as well as any London dispersion forces (see Section 7.4.3) between the ions in the crystal. Details of determining the lattice energy of a salt will be discussed in detail in Chapter 11 and is depicted in Figure 11.6, for the reaction between solid magnesium and molecular oxygen to form MgO. This procedure involves quite a few individual steps for which the energies can be determined experimentally and allows the lattice energy defined above to be derived.

Molten salts are good conductors of electricity, since the ions are free to move. When these mobile ions reach the electrodes, that is, the source of the electric potential, chemical reactions may occur: for example, when positively charged Na^+ ions in a pool of molten sodium chloride reach the electrode, they may pick up an electron and become metallic sodium. In fact, this is the way metallic sodium is produced commercially. However, fairly high temperatures are needed to melt ionic solids due to the strong electrostatic forces that keep the ions in the lattice. Sodium chloride, for example, has a melting point of 801 [°C], similar to that of many metals.

The melting temperature of an ionic compound alluded to above depends on the energy E of interaction between ions in an ionic solid, which is given by (see Eq. 5.17)

$$U = -k \frac{q_1 q_2}{r} \tag{7.2}$$

where q_1 and q_2 are the ionic charges, r is the distance between them, and k is the Coulomb constant. Eq. 7.2 can be obtained by a simple integration of Coulomb's law, which specifies the force F between charged particles as

$$F = k \frac{q_1 q_2}{r^2} \tag{7.3}$$

In order to conform to standard sign conventions used in thermochemistry (see Chapter 11), we drop the negative sign in Eq. 7.2 and consider the energy of interaction between a positive and a negative charge as a negative quantity, lowering the energy of (or stabilizing) the system. Therefore, the lowering in energy of an ionic crystal is larger for ions with higher ionic charge, and lower for larger ions, since the distance r in Eq. 7.2 is larger. Thus, we may predict CaS to have a higher melting point than NaCl (2525 vs. 801 [°C], respectively) since the attractive interaction in CaS are larger (by a factor of 4) than in NaCl. Furthermore, we predict that NaCl has a higher melting point than CsI (801 vs. 621 [°C], respectively) since the ions are much smaller in NaCl than they are in CsI.

7.3 Covalent Solids

There are a few elements and chemical compounds that are described as covalent solids. An example of a covalently bonded molecular solid is quartz, SiO_2, and an example of a covalently bonded element is diamond. In both of these, a network of covalent bonds makes up the entire crystal such that a small diamond actually is one molecule. In diamond, each sp^3-hybridized (see Chapter 6) carbon atom extends four valence orbitals into the corners of a tetrahedron, and therefore has four covalent bonds to its neighboring carbon atoms. This is shown schematically in Figure 7.2. The fact that each carbon atom has four covalent bonds to other carbon atoms in the molecular network of bonds makes diamond the hardest material known.

Figure 7.2 Covalent network found in diamond. The four covalent bonds pointing toward the corners of a tetrahedron are shown for the carbon atom marked with a white dot.

Quartz forms a similar network of bonds, shown schematically in Figure 7.3a. The silicon atoms are surrounded tetrahedrally by four oxygen atoms that themselves are bonded to other silicon atoms. Quartz is one of the most abundant minerals on earth, being composed of the two most frequently found elements in the earth's crust. A quartz crystal, as shown in Figure 7.3b, can be viewed as one large molecule made up of a covalent network of bonds. Such crystals can reach lengths of over a meter.

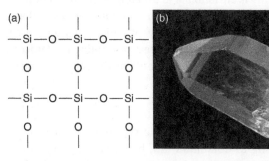

Figure 7.3 (a) Schematic of the covalent network found in quartz. The arrangement of the oxygen atoms around the Si atoms is tetrahedral. (b) Crystalline quartz.
Source: (b) Joe P/Adobe Stock.

When molten quartz is rapidly cooled ("quenched"), it may form an amorphous solid known as "fused silica" or "fused quartz," which is a form of "glass." Glasses, in general, are amorphous, noncrystalline substances that have a fairly broad "melting range," as opposed to the sharp melting temperature of pure, crystalline compounds or metals. In the manufacturing of window glass, for example, impurities such as CaO, B_2O_3, Al_2O_3, or others are added to silicon dioxide as impurities. When a molten mixture of these compounds is cooled, crystallization is prevented by the added impurities and the mixture forms what is referred to as a supercooled liquid. The ability to shape softened, or partially molten, glass into different shapes, such as laboratory beakers, tea pots, etc., is due to the formation of a solid with the properties of a supercooled liquid.

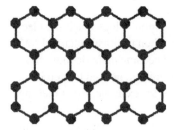

Figure 7.4 Structure of a graphene sheet.

Covalent solids, such as diamond, are very good conductors of heat but very poor conductors of electricity. This can again be understood from the bonding characteristics in diamond. Since the electrons are localized in the covalent network, the electrical conductivity is low, but heat is transported well by the network of bonds: heat transfer, in general, occurs mostly through vibrational energy transfer. The other allotrope of carbon, graphite, which was mentioned at the end of Chapter 6, consists of layers of fused six-member carbon rings as shown in Figure 7.4 that are stacked in the graphite lattice. Due to the delocalized π electrons that span the entire fused ring system, graphite is a good conductor of electricity.

Example 7.1 Distinguish molecular network solids from ionic solids. Give an example of both.

Answer: See answer booklet

Example 7.2 (a) Describe the dominant features of metallic bonding. (b) What distinguishes a metallic conductor from semiconductors? (c) List three characteristic properties of metals.

Answer: See answer booklet

7.4 Intermolecular Forces

Before further discussing solid (and liquid) substances, we need to explore in more detail the forces that act between atoms and molecules. In the previous sections, we discussed metals, ionic solids, covalent solids, and the forces acting between atoms or ions. However, there are three other weaker forces that are important to explain the interactions between molecules and atoms. These are – in decreasing order of strengths – hydrogen bonding, dipole-dipole, and van der Waals interactions that are the most prevalent forces acting between molecules. These forces can be readily explained from our understanding of molecular bonding, structure, and polarity discussed in Chapter 6 and will be presented next.

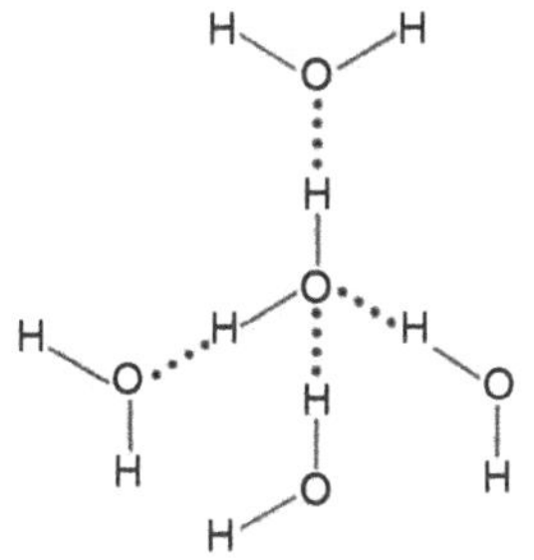

Figure 7.5 Hydrogen bonding (dashed lines) in liquid water.

7.4.1 Hydrogen Bonding

Hydrogen bonding is observed between lone electron pairs on electronegative atoms (such as nitrogen, oxygen, and fluorine) and hydrogen atoms covalently bound to electronegative atoms. A typical example of hydrogen bonding is shown in Figure 7.5, which depicts hydrogen bonding in liquid water. Here, the two hydrogen atoms covalently bonded to oxygen are denuded of electron density due to the large electron withdrawing tendency of oxygen. Thus, the hydrogen atoms attempt to gain electron density from elsewhere, namely from the nonbonding electron pairs from other oxygen atoms in water. These hydrogen bonds are indicated by the dotted "bonds" in Figure 7.5 that coincide with the direction of the two nonbonding electrons on each oxygen atom. Hydrogen bonding is also observed in ammonia and hydrogen fluoride. This is demonstrated by the boiling points observed in these liquids, shown in Figure 7.6. In general, the boiling points of substances increase with increasing molecular mass, as shown for the series CH_4, SiH_4, GeH_4, and SnH_4 in Figure 7.6 (Group 4 hydrides). The reason for this will be discussed in Section 7.4.3. The comparatively high boiling points of ammonia (−33 °C), hydrogen fluoride (20 °C), and water (100 °C) defy this trend and are much higher than expected. This is, of course, due to the hydrogen bonding in these three compounds. Hydrogen bonding is the strongest of the three intermolecular forces. The strength of a hydrogen bond is typically about 10–40 [kJ/mol], weaker than a typical chemical bond (100–400 [kJ/mol]).

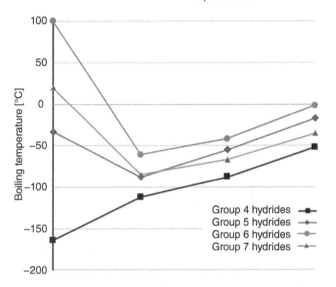

Figure 7.6 Boiling points of hydrides of elements, by groups: Group 4 (CH_4, SiH_4, GeH_4, SnH_4), Group 5 (NH_3, PH_3, AsH_3, SbH_3), Group 6 (H_2O, H_2S, H_2Se, H_2Te), and Group 7 (HF, HCl, HBr, HI).

Hydrogen bonding also occurs within alcohols, where the lone pairs on the oxygen atom of the —OH group interact strongly with H atoms of another molecule. Hydrogen bonds that form between carbonyl groups and amine groups are of great significance, as shown in Figure 7.7a. Proteins are held in their biologically active shapes by hydrogen bonds between the N—H and C=O groups of the amide linkage, Figure 7.7b. Notice here that the hydrogen bond points to the direction expected from the nonbonding oxygen orbitals that form an angle of 120° in a sp^2-hybridized oxygen atom.

The existence of such hydrogen bonds is one of the most important aspects of life, since these hydrogen bonds are most important for the structure and function of deoxyribonucleic acid (DNA), the carrier of biological information. DNA is a macromolecule that consists of a backbone of thousands of repeating deoxyribose (a sugar) and phosphate groups as shown in Figure 7.8. Attached to the deoxyribose molecules are one of four "bases": adenine (A), guanine (G), cytosine (C), or thymine (T). These bases are flat, aromatic (see Section 6.6), nitrogen- and

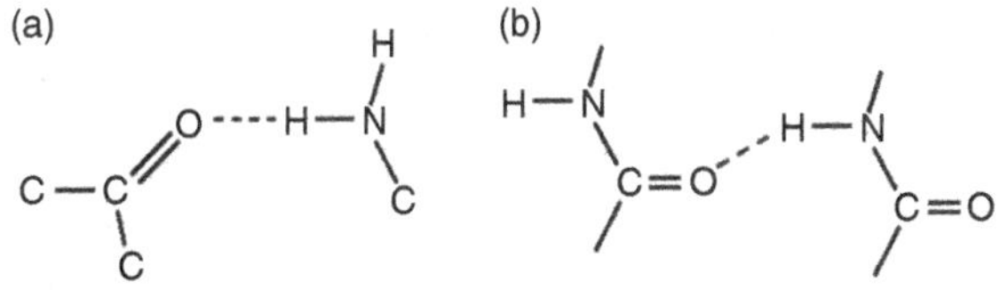

Figure 7.7 (a) Hydrogen bonding between carbonyl carbon and amine hydrogen atoms. (b) Hydrogen bonding between amide linkages in proteins.

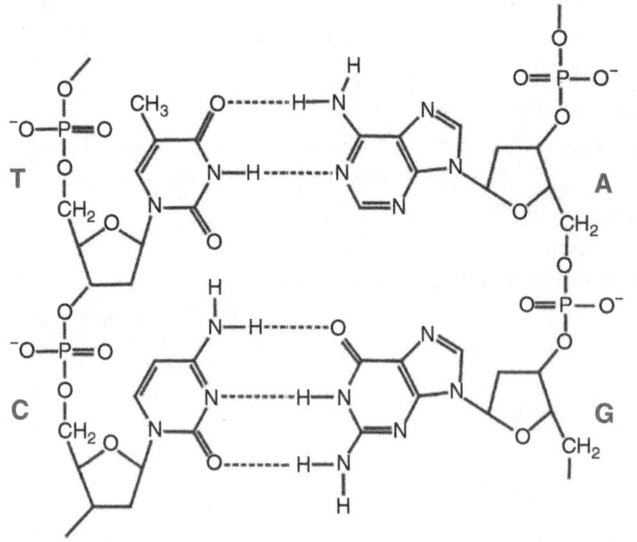

Figure 7.8 Schematic structure of a T–A and a C–G "base pair" in DNA. Remember the comment in Chapter 6 that explained that in organic chemistry and biochemistry, the symbol "C" for carbon atoms is omitted and indicated by the intersection of covalent bonds.

oxygen-containing moieties that can form two or three hydrogen bonds, as shown in Figure 7.8. In the nucleus of a cell, DNA exists as two strands of the polymer forming a double helix with hydrogen bonds between T and A, and between C and G. The fact that the T–A "base pair" contains two and the C–G base pair contains three hydrogen bonds provides selectivity such that C and G can form a base pair, but not C and T or A. Thus, information can be duplicated when DNA is replicated: the double-stranded DNA can form two single-stranded molecules, each of which serves as a template to produce two identical copies of the original DNA. Triads of bases on one strand encode for a given amino acid when a gene, an instruction on how to produce a protein, is transcribed to RNA and used to produce the protein.

It is very interesting that nature chose hydrogen bonds, rather than chemical bonds or any other intermolecular force, to bind the two DNA strands together. This aspect will be discussed next. During cell division, the entire DNA of a cell has to be duplicated. This requires, as indicated above, that regions of the DNA double helix are broken into two individual strands of DNA to create copies of the DNA. Similarly, for transcription of DNA to create an RNA template for protein synthesis, the DNA must be separated into single strands. This separation requires energy, since the interactions between the strands must be broken. This energy is provided by certain enzymes known as helicases that promote unwinding and separation of the DNA strands.

If there were covalent bonds between the strands, the energy requirements to separate the strands would be quite high. If the interaction between strands were based on dipole–dipole forces (see the next section), the energy required would be much less, and the DNA double strand would be much less stable. With the hydrogen bonds between the strands, the interaction between strands is sufficiently strong so that at body temperature (c. 36 [°C]), DNA is stable. However, pure DNA separates ("melts") into single strands at temperature above 50 [°C]. Thus, we see that at body temperature, DNA will not spontaneously separate into two strands; however, the energy required to separate the two strands is only slightly higher than body temperature (for a discussion on how to relate temperature to thermal energy, see Chapters 8 and 11). The hydrogen bonds in DNA, therefore, provide an ideal force between the strands that can be broken in a biological process such as duplication or transcription, but present a stable situation otherwise. In addition, the pattern of hydrogen bonds provides the selectivity or recognition through the base pair specificity.

7.4.2 Dipole–Dipole Interactions

Dipole–dipole interactions are forces that result when polar molecules (see Chapter 6) get in close proximity of each other. These forces are weaker (5–10 kJ/mol) than hydrogen bonds and act on relatively short distances, since the interaction between dipoles depends on $1/r^3$, where "r" is the distance between the dipoles. They are the predominant forces between polar molecules in liquids and solids that do not exhibit hydrogen bonding. An example would be methyl chloride (CH_3Cl), an organic molecule that melts at −97 [°C], boils at −24 [°C], and has a dipole moment along the C—Cl bond. The low boiling point indicates weak intermolecular forces; however, without the dipole–dipole interaction, one would expect an even lower boiling point. This is demonstrated for methane, which has no dipole moment and boils at much lower temperature, −164 °C. This shows that dipole–dipole interactions play an important role as intermolecular forces.

In the liquid and solid state, polar molecules such as CH_3Cl form dynamic structures that can be characterized by the molecules lining up as shown in Figure 7.9, with partially positive and negative portions of the compound in close proximity. This arrangement is found both in the liquid and the solid phase, with the structure in the liquid less well organized than in the solid.

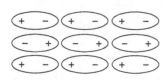

Figure 7.9 Schematic arrangement of polar molecules in the solid state.

7.4.3 London Dispersion Forces (Induced Dipole Forces)

The weakest of the intermolecular forces are induced dipole or "London dispersion forces." These forces act between nonpolar molecules such as solid CO_2, liquid N_2, or even atomic species such as liquid Ar. Induced dipoles arise in molecules such as CO_2 due to the fact that a molecule should not be viewed as a static moiety, but as an entity in which the atoms are in constant motion. This motion, which gives rise to vibrational spectroscopy (see Chapter 15), may distort the molecule as shown in Figure 7.10 from its equilibrium shape (top) to distorted shapes in which it acquires a momentary dipole moment, which may "polarize" other molecules in its vicinity. In addition, the position of electrons, particularly those of larger atoms or double bonds may be subject to perturbations that lead to momentary polarity, for example, a momentary distorted electron cloud. The mobility of the electron cloud is called "polarizability." The polarizability increases from top to bottom in a group of the periodic table. This is one of the reasons for increasing boiling points in the series CH_4, SiH_4, GeH_4, and SnH_4 as pointed out in the previous section. The increased polarizability going down a group is also demonstrated between CO_2 and CS_2, which have similar structures and bonding. Carbon dioxide sublimes at quite low temperatures (-78.5 [°C]), whereas carbon disulfide is a liquid at room temperature and boils at 46 [°C], i.e. it has much stronger induced dipole forces. Even in atomic species, such as liquid argon, there exist dispersion forces due to the polarizability of the atoms. Both London dispersion forces and dipole–dipole interactions are referred to as "van der Waals" forces, named after a Dutch physicist whose name will reappear in our discussions on real gases.

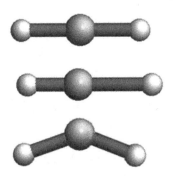

Figure 7.10 Schematic view of vibrational distortions of carbon dioxide (top) that momentarily renders the molecule polar.

Example 7.3 What are the predominant intermolecular forces in solid CO_2, solid PCl_3, solid NH_2OH, and liquid CCl_4?

Answer: See answer booklet

7.5 Macromolecular Solids

The solids formed by small molecules (unless they are ionic) are held together by one of the three intermolecular forces discussed in Section 7.4 and are referred to as molecular solids. Examples of molecular solids are ice, solid carbon dioxide, sugar, caffeine, white phosphorous (P_4), and many others. Ice, for example, is made up of hydrogen-bonded water molecules that form a well-defined lattice. Hence, ice crystals may exist as several well-defined crystal structures. Solids like sugar also incorporate hydrogen bonds between the —OH groups attached to the carbon rings.

Molecular solids that are held together by dipole–dipole or van der Waals forces are less often found in everyday life, since – due to the weak intermolecular interactions – they would be in the gaseous or liquid state at room temperature. Solid CO_2, also known as "dry ice," is a solid where induced dipole forces are the dominant intermolecular forces. Since they are weak, solid CO_2 undergoes a phase transition at very low temperatures (-78.5 [°C]) according to

$$CO_2(s) \rightarrow CO_2(g) \tag{7.4}$$

In contrast, polymeric molecular solids are very common in everyday life. Many solids are composed of polymer chains held together by intermolecular forces such as hydrogen bonding or dipole–dipole interactions. The naturally occurring fiber, cellulose, which constitutes the majority of cotton or wood, is a polymer of sugar rings held together by hydrogen bonds. An artificial polymer, such as polyvinyl chloride (PVC), is composed of chains of repeating units shown in Figure 7.11. The polar C—Cl group will be responsible for dipole–dipole interactions between the chains, whereas the chains themselves, with chain length of thousands of repeating units, are responsible for the strength of the material. Another industrially important polymer is nylon, in which repeating $(-NH-(CH_2)_5-CO-)_n$ chains form hydrogen bonds with neighboring strands as shown in Figure 7.7b.

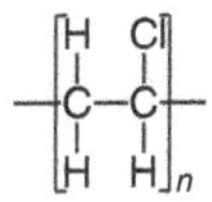

Example 7.4 Discuss the differences in hybridization, structure, and physical properties of the two allotropes of carbon discussed here, diamond and graphite.

Answer: See answer booklet

Figure 7.11 Repeating unit of polyvinyl chloride.

Example 7.5 Define allotropy. What elements exist in allotropic forms?

Answer: See answer booklet

7.6 Liquids and Solutions

The liquid state is intermediate between the solid and gaseous phases in many properties. The density of liquids is generally less than that of solids, but still much higher than that of gases (see Chapter 8). The exception to this statement is water: the density of ice is less than the density of liquid water; hence, ice forms on top of liquid water. The reason for this is the special, hydrogen-bonded structure of ice.

The order of molecules in liquids is lower than in solids, but the same intermolecular forces found in solids act between the molecules in the liquid phase. Liquids conform to the shape of their containers and form horizontal surfaces, except at the contact points with the container, where the surface tension distorts the flat surfaces. The surface tension, which is responsible, for example, for the pearling of water droplets off hydrophobic surfaces, is actually a quite difficult subject that results from the fact that molecules at the surface are not surrounded uniformly by the intermolecular forces acting on a molecule deep inside the liquid. Thus, the surface molecules are pulled into the liquid phase by the intermolecular forces. Consequently, the surface tension tends to minimize the surface area.

Another property of liquids is the tendency of molecules at the surface of a liquid to escape the surface and establish what is referred to as the vapor pressure of the liquid. This tendency of escaping the liquid increases exponentially with increasing temperature, since molecules at the surface will have higher kinetic energy to overcome the intermolecular forces that exist in the liquid. The vapor pressure is a typical dynamic equilibrium situation where, at equilibrium, the number of molecules condensing into the liquid equals the number of molecules vaporizing from the surface into the gaseous phase. We shall revisit this subject again in Chapter 9 on Chemical Equilibria.

7.6.1 General Aspects of Solutions and Solvation

At this point, one aspect of many liquids will be introduced, namely their tendency to act as solvents for other compounds. Solutions and some of solution properties have been introduced in Chapter 4, in particular the definition of concentration units, dilution, and so forth. Also introduced was the concept of ionic compounds dissolving in a polar solvent, such as water, as shown in Figure 4.1.

Since most biological and biochemical processes occur in aqueous solution, let us concentrate for the remainder of this chapter on some of the principles of aqueous solutions. As indicated in Chapter 4, enormous changes occur to both the solute and the solvent when an ionic compound dissolves in water. As shown in Figure 4.1, the ions formed are totally solvated (i.e. surrounded) by water molecules. The energy of attraction between the ionic charge and the dipole moment of the water molecule often exceeds the lattice energy; therefore, many ionic compounds readily dissolve in water. For limitations on the solubility of compounds in water, please refer to Section 9.9, Solubility and Solubility Product. The process of dissolving a compound also has a profound effect on the structure of liquid water, which can be readily observed by the contraction of the molar volume of water when mixed with methanol or ethanol, see Section 4.7. Water also changes its structure in response to nonpolar solutes; in particular, water molecules may form a clathrate, or cage structure, around nonpolar moieties. In either case – the direct solvation or the formation of a cage structure – the original water structure is perturbed, which results in a reduction of the aforementioned vapor pressure of water, that is, the tendency of water molecules to escape from the liquid into the gaseous phase. A detailed discussion of water vapor pressure is presented in Chapter 9; however, some aspects will be presented here to explain what is known as the colligative properties of solutions in the next section.

7.6.2 Colligative Properties

Freezing point depression, boiling point elevation, and osmotic pressure are three colligative properties. These effects depend on the number of solute particles (molecules or ions) and not on their nature (as long as they are nonvolatile). Thus, the freezing point depression of water is the same for a 1 molal solution of sucrose in water or a half-molal solution of sodium chloride in water since 1 mole of NaCl dissociates into 2 moles of ions, and the freezing point depression depends only on the concentration not the nature of the solute. The best way to understand both freezing point depression and boiling point elevation is by referring to Figure 7.12 (solid trace). For water, the vapor pressure is 760 [mm Hg] or 1 [atm] at 100 [°C]. This implies that in boiling water at 100 [°C] and 1 [atm] pressure the vapor pressure inside the liquid equals the external pressure, and bubbles form within the liquid. At 0 [°C], the water vapor pressure is about 4.6 [mm Hg]; this vapor pressure explains why snow and ice disappear in winter even at temperatures at (or below) the freezing point of water and why a puddle of water dries up even at temperatures below the boiling point of water. The former process is called sublimation, whereas the latter is generally referred to as evaporation or vaporization. Figure 7.12 also explains why at higher elevation, water boils at a lower temperature: since the external atmospheric pressure is lower, water vapor can form bubbles inside the liquid at temperatures below 100 [°C]. The (exponential) dependence of the vapor pressure on the temperature will be discussed in more detail in Chapter 9. The vapor pressure of solid water, below 0 [°C], also obeys an exponential function.

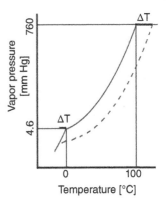

Figure 7.12 Schematic representation of the vapor pressure curve of water (see also Figure 9.4). Solid trace: vapor pressure of pure water; dashed trace: vapor pressure over an aqueous solution.

The effect of the solute is to reduce the concentration of the solvent at the surface of the solution. In addition, the solute exerts forces on the solvent molecules, which makes their escape from the surface more difficult. This lowers the vapor pressure of the solvent according to Raoult's law:

$$p_{vap}^{solution} = p_{vap}^{solvent} \cdot X^{solvent} \tag{7.5}$$

where $p_{vap}^{solution}$ is the vapor pressure over the solution, $p_{vap}^{solvent}$ is the vapor pressure of the pure solvent, and $X^{solvent}$ is the mole fraction of the solvent. The mole fraction is a concentration unit that was introduced in Chapter 4 and is defined as

$$X^{solvent} = \frac{n_{solvent}}{n_{solvent} + n_{solute}} \tag{7.6}$$

$$X^{solute} = \frac{n_{solute}}{n_{solvent} + n_{solute}} \tag{7.7}$$

The mole fraction is indicative of the fact that in a solution, there are just fewer solvent molecules at the surface; thus, the vapor pressure of the solvent is lowered. Since the number of moles of solvent plus the number of moles of solute equals the mass of the solution, we see that the mole fraction of the solute is directly proportional to the molality of the solution. Furthermore,

$$X^{solvent} + X^{solute} = 1 \tag{7.8}$$

Thus, we see that the vapor pressure of the solution decreases linearly with the mole fraction of the solvent. How far the vapor pressure curve (dashed line in Figure 7.12) is offset from the vapor pressure curve of the pure solvent depends on the mole fraction (or molality) of the solution. As pointed out before, the molality and the mole fraction are proportional to each other; hence, both of them could be used to express the boiling point elevation of a solution. However, both the boiling point elevation and the freezing point depression are usually expressed in molal units.

The change in temperature for the boiling or freezing point of a solution is indicated by the heavy black lines marked ΔT in Figure 7.12.

The boiling point elevation, ΔT_{pb}, is given by

$$\Delta T_{bp} = K_{bp} \cdot C_m \tag{7.9}$$

where C_m is the molal concentration (see Eq. 4.8) of the solute and K_{bp} is the boiling point elevation constant of the solvent. K_{bp} for water is 0.512 [°C/molal]. For a nondissociating compound such as sucrose, a 1 molal solution in water has a boiling point of 100.51 [°C]. A 1 molal solution of NaCl has a boiling point of 101.2 [°C] since this ionic compound produces 2 moles of dissolved ions. The equations for the freezing point depression of water mirrors Eq. 7.9:

$$\Delta T_{fp} = K_{fp} \cdot C_m \tag{7.10}$$

K_{fp} for water is 1.86 [°C/molal]. Notice that ΔT in both Eqs. 7.9 and 7.10 is defined as the difference in freezing/boiling temperature of the solutions, as compared to the pure solvent.

Example 7.6

(a) What is the freezing point of an aqueous solution of 20 [g] of NaCl in 800 [g] of water?
(b) Repeat the calculations for a solution of 20 [g] of CaCl$_2$ in 800 [g] of water.

Answer: (a) −1.59 [°C]; (b) −1.26 [°C]

Example 7.7 An aqueous solution of a certain sugar has a boiling point of 100.299 [°C]. The solution was prepared by dissolving 10.0 [g] of the sugar in 50 [g] of water. What is the gram molecular mass $\mathcal{M}$ of the sugar?

Answer: 342 [g/mol]

Finally, another colligative property will be discussed that is known as the "osmotic pressure" of a solute/solvent system, which is associated with the process of osmosis. One of the best ways to think of this phenomenon is *via* an example of enormous importance to all of us: the cells in our body, in particular the red blood cells known as erythrocytes. Erythrocytes contain hemoglobin, the carrier of oxygen in blood. An erythrocyte is enclosed by a cell membrane that allows the solvent (water) to permeate it freely but does not allow the components inside the cell to leave it. Such a membrane is commonly referred to as a "semipermeable" membrane.

In a system where a solution at high concentration is separated from a pure solvent or a less concentrated solution by a semipermeable membrane, there is a tendency of the system to reach equal concentrations by solvent molecules from the lower concentration or pure solvent side to migrate to the side of higher concentration. In this process, the high concentration side will be diluted, and the concentration in the low concentrating side will increase. This process is known as osmosis. In human blood, therefore, the concentration of salts and all other compounds must equal the concentration of components inside the erythrocyte. Otherwise, water would migrate through the cell membrane to equalize the concentrations inside and outside the cell, with an increase of the volume of the erythrocyte that would eventually burst. Thus, whenever liquid is added to blood, such as in an intravenous (IV) drug delivery or transfusion, the liquid added must be "isotonic," i.e. must have the same concentration of solutes as the erythrocyte.

Similar to the freezing point depression and boiling point elevation discussed earlier, the nature of the solute does not matter in osmosis; only the concentration of all solute components matters. Thus, an isotonic solution for IV transfusion could consist of 0.9% solution of sodium chloride (known as a "normal saline" solution), or what is known as "Lactated Ringer's" solution that contains 6.0 [g] of sodium chloride, 3.10 [g] sodium lactate, 0.3 [g] potassium chloride, and 0.2 [g] calcium chloride per liter. If sugar (dextrose) is added to the IV solution, all other concentrations need to be reduced to keep the total concentration isotonic.

The osmotic pressure, Π, is the pressure, typically expressed in units of [atm], which is needed to prevent the osmosis, i.e. the migration of solvent molecules through the membrane toward the side of higher concentration. This pressure is defined by an equation that looks very much like the ideal gas law:

$$\Pi = i\, C_M\, RT \tag{7.11}$$

where C_M is the molar concentration of (all) solute species (see Eq. 4.9), and R and T have their usual meaning. i is known as the van't Hoff index that takes into account that ionic compounds may dissociate into several ions in solution. However, this factor also takes into account the fact that the same solute species may interact in solution and produce solutions whose apparent concentration is slightly less than the concentration expected by merely counting the ions produced.

Example 7.8 Calculate the osmotic pressure at room temperature of a solution of 1.25 g of glucose ($C_6H_{12}O_6$) dissolved in 10.0 g of diethyl ether, if the density of the solution is 1.30 g/cm^3.

Answer: 19.6 [atm]

Example 7.9   Which of the following ionic compounds has the highest melting point?

(a) SrS
(b) NaF
(c) AlN
(d) Cannot tell without further information

Answer: (c) Details in answer booklet

Example 7.10   Silicon carbide (also known as carborundum) with the chemical formula SiC is one of the hardest materials known. Therefore, its inter/intramolecular forces are most likely

(a) van der Waals forces
(b) dipole–dipole forces
(c) ionic forces
(d) covalent bonds

Answer: (d) Details in answer booklet

Further Reading

Bonding in Semiconductors: https://chem.libretexts.org/Bookshelves/General_Chemistry/Chem1_(Lower)/09%3A_Chemical_Bonding_and_Molecular_Structure/9.11%3A_Bonding_in_Semiconductors.
H-bonds and DNA: https://portlandpress.com/biochemist/article-pdf/41/4/38/856222/bio041040038.pdf.
Vapor pressure lowering: https://www.chem.purdue.edu/gchelp/solutions/colligv.html.
Colligative properties: https://byjus.com/jee/colligative-properties/.

8

The Gaseous State

Although we are surrounded by gas (to be more precise, a mixture of gases) in everyday life, it took scientists quite a while to understand many of the properties of gases that we take for granted nowadays. We now know that gases are a form of matter, have mass, occupy a volume, and exert pressure. Gases have a temperature as well, and in this chapter, we concern ourselves with the relationship of pressure, volume, mass, and temperature of gases. Furthermore, gases exhibit properties similar to flowing liquids referred to as aerodynamics, which, of course, are related to flight. Compared to the solid and liquid phases, the gaseous state is the most disordered and least dense, with gaseous atoms or molecules moving totally randomly, with relatively high velocity, and with minimal interaction between them.

The gaseous state of matter is reached when the atoms or molecules in a substance have sufficient thermal or kinetic energy to overcome the attractive (i.e. intermolecular) forces that hold them in the liquid or solid phase. These forces were discussed before and include – in order of decreasing strength – metallic bonding, hydrogen bonding, dipole–dipole, and induced dipole forces. This explains why with increasing temperature, all substances turn into the gaseous state or decompose into several gases. This process can occur over huge temperature ranges for different materials: the metal tungsten has a boiling point of over 5000 [K], water boils at 373 [K], and liquid helium vaporizes (boils) at 4 [K]. These temperatures may be viewed as a direct measure of the intermolecular forces that need to be overcome in the process of turning a liquid into a gas.

Empirical observations of the behavior and properties of gases not only led to the formulation of the gas laws to be discussed next but also provided a deeper understanding of the principles of matter, conservation of mass and energy, and the formulation of atomic theory. These observations paved the way for the development of the scientific method: detailed, reproducible measurements are explained by theoretical models that are refined as new observations are made.

8.1 General Properties of Gases

As compared to the physical states discussed in the last chapter, gases have low density (typically 0.001 [g/mL]) as compared to solids (3–15 [g/mL]) or liquids (1–3 [g/mL]). Whereas one finds dense and highly ordered packing of atoms or molecules in the solid phase and less ordered but still quite dense packing in the liquid phase, gases consist mostly of empty space with atoms or molecules moving freely, completely randomly, and with high velocity, typically hundreds of meters per second at ambient temperature (see Section 8.5). Nevertheless, the gaseous state shares the concepts of mass, temperature, and volume with the other phases, yet with a few differences.

The mass of gases, in particular air, can easily be demonstrated by an experiment shown in Figure 8.1. A 1 L round-bottom flask with a stopcock filled with air at ambient pressure and temperature can be weighed on an analytical balance with an accuracy of about 0.1 [mg]. After evacuation with a vacuum pump that removes most of the air (more than 99.99%), it is found that the weight decreased by about 1 [g]. If the volume of the flask is accurately known, the density of air can be determined and is $d = 1.18$ [kg/m^3 or g/L] at room temperature and standard pressure (see later). This is about 800 times less than the density of water (1.0 [g/mL]); hence, air bubbles released underwater rapidly rise to the surface of water. The volume of a gas can be

Figure 8.1 A round-bottom flask with a stopcock can be used to determine the mass of a given volume of a gas such as air.

Understanding Essential Chemistry, First Edition. Max Diem.
© 2025 John Wiley & Sons, Inc. Published 2025 by John Wiley & Sons, Inc.

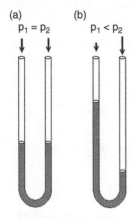

Figure 8.2 Schematic and function of a U-tube manometer.

determined from the geometry of its containers since gases take up the entire volume of a container and conform to the shape of the container, in contrast to solids. Liquids, as discussed earlier, tend to fill a container to a certain level and form horizontal surfaces or interfaces. Gases can be heated or cooled, as we know from cold winter air and hot summer air. This temperature can conveniently be determined using a thermometer.

In addition, gases exhibit a property that is unique to them, namely pressure. Gas pressure is the force of the collisions the gas particles exert on the walls of the container and is defined as the force per area: The pressure inside the tire of a car is expressed in pounds per square inch (psi). The pressure of gases can be measured *via* a "manometer," the simplest of which is the U-tube manometer, shown in Figure 8.2. It consists of a U-shaped glass tube filled with a liquid (preferably one with low vapor pressure and high density, see Chapter 9). Typically, mercury is used. When the pressure in both legs of the manometer is equal, the mercury level will be the same in both legs (Figure 8.2a). If the pressure in one leg is less than that in the other (Figure 8.2b), the mercury column in the leg with lower pressure will rise higher than in the leg with higher pressure. When all the air is removed over one of the legs *via* a vacuum pump (again, a mechanical vacuum pump can remove more than 99.99% of the air), the mercury in this leg rises to 760 [mm] (if you carry out this experiment at sea level – see below). This is shown in Figure 8.2b. Since the pressure p_1 in this figure is zero (or very close to it), the difference in heights of the mercury represents the mass of the air column over the right leg, which is balanced by the mass of the mercury column. Thus, we may estimate the pressure (= force/area) the air column exerts on each unit area at the earth's surface, as follows. Let us assume that the area of the mercury surface in each of the legs of the manometer is exactly 1.0 [cm^2]. Then the volume of the mercury column is 76 [cm] × 1 [cm^2] = 76 [cm^3]. Since mercury has a density of 13.5 [g/cm^3], the mass of the mercury in balance with the air pressure is 1026 [g] or 1.026 [kg]. We call this pressure the air column exerts on each square centimeter of surface 1 atmosphere [atm] or 760 [mm Hg].

Example 8.1 To demonstrate that it is really the mass of the air resting on one leg of the manometer that drives up the mercury in the other leg of the manometer, let us carry out the following thought experiment: what will happen if we evacuate both legs of the manometer?

Answer: See answer booklet

Although the use of atmospheres and [mm Hg] as the unit of pressure is widespread in Chemistry, these units are not SI units, and different branches of science and engineering use quite different pressure units. In meteorology, the metric unit of pressure, the "bar" is used with subdivisions of "millibar" [mbar]. It is based on the SI unit of pressure, which is the "Pascal," abbreviated as Pa. As pointed out above, pressure is the force exerted by a gas on an area; thus, Pascal is defined as the force of 1 [N] (Newton) per square meter:

$$1 \, [\text{Pa}] = 1 \frac{[\text{N}]}{\text{m}^2} \tag{8.1}$$

Within the metric system, 1 bar is defined as

$$1 \, [\text{bar}] = 100\,000 \, [\text{Pa}] \tag{8.2}$$

and is roughly equal to 1 [atm]:

$$1 \, [\text{bar}] \approx 0.987 \, [\text{atm}] \tag{8.3}$$

$$1 \, [\text{atm}] = 760 \, [\text{mm Hg}] \tag{8.4}$$

Thus, the standard atmospheric pressure of 1 [atm] corresponds to about 1.013 [bar]. You may have heard in the news during hurricane season that the center (eye) of a strong hurricane exhibits a pressure of less than 900 [mbar]; thus, strong winds (with speeds up to 300 [km/h]) will flow toward the center of the storm to increase the pressure up to standard atmospheric pressure.

Example 8.2 Express the pressure in the eye of a hurricane (900 [mbar]) in units of [mm Hg], atmospheres [atm], and [Pascal].

Answer: 0.900 [bar], 0.888 [atm], 675 [mm Hg], 90 000 [Pa]

The remainder of this book will use the standard atmosphere, subdivided into 760 [mm Hg], as a unit of pressure, since this is in line with commonly used chemistry literature. Any pressure conversions can be carried out easily using Eqs. 8.1 through 8.4. In engineering, the American pressure unit of pound (or force) per square inch [psi] is used. The conversion between [atm] and [psi] is 1 [atm] = 14.7 [psi].

8.2 Empirical Gas Laws

As so often in science, discoveries are made when technology is developed to an extent that reliable and reproducible observations can be made. In the case of the gas laws, this required having equipment to measure gas pressure – a "manometer" or "barometer" – and flasks with sufficiently good seals to allow changes in the volume of the flask without the gas escaping. One of the earlier gas laws discovered was Boyle's law, which states that for a given amount of gas at constant temperature, the product $p \cdot V$ is constant:

$$p \cdot V = \text{constant} \tag{8.5}$$

This result was found empirically, using an apparatus similar to one you may encounter in an experiment in Chemistry laboratory, shown in Figure 8.3a. Here, you have a syringe connected to a simple manometer. When you change the volume by moving the plunger of the syringe, the pressure of the gas trapped in the syringe changes, and the liquid column in the manometer rises or falls. When you plot the volume (in appropriate units) vs. the pressure (in appropriate units), you will get a plot as shown in Figure 8.3b. The resulting curve appears to be a hyperbola; thus, we try to plot the pressure against the inverse of the volume and obtain a straight line (Figure 8.3c).[1] From this, we conclude that the pressure is inversely proportional to the volume, or

$$p \propto \frac{1}{V} \quad \text{or} \quad p \cdot V = \text{constant} \tag{8.6}$$

from which Eq. 8.5 follows. Eq. 8.6 is known as Boyle's law. The plot indicates that for very large volumes, when $1/V$ approaches zero, the pressure also approaches zero. This observation makes sense: for a fixed amount of gas at constant temperature, we assume that the gas pressure gets very small (approaches zero) when the volume gets infinitely large, and $1/V$ approaches zero.

Example 8.3 Since mercury vapor is quite poisonous, the laboratory experiment shown in Figure 8.3a may be carried out with water in the manometer. Will the experiment still work? In order to obtain a linear p vs. $1/V$ plot, do you need to convert the pressure to [atm] or [mm Hg] units, or can you use "mm H_2O" as pressure units?

Answer: Still works. For more details, see answer booklet.

An experiment to determine the dependence of the pressure with temperature can be carried out by attaching a manometer to a 1.00 [L] flask shown in Figure 8.1 and immersing the flask in baths of different, readily achievable temperatures, for example in a water/ice mixture (0 [°C]) or boiling water (100 [°C]) as shown in Figure 8.4a. We assume that the volume

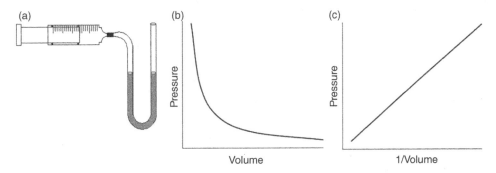

Figure 8.3 (a) Simple apparatus to determine the relationship between volume and pressure of a gas. (b) Plot of observed pressure vs. volume data. (c) Plot of observed pressure vs. inverse volume.

---

1 You may want to consult Chapter 1 if you need to review concepts of proportionality, proportionality constants, equations, and graphs.

of the flask does not change with temperature. In such an experiment, we find that the pressure is directly proportional to the temperature:

$$p \propto T \quad \text{or} \quad \frac{p}{T} = \text{constant} \tag{8.7}$$

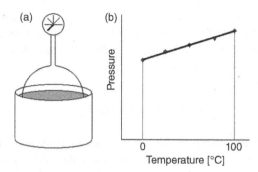

Figure 8.4 (a) Simple apparatus to determine the relationship between pressure and temperature of a gas. (b) Pressure vs. temperature plot.

When the pressure is plotted against the temperature of the gas, we obtain a graph as shown in Figure 8.4b. Eq. 8.7 is known as Gay-Lussac law. The linear plot confirms the direct proportionality between pressure and temperature, but at zero [°C] the pressure of a gas it not zero. If we extrapolate the straight line to the left to a point where it intersects the abscissa, we find that at −273 [°C], the pressure of an "ideal" gas approaches zero. We call the temperature scale in which either pressure or volume become zero with the temperature approaching zero, the absolute temperature scale and express it in degree Kelvin [K], with the relationship

$$T_K = T_C + 273 \quad \text{or} \quad T_C = T_K - 273 \tag{8.8}$$

In Eq. 8.8, T_C denotes the temperature expressed in centigrade units and T_K the temperature expressed in [K]. (However, keep in mind that an "ideal" gas does not exist and all "real" gases liquefy at sufficiently low temperatures, and that the temperature of −273 [°C] cannot be achieved experimentally.) You will find more on the concepts of real and ideal gases later in this chapter.

A somewhat more difficult experimental setup demonstrated that for a given amount of gas, the volume of the gas is directly proportional to the temperature as well:

$$V \propto T \quad \text{or} \quad \frac{V}{T} = \text{constant} \tag{8.9}$$

Eq. 8.9 is known as Charles' law. Combining Eqs. 8.6, 8.7, and 8.9, an empirical relationship or an "empirical gas law" is obtained:

$$\frac{p \cdot V}{T} = \text{constant} \tag{8.10}$$

for a given amount of gas, or

$$\frac{p_1 \cdot V_1}{T_1} = \frac{p_2 \cdot V_2}{T_2} \tag{8.11}$$

Here, the subscripts 1 and 2 denote different conditions. It is important to notice that Boyle's law, Eq. 8.6, holds for whatever units of volume and pressure are used, since the two variables, p and V are inversely proportional. This implies, for example, that whenever p goes to zero (in any arbitrary units), the volume goes to infinity, and *vice versa*. However, Eqs. 8.7 and 8.9 only hold if the temperature is expressed in the absolute temperature scale (the Kelvin scale, T_K). This is because pressure or volume is directly proportional to the temperature; thus, for either pressure or volume to become zero with the temperature approaching zero, it has to be expressed in degree Kelvin [K].

Example 8.4 A gas is collected at room temperature (25.0 [°C]) and atmospheric pressure and occupies a volume of 3.00 L. What will be the pressure of this gas at 0.00 [°C] when it occupies a volume of 10.0 [L]?

Answer: 0.275 [atm]

Example 8.5 A car tire is pumped up to 38 [psi] at 34 [°F]. A sudden heat wave arrives with a summerly temperature of 76 [°F].

(a) What is the pressure in the tire at the new temperature? Assume that the volume of the tire does not change.
(b) If your answer to part (a) was 84.9 [psi], what mistake did you make?

Answer: (a) 41.2 [psi]. (b) You used the temperature in degrees Fahrenheit.

The last of the empirical gas laws is Avogadro's law that states that at the same temperature and pressure, the volume of a gas is directly proportional to the number of moles, n, (as defined in previous chapters) of the gas:

$$V \propto n \tag{8.12}$$

Eq. 8.12 can also be stated as follows: equal volumes of gases contain the same number of moles of the gas, and therefore, the same number of particles (atoms or molecules). At first glance, this law appears somewhat trivial, but its acceptance by the middle of the nineteenth century paved the way for the definition of the mole, the concept of Avogadro's number, and atomic and molecular masses. We shall discuss more of this discovery process in Section 8.6 when the kinetic theory of gases will be introduced.

8.3 The Ideal Gas Law

We can combine Eqs. 8.6, 8.7, 8.9, and 8.12 into one equation that is known as the ideal gas law:

$$\frac{p \cdot V}{n \cdot T} = \text{constant} \tag{8.13}$$

Since the expression of the left-hand side of Eq. 8.13 is constant, we may call it the "gas constant" and designate it by the symbol R:

$$pV = nRT \tag{8.14}$$

and call Eq. 8.14 the "ideal gas law." The numerical value of R, the gas constant can be determined experimentally as follows: for a set of well-defined conditions, namely at a temperature $T = 0$ [°C] = 273 [K] and a pressure $p = 1.0$ [atm], the volume of 1 [mol] ($n = 1$) of gas was found to be 22.4 [L]. Thus,

$$R = \frac{pV}{nT} = \frac{1 \cdot 22.4}{1 \cdot 273} = 0.082 \left[\frac{\text{L atm}}{\text{K mol}}\right] \tag{8.15}$$

The gas constant happens to be one of the most important constants, as will be discussed next. We shall see later that the product "pV" has units of energy or work (for example, volume work or the compression or expansion of a gas, see Chapter 11); thus, R often is expressed in units of energy [Joule] per degree Kelvin and moles:

$$R = 8.314 \left[\frac{\text{J}}{\text{K mol}}\right] \tag{8.16}$$

This constant relates energy, in particular the kinetic energy of gaseous molecules, to temperature. (More on this in Section 8.6). Thus, we may say that the thermal energy E_{therm} is proportional to the absolute temperature:

$$E_{\text{therm}} \propto T \tag{8.17}$$

with the proportionality constant R. At room temperature (25 [°C] or 298 [K]), the thermal energy E_{therm} is about

$$E_{\text{therm}} = 8.3 \cdot 298 \approx 2.5 \left[\frac{\text{kJ}}{\text{mol}}\right] \tag{8.18}$$

Thus, the phrase "... at the energies of chemical and biochemical reactions, and in everyday life, the view of atoms consisting of protons, neutrons and electrons is adequate..." in Section 2.1 implies an energy of a few [kJ] in our everyday life. At extreme conditions, for example, inside the sun, the temperature exceeds 10^7 [K], corresponding to energies in the hundreds of mega joules, and in modern particle accelerators, temperatures in excess of 10^{12} [K] have been achieved, corresponding to energies in the 10^4 [GJ] range. At these temperatures, atoms are found to be no longer stable and fall apart into high energy nuclear particles.

Eq. 8.14 allows the calculation of temperature, pressure, and volume of any gas under ideal gas conditions, since – as compared to Eq. 8.10 – the standard state, 273 [K], 1 [atm] pressure, and 22.4 [L] volume are contained in the gas constant R. Below are a few examples to demonstrate the usefulness of the ideal gas equation.

Example 8.6 Calculate the volume of 1.00 [mol] of an ideal gas at room temperature (25 [°C]) and a pressure of 1.00 [atm].

Answer: 24.4 [L]

Example 8.7 Why is it, as insinuated in Examples 8.3 and 8.5, that pressure and volume can be expressed in any units in Eq. 8.6, while the temperature has to be converted to Kelvin units in Eqs. 8.7 and 8.8?

Answer: See discussion in answer booklet

A further example of the utility of the ideal gas law is the inclusion of a gas density in the equation. The density d (of any material) is given by the ratio of mass over volume:

$$d = \frac{m}{V} \tag{8.19}$$

For gases, the units are typically [g/L] or [kg/m^3] (see Section 8.1). Using the relationship between mass, molecular mass, and number of moles (Eq. 3.6), Eq. 8.14 can be rewritten as

$$pV = \left(\frac{m}{\mathcal{M}}\right)RT \tag{8.20}$$

from which follows

$$p = \frac{d\,RT}{\mathcal{M}} \quad \text{or} \quad d = \frac{p\,\mathcal{M}}{RT} \tag{8.21}$$

For an elementary gas, such as the noble gases, the atomic mass $\mathcal{A}$ is used instead of the molecular mass. From Eq. 8.21, we see that the density of a gas is directly proportional to the molecular mass and indirectly proportional to the temperature. Thus, hot gases have a lower density than cold gases, which explains the phenomenon of hot air rising. It also explains that gases with higher molecular mass have a higher density than gases with lower molecular mass. Thus, CO_2 will accumulate in the bottom of mine shafts, hence the famous canary in the coal mine (look it up!).

Example 8.8 (a) Estimate the average gram molecular mass of an "air molecule" at room temperature (25 [°C]) and 1.00 [atm] pressure from the density of air given in Section 8.1.
(b) From the results of part (a), estimate the percentages of O_2 and N_2 in air, assuming oxygen and nitrogen are the only constituents of air.

Answer: (a) $\mathcal{M}$ = 28.8 [g/mol]. (b) The percentage of oxygen is 20%. See answer booklet.

Example 8.9 Consider a hot-air balloon with a volume of 2000 [m^3], corresponding to a radius of the balloon of just under 8 [m] (verify the volume calculation). What is the lift of this balloon at atmospheric pressure if the air inside is heated to 80 [°C], and the outside air temperature is 25 [°C]. Use the density of air at 25 [°C] d_{298} = 1.19 [g/L = 1.19 [kg/m^3]].

Answer: 380 [kg]

Example 8.10 (a) Calculate the density of He(g) at 1 [atm] and 25 [°C].
(b) Repeat the calculations for the lift of the balloon if He is used, rather than hot air, to fill the balloon.

Answer: (a) d_{He} = 0.163 [kg/m^3]. (b) The lift of the balloon is 2060 [kg].

8.4 Real Gases

The ideal gas law, as discussed in Section 8.3, is a really useful method to predict the relationship between amount, pressure, temperature, and volume of gases (and as we shall see in the next section), as well as gaseous mixtures. However, it is a model for a gas that does not exist: all gases in nature are "real" gases, which means that the gaseous particles do have a finite, albeit small, volume and that interactions between these gaseous particles exist that eventually lead to condensation of the gas if the temperature is sufficiently low and/or the pressure is sufficiently high.

Thus, the ideal gas law really never holds exactly; however, a gas far away from the temperature at which it condenses will exhibit properties very close to the ideal gas law. Consider helium, for example. Its normal boiling point is about −269 [°C] or 4 [K]. Thus, at 273 [K], it is very far away from its boiling/condensation temperature, and it behaves nearly exactly like an ideal gas. This can be seen from the fact that at 273 [K], its volume is about 22.38 [L], as compared to the ideal gas volume of 22.40 [L]. The discrepancy is due to two factors: first, the He atoms do occupy a small volume, whereas in the description of an ideal gas, it is assumed that the gaseous particles have no volume. Furthermore, there exist very small, but finite, intermolecular attractions (Dispersion forces, see Section 7.4.3) between the He atoms that make the gas condense at very low temperature and tend to reduce the volume the gas occupies at higher temperatures.

In order to describe the behavior of real gases, one amends the ideal gas law by two aspects that were omitted when defining the ideal gas law: the residual volume of the atoms or molecules of the gas and the fact that the gaseous particles do exert attractions to each other. This is manifested in the so-called van der Waals equation

$$\left(p + \frac{an^2}{V^2}\right)(V - nb) = nRT \tag{8.22}$$

where "a" and "b" are material-specific constants determined to fit experimental data. Interestingly, when J. van der Waals, a Dutch physics professor and Nobel laureate, performed experimental and theoretical work on intermolecular forces and nonideal behavior of gases in the last decades of the nineteenth century, the concept of molecules was still debated, and it was not generally accepted that the liquid and gaseous states of a given substance consist of the same "molecules." Thus, the phase transitions and the existence of the same compound in different phases, as discussed at the beginning of Chapter 7, were not yet understood.

In summary of this section, it should be emphasized that the ideal gas law will allow us to predict the properties of gases at conditions far removed from the temperature and pressure conditions where condensation occurs. Thus, air can be modeled by this law very well under standard conditions. At conditions closer to liquefaction, modifications to the ideal gas law are introduced that take into account the nonideal behavior of real gases.

8.5 Gaseous Mixtures and Partial Pressures

In the previous discussion, we mostly have assumed that the gas was composed of one compound. However, the gas we are most familiar with – air – is a mixture of several gaseous components: approximately 80% N_2, about 18% O_2, a little CO_2, H_2O, and Ar. The concentrations of CO_2 and H_2O vapor may vary a bit. What we know from first experience is that each gaseous component occupies the total gas volume. When we mix pure oxygen and nitrogen gases, they both nearly immediately fill the entire container. This is quite important for everyday life because we would be extremely uncomfortable if all the oxygen and nitrogen molecules were to segregate into different zones. We also assume (and show in the next section on the kinetic theory of gases) that all components of a gaseous mixture have the same temperature, because gaseous molecules undergo huge numbers of collisions each second, and we can assume that these collisions equalize the kinetic energies of all gaseous particles.

Thus, since we know that the temperature of the gaseous particles is the same, and that the volume of each gaseous component is the total volume V_T of the gas, we may define the "partial pressure" of a gaseous component A in a mixture of gases as

$$p_A = \frac{n_A R T}{V_T} \tag{8.23}$$

and the total pressure of a mixture of gases as

$$p_T = p_A + p_B + \ldots \tag{8.24}$$

Example 8.11 Calculate the partial pressure of oxygen and nitrogen of a 100 [g] sample of air at 25 [°C] in a 10.0 [L] container using the percentages of oxygen and nitrogen calculated in Example 8.8. What is the total pressure p_T in the container?

Answer: $p(N_2) = 6.98$ [atm]; $p(O_2) = 1.53$ [atm]; $p_T = 8.51$ [atm]

Thus, we see that the pressures and number of moles in a gaseous mixture are proportional and refer to each component of the mixture, whereas the volume and temperature are properties of the mixture.

8.6 Kinetic Theory of Gases

In the second half of the nineteenth century, the theory of gases based on the collisions of gaseous molecules with themselves and the walls of the container in which the gas is confined was developed by leading physicists of the era, such as R. Clausius, J. Maxwell, and L. Boltzmann. The resulting model is quite simple and provided a theoretical framework for the ideal gas law, and was later refined to include many aspects of modern statistical thermodynamics.

The kinetic theory of gases is based on the assumption that the atoms or molecules in a gas move randomly at a relatively high velocity $\mathbf{v}$ and have a momentum associated with their velocity given by

$$\boldsymbol{p} = m\,\mathbf{v} \tag{8.25}$$

In the following discussion, we use the lower-case letter $\mathbf{v}$ for the velocity of a gaseous particle, and the upper-case letter V for the gaseous volume. Furthermore, we denote the momentum a gaseous particle possesses by the script symbol $\boldsymbol{p}$. Letters in bold typeface refer to vectorial quantities (notice that in Section 8.6, symbols are used that may have different meanings in other chapters). The gaseous particles undergo frequent elastic collisions with the walls of the container to which the gas is confined. The momentum transferred to the walls by these collisions is what we observe as the pressure of the gas. Let us assume, for simplicity, that the gas consists of atomic species, such as Ar atoms, and let us start by considering the gas to contain just one Ar atom. The momentum $\boldsymbol{p}$ of such a moving Ar atom is given in classical mechanics by Eq. 8.25, where m is the mass of the atom. The total momentum $\boldsymbol{p}_t$ transferred by collisions of the one Ar atom to the container wall would be

$$\boldsymbol{p}_t = m \cdot \mathbf{v} \cdot C \tag{8.26}$$

where C is the number of collisions per second of the atom with the walls. This number C can be determined as follows. Let us assume that the atom can move only in the x direction, and that its velocity in the x direction is given by v_x. Thus, Eq. 8.26 simplifies to

$$p_x = m \cdot v_x \cdot C \tag{8.27}$$

If the walls are separated by distance x, it will take the atom a time t to fly from wall to wall:

$$t = \frac{x}{v_x} \tag{8.28}$$

Thus, the frequency of collisions C of one particle with the two walls is

$$C = \frac{1}{t} = \frac{v_x}{x} \tag{8.29}$$

Thus, in the one-dimensional case, the momentum transferred to the walls is

$$p_x = \frac{m \cdot v_x \cdot v_x}{x} \tag{8.30}$$

Next, a sample of gas certainly does not consist of one atom only. The momenta transferred by N atoms then would be given by

$$p_x = \frac{N \cdot m \cdot v_x^2}{x} \tag{8.31}$$

However, Eq. 8.31 assumes that all gaseous atoms exactly have the same velocity (component). In reality, the gaseous particles do not all have to have the same velocity, but they have a distribution of velocities (see later). Thus, we need to average the velocity of the gaseous particles over the distribution of velocities, and we substitute the mean square velocity $\overline{v_x^2}$ in Eq. 8.31:

$$p_x = \frac{N \cdot m \cdot \overline{v_x^2}}{x} \tag{8.32}$$

In a three-dimensional container with a volume $V = xyz$, the particle can move in all three dimensions with velocity components $\overline{v_x^2}, \overline{v_y^2},$ or $\overline{v_z^2}$. According to the Pythagorean theorem, the mean square velocity of a particle will be

$$\overline{v^2} = \overline{v_x^2} + \overline{v_y^2} + \overline{v_z^2} = 3\,\overline{v_x^2} \tag{8.33}$$

since the three velocity components $\overline{v_x^2}$, $\overline{v_y^2}$, and $\overline{v_z^2}$ can be assumed to be equal. This is because there is no preferred direction in the random motion of the atoms. Notice that in the term $\overline{v^2}$, the velocity was first squared and then averaged. If one averaged the velocities first and then squared the result, i.e. if one had calculated $\bar{v}^2$, one would obtain quite different a result: the average velocity of the gas particles would be zero, since the probability of a particle moving in the opposite direction is equal for a gas confined in a container. Thus, averaging the velocities would produce a zero result and squaring this result would still be zero. However, squaring the velocity (components) first gives a nonzero value, which subsequently is averaged. Therefore, $\overline{v^2}$ is known as the mean square velocity.

We may write Eq. 8.33 in all three directions as

$$p_t = \frac{N \cdot m \cdot 3\,\overline{v_x^2}}{xyz} = \frac{N \cdot m \cdot 3\,\overline{v_x^2}}{V} \tag{8.34}$$

If we equate the total momentum p_t transferred to the wall to the pressure p of the gas, we obtain

$$p = \left(\frac{1}{3}\right) \frac{N \cdot m \cdot \overline{v^2}}{V} \tag{8.35}$$

Since $p \cdot V = nRT$,

$$nRT = \left(\frac{1}{3}\right) N \cdot m \cdot \overline{v^2} \tag{8.36}$$

With $N = n\,N_A$, $\tag{2.9}$

we obtain

$$nRT = \left(\frac{1}{3}\right) n\,N_A \cdot m \cdot \overline{v^2} \quad \text{or} \tag{8.37}$$

$$3RT = N_A\,m\,\overline{v^2} \tag{8.38}$$

Eq. 8.38 implies that the temperature of a gas is proportional to the mean square velocity of the gaseous particles. The term $N_A \cdot m$, Avogadro's number multiplied by the mass of one atom, is just the atomic mass $\mathcal{A}$ defined before (Eq. 2.8). Thus,

$$\overline{v^2} = \frac{3RT}{\mathcal{A}} \tag{8.39}$$

The square root of the square mean velocity is known as the root mean square velocity v_{rms}, which is then

$$v_{\text{rms}} = \sqrt{\left(\frac{3RT}{\mathcal{A}}\right)} \tag{8.40}$$

When dealing with molecular gases, rather than atomic gases, one would use the molecular mass $\mathcal{M}$, rather than the atomic mass $\mathcal{A}$ in Eqs. 8.39 or 8.40.

It is useful to look at an example to get a feeling for magnitudes and units. Let us calculate the root mean square velocity of an oxygen molecule at room temperature, using Eq. 8.40:

$$\begin{aligned} v_{\text{rms}} &= \sqrt{\left(\frac{3RT}{\mathcal{M}}\right)} \\ &= \sqrt{\frac{3 \cdot 8.314 \cdot 298}{0.032}} \left[\frac{\text{J K}}{\text{kg}} = \frac{\text{kg m}^2}{\text{kg s}^2}\right] = \sqrt{232272} = 480 \left[\frac{\text{m}}{\text{s}}\right] \end{aligned} \tag{8.41}$$

Notice that we expressed the gas constant R in units of [J]; therefore, we had to enter the molecular mass $\mathcal{M}$ in units of [kg]. This result implies that the oxygen molecule zips around at very high speed, covering the distance in a large lecture hall from wall to wall in fractions of a second. In reality, however, each gaseous molecules, at 1 [atm] pressure, undergoes about 10^{10} collisions with other gaseous molecules per second, such that the mean free path, i.e. the mean distance between collisions, is about $480/10^{10}$ or about 48 [nm]. However, if the pressure of the gas is lowered to ultra-high vacuum levels, the mean path between collisions increases to kilometers.

The result presented in Eq. 8.40 ties into the discussion earlier in this chapter (Eq. 8.17) that related the kinetic energy of a gaseous particle directly to the temperature. Since the kinetic energy of a moving particles is given by

$$E_{\text{kin}} = \frac{1}{2} m v^2 \tag{8.42}$$

It follows, using Eq. 8.38 that

$$E_{\text{kin}} = \frac{3}{2} k T \tag{8.43}$$

where k is known as the Boltzmann constant, $k = R/N_A$. This constant, with a value of

$$k = 1.38 \cdot 10^{-23} \left[\frac{J}{K}\right] \tag{8.44}$$

may be considered the gas constant per atom, rather than per mole. This constant is as important as the gas constant in a field known as statistical thermodynamics where the macroscopic properties of substances (i.e. the distribution of energy among atoms and molecules) are evaluated on a microscopic level. Eq. 8.43 gives the exact form of the relationship insinuated by Eq. 8.17, namely that the temperature of a gas is a measure of the kinetic energy of the individual gas particles. Eq. 8.40 also indicates that lighter particles must move faster than heavier particles to have the same kinetic energy at the same temperature. Thus, the root mean square velocity of a hydrogen molecule is four times the root mean square velocity of an oxygen molecule at the same temperature.

Example 8.12 (a) Verify the statement contained in the last sentence.
(b) Calculate the root means square velocities of H_2 and N_2 molecules at room temperature.

Answer: (a) See answer booklet; (b): $1926 \left[\frac{m}{s}\right]$ $515 \left[\frac{m}{s}\right]$

The discussion of the kinetic theory of gases has provided us with a more detailed understanding of many properties of gases. First, the concept of gas pressure can be described as the force of the momentum transfer of the gaseous molecules with the walls of the container. Second, since the temperature is a measure of the kinetic energy of the gaseous molecules, an increase in temperature will increase the momentum (and kinetic energy) of the gas particles, thereby increasing the pressure, as formulated in Eq. 8.7. We now can see the effect of lowering the temperature to the point where gaseous molecules condense: their kinetic energy becomes insufficient to overcome the intermolecular interactions. Third, the motion of gaseous particles, described by high velocities and frequent collisions with other gaseous particles, leads to the concepts of diffusion and effusion of gases, discussed in the next section. Finally, the kinetic theory demonstrates very elegantly how scientific model building leads from very simple physical considerations – the collisions of the gaseous particles with the walls of the container – to fundamental insights into the microscopic behavior of matter. Very few assumptions were needed to produce these results: the major restriction in this model is the assumption that the collisions between gaseous particles with each other and the walls of the container are elastic, i.e. no energy is lost in these collisions.

8.7 Diffusion and Effusion of Gases

The results obtained in the previous sections provide us with an explanation of two phenomena of gases that will be discussed next: diffusion and effusion. Both these processes are due to the random motion of gaseous particles and their frequent collisions with each other. Diffusion is a process that describes the dispersion of a gas into the total volume available to the gas, regardless of the presence of another gas. Consider the situation shown in Figure 8.5a in which a gas is contained in one container, with the other evacuated. When opening the stopcock between the two halves, the gas will immediately disperse into the entire volume as shown in Figure 8.5b. Eventually, an equilibrium situation

Figure 8.5 Diffusion of a gas (dark gray dots, panel a) into a vacuum. A short time after opening the stopcock, gaseous particles will have diffused into the empty space (panel b), eventually reaching an equilibrium with equal pressures in both containers. Panel (c): diffusional mixing of two gases (gray and white spheres) before (panel c) and after the stopcock has been opened (panel d). *Source*: Adapted from UCF Pressbooks, University of Central Florida.

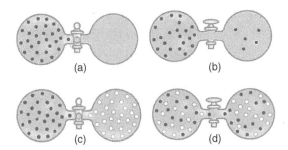

will be reached in which the concentration of gaseous particles in both containers is equal. From the discussion of the velocity of gaseous particles above, it is clear why this process occurs: the gaseous particles moving with high velocity will encounter few collisions with other gaseous particles when expanding into the evacuated part and, therefore, will very quickly fill the total volume.

In the presence of another gas, as shown in Figure 8.5c, both gases will fill both halves of the container, since the motion of gaseous particles is completely random. However, due to the large number of collisions the particles undergo, the mixing process is slower and is determined by the rate of diffusion, which itself depends on the concentration gradient of the components, the area along which the diffusion occurs, and the root mean square velocity of the gaseous particles, and thereby on the atomic or molecular mass of the gaseous particles. Since the root means square velocity of a gas is inversely proportional to the molecular (or atomic) mass, we see that lighter particles will diffuse more rapidly than heavier ones. This is expressed in Graham's law

$$\frac{r_1}{r_2} = \sqrt{\frac{\mathcal{M}_2}{\mathcal{M}_1}} \tag{8.45}$$

where r_1 and r_2 are the rates of diffusion (or effusion, see later) of gases 1 and 2, respectively, and $\mathcal{M}_1$ and $\mathcal{M}_2$ are their molecular masses. A visualization of the random motions of gaseous particles during the diffusion process can be found at https://www.sciencephoto.com/media/559327/view/gas-diffusion-animation.

Effusion is a similar process where gaseous particles escape through tiny holes (with a size of approximately the mean free path of particles between collisions) into a vacuum or another gas. Again, the effusion rate is governed by Graham's law.

The term "diffusion" is also used for similar processes that occur in condensed phases. When adding a drop of dye into water, one observes that the initial deep color of the drop of dye eventually spreads out throughout the aqueous phase and dilutes the color of the dye throughout the water. This process is also driven by the random motion of the dye molecules, but their mean square velocity is much lower than that of gas phase molecules due to larger intermolecular forces. Also, due to the higher density of liquids than gases, the number of collisions each particle experiences is larger than in the gaseous state; consequently, the free path between collisions is even smaller. Therefore, a diffusion process in the liquid phase is much slower than in the gaseous state.

Example 8.13 The separation of isotopes can be carried out by gas effusion of isotopic species. Calculate the ratio of effusion rates of $^{32}SO_2$ to $^{34}SO_2$ at the same temperature.

Answer: 1.0155. The lighter isotopic species effuses 1.016 times faster than the heavier species.

Example 8.14 An unknown gas effuses 1.05 times faster than NF_3 at the same temperature and pressure. The molar mass of the unknown gas is approximately

(a) 64.4 [g/mol]
(b) 67.6 [g/mol]
(c) 78.1 [g/mol]
(d) None of the above

Answer: (a) Details in answer booklet

Example 8.15 In outer space, the temperature is about 3 [K], and the average number of particles found is about 1000 [molecule/L]. The ideal gas pressure under these circumstances is about

(a) $2.46 \cdot 10^{-22}$ [atm]
(b) $4 \cdot 10^{-22}$ [atm]
(c) 246 [atm]
(d) None of the above

Answer: (b) Details in answer booklet

Further Reading

https://en.wikipedia.org/wiki/Kinetic_theory_of_gases.

https://chem.libretexts.org/Bookshelves/General_Chemistry/General_Chemistry_Supplement_(Eames)/Gases/Diffusion_and_Effusion.

9

Chemical Equilibrium

9.1 What Is a System "at Equilibrium"?

When we discussed chemical stoichiometry in Chapter 4, we assumed that a reaction goes completely in the direction indicated by the arrow. For example, in the thermite reaction

$$Fe_2O_3(s) + 2Al(s) \rightarrow Al_2O_3(s) + 2Fe(l) \tag{4.3}$$

we assumed that two moles of aluminum react with one mole of iron III oxide to produce one mole of aluminum oxide and two moles of iron. For this reaction, this is more or less the case, but this is not true for all chemical reactions. For example, if you mix hydrogen gas and fluorine gas at an elevated temperature, you will find that these two gases react to produce hydrogen fluoride gas. However, an analysis of the gaseous components reveals that the reaction does not go to completion, but that, once the reaction is finished, there is hydrogen and fluorine left, in addition to the hydrogen fluoride. We call this mixture of reactants and products an equilibrium mixture, and write the chemical equation as

$$H_2(g) + F_2(g) \rightleftarrows 2HF(g) \tag{9.1}$$

Here, the double arrow indicates that hydrogen and fluorine react to give HF, but that HF also falls apart to give us back the starting materials. This is what we refer to as a dynamic equilibrium. Once a dynamic equilibrium is established, it appears that nothing is happening, but on a microscopic level, we see that both the forward and reverse reactions occur, but that the concentration of reactants and products stay unchanged. This appears somewhat strange, so let us look at a macroscopic analog to this situation.

In Figure 9.1a, we see a dedicated, but not too smart person starting to dig a hole. He uses his shovel to throw the dirt he digs out over his shoulder, and makes good progress with his ditch (Figure 9.1b). As the pile of dirt he excavates grows, and the ditch gets deeper, some of the dirt slides back into the ditch. The guy is not too bright, and does not notice the dirt sliding back into the ditch. After a while, the ditch is sufficiently deep, and the pile of dirt sufficiently high, that all the dirt he excavates and throws over his shoulder slides right back into the ditch (Figure 9.1c). Although he keeps on digging, the ditch is not getting any deeper, since the *forward* reaction (namely, the dirt being thrown out of the ditch) is exactly as fast as the *reverse* reaction, namely the dirt sliding back into the ditch.

This is an example of what we call a dynamic equilibrium. Although the poor guy keeps working, the ditch is not getting any deeper. From the point of view of the employer, nothing seems to happen: no further progress is being made toward deepening the ditch. This is one of the important criteria of a dynamic equilibrium: there seems to be no progress toward the side of the reaction or the products. This is, since the forward reaction (the deepening of the ditch, or the formation of HF) with the same rate as the reverse reaction (the dirt sliding back into the ditch, or the reformation of H_2 and F_2 from HF). Thus, nothing seems to happen, although on the microscopic level, the reaction occurs in both directions.

This is a common situation in Chemistry. Most chemical reactions proceed toward equilibrium. The first example we shall discuss in detail, namely the vaporization of a liquid to establish its vapor pressure, is a typical example of a dynamic equilibrium. This subject is contained in most textbooks in the chapters on intermolecular forces, but will be discussed in this text as an introduction to equilibria.

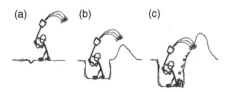

Figure 9.1 Example of a dynamic equilibrium: the rate of excavation is the same as the rate of dirt sliding back into the ditch.

9.2 Liquid–Vapor Phase Equilibrium: Vapor Pressure

We are familiar with the fact that liquids have a tendency to vaporize. If you open a bottle of perfume, or a bottle of whiskey, your nose tells you that molecules of the respective liquids have escaped from the liquid and entered the gas phase; hence, your nose detects them. Thus, we say that a liquid has the tendency to "vaporize" or evaporate, that is, to establish gaseous molecules over the surface of the liquid. This "vaporization" is actually quite complicated a process. How does this occur, and how does this process differ from the process that occurs when you boil a liquid?

First of all, let us take a look at a liquid. All real atoms or molecules – as opposed to the "ideal" molecules we discussed in the context of ideal gases – experience intermolecular forces: hydrogen bonding, dipole–dipole or van der Waals forces. These forces, which hold the molecules in the liquid phase, may be very small, or quite large. In liquid helium, they are so small that liquid helium boils at $-269\,[°C]\,(4\,[K])$, whereas in water, the hydrogen bonds are so strong that water does not boil until $100\,[°C]$ or $373\,[K]$. When a compound evaporates, these intermolecular forces are broken. Thus, a vaporization process, such as

$$H_2O_{(liquid)} \rightleftarrows H_2O_{(vapor)} \tag{9.2}$$

does not involve the breaking of chemical bonds (the bonds bolding hydrogen and oxygen together), but the hydrogen bonds and dipole–dipole interactions that act between individual water molecules.

Before we look at the equilibrium aspects of this process, let us define a few terms. We call the forward process given by Eq. 9.2 "vaporization" and the reverse process "condensation." For all known liquids, heat is required for the vaporization process. This heat of vaporization is also referred to as enthalpy of vaporization, and given the symbol ΔH_{vap}. In Chapter 11, we shall discuss the terms heat and enthalpy in more detail. For the time being, let us use heat and enthalpy interchangeably. For water, the heat of vaporization is large,

$$\Delta H_{vap} = 40.66\,[kJ/mol] \text{ at } 100\,[°C] \tag{9.3}$$

This value is for the vaporization of water at $100\,[°C]$ and is dependent on temperature. At room temperature, the value is $44.0\,[kJ/mol]$). If heat or enthalpy is positive ($\Delta H_{vap} > 0$), we call the process "endothermic." The reason that vaporization is endothermic is obvious: since intermolecular forces need to be broken for a molecule to escape from the liquid to the gaseous state, this energy must be supplied in the form of heat. For water, the heat of vaporization is large, since hydrogen bonds are the strongest of the intermolecular forces.

Thus, heating a liquid promotes vaporization, and the higher the temperature of the liquid, the more particles of the liquid will possess sufficient energy to break the intermolecular forces and escape the liquid.

The reverse process, the condensation of water at $100\,[°C]$, is exothermic ($\Delta H_{vap} < 0$) by the same amount:

$$\Delta H_{cond} = -40.66\,[kJ/mol] \tag{9.4}$$

that is, heat is released when water vapor condenses. For this reason, steam burns are particularly damaging to human tissue: not only is the steam very hot, but when the stem condenses to water, additional heat is released.

Back to our discussion of the vaporization of a liquid, and the dynamic aspects of this process. Let us carry out a simple thought experiment. Imagine we have an evacuated container, shown in Figure 9.2a, into which we introduce a dish of water. According to our previous discussion, some molecules of the liquid water will have sufficient kinetic energy to overcome the intermolecular forces and escape into the evacuated flask. Thus, vaporization will occur, as indicated by the up arrow in Figure 9.2a. Now, as vaporization proceeds, more and more water molecules whiz around in the container. Inevitably, some will collide with the surface of the water. If this collision occurs at low velocity of the gaseous water molecule, the surface (i.e. the intermolecular forces acting at the surface) may trap the gaseous molecule and pull it back into the liquid. The more molecules are in the gaseous state, the more likely it is that a molecule gets trapped back into the liquid. After a while, the number of molecules leaving the liquid (i.e. the rate of evaporation) equals the number of molecules being trapped (i.e. the rate of condensation), as shown in Figure 9.2b and indicated by the two arrows being equal in length. This is, of course, another example of a dynamic equilibrium: after a while, the amount of water vapor in the container does not change any more, although evaporation and condensation keep occurring.

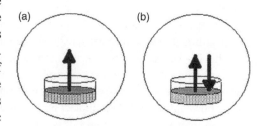

Figure 9.2 Establishing dynamic equilibrium between liquid and gaseous water. (a) Evaporation and (b) evaporation and condensation.

We call the equilibrium concentration of water vapor that is established in equilibrium with liquid water the water "vapor pressure," and designate it p_{vap}. For water at room temperature (25 [°C] or 298 [K]), this equilibrium pressure is 24 [mm Hg] (= 0.032 [atm]), and depends on the temperature. It does not depend on the presence of another gas, as discussed below.

Now, let us look at the implication of this dynamic equilibrium process. As long as we have liquid water in a container – regardless of the volume of the container – the water vapor pressure will be 24 [mm Hg], if we keep the temperature at 25 [°C]. What happens if we double the volume of the container, after the equilibrium pressure (p_{vap}) was established, as shown in Figure 9.3? We know what happens for an ideal gas when we double the volume: the pressure decreases by a factor of two, since the product $p \cdot V$ = constant. For the situation shown in Figure 9.3, however, we have a totally different scenario: we are not dealing with an ideal gas (remember, an ideal gas cannot be liquefied, since there are no intermolecular forces), but with water vapor, which is a very real gas. So, when we increase the volume of the container, the gaseous water molecules have more space to occupy. Thus, the chance of them bumping into the surface of the liquid will be reduced, as will be the rate of condensation. The rate of evaporation, however, stays the same. As a consequence, more water will evaporate, and the reaction, given by Eq. 9.2, will occur in the forward direction until the water vapor pressure is reestablished. This is shown in Figure 9.3: the vapor pressure is the same, although the volume is twice as large. This is another important feature of equilibria: when condition changes, such as the volume in this example, the reaction (here, the evaporation) will occur in order to reestablish the equilibrium. This will happen over and over, if we repeatedly change the volume. The only exception here is if we make the volume so large that all the liquid water evaporates. Then we have reached an altogether different situation: we are no longer dealing with a liquid/vapor equilibrium, but with a vapor only. Thus, a different law applies, namely that for one-phase, gaseous components: the gas law. More on this below.

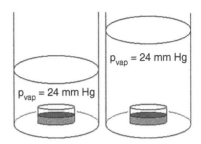

Figure 9.3 The vapor pressure of water is independent of the volume, as long as there is sufficient liquid water.

One of the intriguing aspects of the vapor pressure of a liquid is the fact that the vapor pressure is established regardless of the presence of other gases. Thus, water at room temperature will establish 24 [mm Hg] vapor pressure if the container shown in Figure 9.2 is originally evacuated, or if it contains 760 [mm Hg] of dry oxygen, nitrogen, or air. If you fill an empty container, at room temperature, with 760 [mm Hg] of dry air (80%, or 608 [mm Hg] nitrogen, and 20%, or 152 [mm Hg] of oxygen, see Example 8.11), and introduce a dish of liquid water, and wait for a few minutes, the liquid water will establish its vapor pressure of 24 [mm Hg], and the total pressure inside the flask is now 760 + 24 = 784 [mm Hg]. This latter aspect accounts for the fact that the atmospheric pressure at sea level can vary quite a bit, depending on its water vapor content.

Let us pursue the aspect of when to use the gas law, and when to use the vapor pressure formalism. Let us go back to the experiment in Figure 9.3, but treat it quantitatively. Let us make the volume of the container 10.0 [L] and fill it with completely dry air, i.e. the initial water vapor pressure $p_{vap} = 0$. Then, we fill a dish with 1.0 [g] of liquid water, and introduce the dish of liquid water into the container and wait a bit for equilibrium to establish.

At equilibrium, we find 24 [mm Hg] of water vapor in the 10.0 [L] volume of the container. If we *just consider the gaseous phase*, we can calculate the number of moles and the mass of water in the gas phase:

$$n = \frac{p \cdot V}{R \cdot T} = \frac{\left(\frac{24}{760}\right) \cdot 10.0}{0.082 \cdot 298} \left[\frac{\text{atm L}}{\frac{\text{L atm K}}{\text{K mol}}}\right] = 1.29 \cdot 10^{-2} \, [\text{mol}] \text{ or } 0.23 \, [\text{g}] \text{ of water} \quad (9.5)$$

Thus, we see that 0.23 [g] of water will evaporate to establish the vapor pressure of 24 [mm Hg], and 1.0 − 0.23 = 0.77 [g] of liquid water will remain. Now, let us look at the same problem, with slightly different conditions. We start out with only 0.20 [g] (just a few drops) of liquid water in our dish, and with an original volume of dry air of 50.0 [L], still at 25 [°C]. What is going to happen?

The amount of water required to build up the vapor pressure is:

$$n = \frac{p \cdot V}{R \cdot T} = \frac{\left(\frac{24}{760}\right) \cdot 50.0}{0.082 \cdot 298} \left[\frac{\text{atm L}}{\frac{\text{L atm K}}{\text{K mol}}}\right] = 6.46 \cdot 10^{-2} \, [\text{mol}] \text{ or } 1.2 \, [\text{g}] \text{ of water} \quad (9.6)$$

Thus, 1.2 [g] of water would be required to fill the 50.0 [L] container to the equilibrium vapor pressure, 24 [mm Hg]. But we only started with 0.20 [g] of liquid water! So, we do not have a sufficient amount of water to reach the allowed pressure, and all water will evaporate without ever reaching the equilibrium pressure.

What will be the final partial pressure of water in this example? Since all water evaporates, and no liquid water is left, we may use the gas law to calculate the final pressure. Since we started with 0.20 [g] of water, we know we had

$n = 0.20/18 = 0.0111$ [mol] of water. Thus, the partial pressure of water will be

$$p = \frac{nRT}{V} = \frac{0.0111 \cdot 0.082 \cdot 298}{50.0} \left[\frac{mol\left(\frac{L\,atm}{K mol}\right)K}{L} \right] 5.42 \cdot 10^{-3} \text{ [atm]} \tag{9.7}$$
$$= 4.1 \text{ [mm Hg]}$$

What we have discussed so far for water is true for all liquids: they all exhibit vapor pressure, that is, the tendency for molecules to escape from the liquid. Liquids that have a high vapor pressure at room temperature are called volatile. Examples are the main ingredient of gasoline (octane), or the liquid found in a cigarette lighter, butane. A liquid with low vapor pressure is, for example, mercury. However, as the temperature increases, the vapor pressure of any liquid increases as well. This will be discussed in the next section.

Example 9.1 Consider the liquefied gas butane in a cigarette lighter. As purchased, the lighter contains 5 [g] of butane in a volume of 10 [cm^3]. The vapor pressure of butane is 2.41 [atm] at room temperature and its molecular mass $\mathcal{M}$ is 58 [g/mol].

(a) Calculate the pressure inside the lighter if all butane were in the gaseous phase.
(b) What is the pressure of the butane in the cigarette lighter?

Answer: (a) 211 [atm]; (b) 2.41 [atm], see answer booklet for details

9.3 Temperature Dependence of Vapor Pressure

As pointed out previously, the vapor pressure of a liquid depends on temperature. As temperature increases, the vapor pressure increases as well until the liquid starts to boil. Before any further discussion, we need to clarify one major problem. What is the difference between the evaporation process we discussed in the previous section, and the boiling process mentioned in the previous sentence. Are not these identical processes?

Let us look at water again, because we are most familiar with this liquid. As we discussed, water evaporates at room temperature to establish its vapor pressure. This is a process we are thoroughly familiar with: after all, a puddle of water after a rain storm dries up (that is, the water evaporates) although the temperature is well below the boiling point of water, which is 100 [°C].

Then what is the difference between this evaporation and the boiling process? If you watch water boil, you will notice that water vapor forms throughout the liquid and escapes in large bubbles, which makes the liquid "boil" and roll. The evaporation that occurs at temperatures below the boiling point only occurs at the surface of the liquid; remember, we described it as a process of liquid molecules overcoming the intermolecular forces and escaping into the gas phase.

What conditions are necessary for the "boiling" process to occur; i.e. for vapor forming throughout the liquid? Very simple: *the vapor pressure of the liquid must be at least equal to the ambient (atmospheric) pressure.* So, we say that water boils when its vapor pressure exceeds the external pressure of 1 [atm]. This was introduced before, see Figure 7.12 in Chapter 7, when we discussed colligative properties. This, of course, implies that if you lower the atmospheric pressure (for example, by moving to Denver, the "mile high city") water boils at a lower temperature.

Thus, in order to predict at which temperature a liquid will boil (at standard pressure), we need to establish the relationship between vapor pressure and temperature. Figure 9.4 shows experimental results of the dependence of vapor pressure on temperature for water and acetone. We see that the overall shape for these two curves is similar, but the mathematical function appears to be complicated. The shape of these curves suggests that the vapor pressure depends exponentially on the temperature (for an introduction to exponential functions, see Chapter 1). When the logarithm of the vapor pressure, $\log(p_{vap})$ is plotted against $1/T$ (in degree Kelvin), we obtain a straight line, Figure 9.5. Since this plot is linear, we conclude that the vapor pressure depends on the temperature as follows:

$$\log(p_{vap}) \propto -1/T \tag{9.8}$$

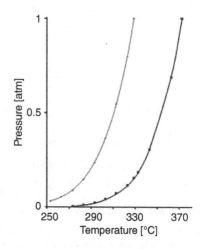

Figure 9.4 Plot of the observed vapor pressure p_{vap} vs. temperature for water (black) and acetone (gray).

since the slope of the line is negative, or

$$p \propto 10^{-(1/T)} \tag{9.9}$$

Thus, the vapor pressure depends exponentially on the negative inverse absolute temperature.

A thermodynamic derivation that is beyond the level of discussion here shows that the exponent in Eq. 9.11 is

$$p_{vap} \propto 10^{-\frac{\Delta H_{vap}}{2.303RT}} \tag{9.10}$$

or

$$\log(p_{vap}) \propto -\Delta H_{vap}/2.303RT \tag{9.11}$$

where ΔH_{vap} is the enthalpy of vaporization of the liquid in units of [kJ/mol] and the factor 2.303 results from the conversion of natural to decadic logarithms. Thus, the slope of the line in Figure 9.5 equals $-\Delta H_{vap}/R$. Before continuing with the mathematical aspects of Eqs. 9.10 and 9.11, a word on the expression $\Delta H_{vap}/RT$ is in place. In the discussion of the kinetic theory of gases, we derived that RT is a measure of the kinetic energy of the gaseous molecules. Thus, the expression $\Delta H_{vap}/RT$ compares the average kinetic energy of a particle to the energy required to leave the surface of the liquid.

Eqs. 9.10 and 9.11 were written as proportionalities because their forms as real equations need some discussion. The right-hand side of these equations is a unitless quantity, since

$$\frac{\Delta H_{vap}}{RT}\left[\frac{\frac{J}{mol}}{\frac{J}{mol\,K}K}=1\right] \tag{9.12}$$

The left-hand side of Eq. 9.11 has the logarithm of a quantity that has units of pressure, and that is a mathematical no-no. You cannot take the logarithm of a number with units, and since the right-hand side of the equation is unitless, something is amiss. As it turns out, Eq. 9.11 is derived by an integration of an expression (see Chapter 1.8)

$$\int_{p_1}^{p_2}\frac{1}{p}dp \tag{9.13}$$

that gives

$$\ln\frac{p_2}{p_1} \tag{9.14}$$

or the natural logarithm of the ratio of two pressures. The ratio is unitless, so we are allowed to take the logarithm. The final dependence of the vapor pressure is given by Eq. 9.15, which is known as the Clausius–Clapeyron equation. In this equation, both pressures are vapor pressures, but the subscript "vap" has been dropped for better readability):

$$\ln\frac{p_2}{p_1}=\frac{\Delta H_{vap}}{R}\left(\frac{1}{T_1}-\frac{1}{T_2}\right) \tag{9.15}$$

Eq. 9.15 can be written using the decadic, rather than the natural logarithm, using the conversion factor 2.303 (see Chapter 1)

$$\log\frac{p_2}{p_1}=\frac{\Delta H_{vap}}{2.303\,R}\left(\frac{1}{T_1}-\frac{1}{T_2}\right) \tag{9.16}$$

These equations permit the computation of the enthalpy of vaporization from vapor pressure and temperature data, or vapor pressures from temperature data and the enthalpy of vaporization. This is demonstrated in the following example. There are two pressures and two temperatures, as well as ΔH_{vap} in the Clausius–Clapeyron equation. Thus, it will be possible to compute ΔH_{vap} if two points on the vapor pressure curve in Figure 9.5 are given. Let us take the example of water: we have discussed that the vapor pressure is 24 [mm Hg] at 25 [°C]. Furthermore, we know that the vapor pressure is 760 [mm Hg] at 100 [°C]. How do we know this?

We use 9.16 and substitute the given pressures and temperatures

$$\log\frac{760}{24}=\frac{\Delta H_{vap}}{2.303\cdot 8.3}\left(\frac{1}{298}-\frac{1}{373}\right) \tag{9.17}$$

Notice that we may substitute the pressure in whatever units, as long as the units in the fraction are the same. The temperature has to be entered in Kelvin, and R in units of [J/(K mol)]

$$1.5006\cdot 2.303\cdot 8.31=\Delta H_{vap}(0.0033557-0.0026809) \tag{9.18}$$

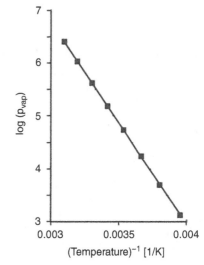

Figure 9.5 Plot of $\log(p_{vap})$ vs. $1/T$ for acetone.

Notice that we carry more significant figures during the computations and round off afterward. This is necessary for exponential or logarithmic computations

$$\Delta H_{vap} = 28.68/0.00067474 = 42511 \approx 42.5 \, [\text{kJ/mol}] \tag{9.19}$$

in reasonable agreement with the value given in Eq. 9.3 and the comment following this equation.

As a final example, let us look at the water vapor pressure at 110 [°C]. Why would we want to know the water vapor pressure at a temperature above the boiling temperature of water? Imagine you heat water in a closed container. At 100 [°C], the vapor pressure reaches 760 [mm Hg], or 1 [atm]. But if the vapor is not allowed to escape (i.e. confined in a closed container), the vapor pressure in equilibrium with the liquid water will keep increasing, and water will boil at a temperature higher than 100 [°C]. This is opposite to the lowered boiling point in the "mile high city" discussed before, and is, of course, the principle behind pressure cookers, steam engines, steam turbines, and all the good stuff of nineteenth-century engineering!

Back to the problem at hand. We again use Eq. 9.16 in a slightly modified form, and ΔH_{vap} of 40.66 [kJ/mol] at 100 [°C]

$$\log p_2 - \log p_1 = \frac{40660}{2.303 \, R} \left(\frac{1}{T_1} - \frac{1}{T_2} \right) \tag{9.20}$$

Here, we have used the identity

$$\log(a/b) = \log a - \log b \tag{9.21}$$

Just make sure that the two pressures have the same units! Then,

$$\log p_2 - \log(760) = \frac{40660}{2.303 \, R} \left(\frac{1}{373} - \frac{1}{383} \right) \tag{9.22}$$

$$\log p_2 - 2.8801 = 2127.14 \, (0.002680965 - 0.0026109660) \tag{9.23}$$

$$\log p_2 - 2.8801 = 0.14890 \tag{9.24}$$

$$\log p_2 = 3.0290 \tag{9.25}$$

$$p_2 = \text{anti} \log(3.0290) = 10^{3.0290} = p_2 = 1069 \, [\text{mm Hg}] \text{ or about } 1.41 \, [\text{atm}] \tag{9.26}$$

So, the water vapor pressure increases rapidly at elevated temperatures, and such pressurized water vapor is used in steam engines and steam turbines since the expansion of this vapor to atmospheric pressure can perform work.

Example 9.2 Using information from Example 9.1 and the enthalpy of vaporization of butane, 22.4 [kJ/mol], calculate the vapor pressure of butane at 200 [°C].

Answer: $p_{473} = 68.7$ [atm]

Example 9.3 The propellant gas in spray paint cans is a mixture of liquefied gases that may be approximated by the properties of butane.

(a) Why do you think a liquefied gas, such as butane, is used as a propellant, rather than a compressed gas such as N_2?
(b) Spray paint cans bear the warning not to expose them to heat or fires. Based on your answer in Example 9.2, what would happen if you ignored this warning?
(c) After prolonged use of a spray paint can, you may realize that the can cools down significantly. Explain.

Answer: See answer booklet

The amount of water vapor in the atmosphere depends on the temperature as given by Eq. 9.16, and we can assume that the water vapor is additive to the total air pressure, as indicated by the law of partial pressures (Eq. 8.24). This water vapor pressure affects us quite a bit in terms of our well-being: on a hot, muggy day the water vapor pressure can reach its saturation value given by Eq. 9.16. Thus, a weather forecast of 100% humidity at 100 [°F] refers to the vapor pressure of water at this temperature, which is c. 54 [mm Hg]. The relative humidity you hear about in the weather report is given by the measured humidity multiplied by 100, divided by the vapor pressure of water at a given temperature.

Example 9.4 Calculate the water content of the atmosphere on a cold, crisp winter day when the temperature is 4 [°C] and the relative humidity is 10%. Use $\Delta H_{vap} = 44$ [kJ/mol] and $p_{vap}(25 [°C]) = 24$ [mm Hg].

Answer: $p_{\text{water vapor}} = 0.62$ [mm Hg]

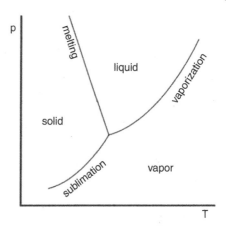

Figure 9.6 Schematic phase diagram of water.

Solids also exhibit vapor pressure. For most solids, for example metals, the value of the vapor pressure is so small that we may say that vaporization is negligible. However, solid water (ice) exhibits a significant vapor pressure that is responsible for the sublimation of snow and ice, even below the melting point of water. This explains the slow disappearance of a snow pack even if the temperature never reaches the melting temperature. The vapor pressure of a solid is governed by an equation similar to the Clausius–Clapeyron equation (Eqs. 9.17 or 9.18) where the heat of vaporization, ΔH_{vap}, is replaced by the heat of sublimation, ΔH_{sub}. The sublimation curve, along with the vaporization curve, is shown in the phase diagram of water, depicted in Figure 9.6. This figure shows the intersection point of the vaporization curve, the sublimation curve, and the melting curve in a pressure vs. temperature diagram. The point where all three phases can coexist (the aforementioned intersection of the sublimation, vaporization, and melting curves) is known as the "triple point." For water, the triple point occurs at about 0.01 [°C] and 4.6 [mm Hg].

This diagram can be read as follows. We know that at very high temperature, water exists as vapor only. At intermediate temperatures, between 0 [°C] and a few hundred degrees Centigrade, a liquid water/water vapor equilibrium exists. This is shown by the curve marked as "vaporization." At a temperature below the freezing point (which is close to the triple point), water will be a solid, in equilibrium with vapor, indicated by the sublimation curve. The shapes of the vaporization and sublimation curves reflect the exponential dependence of the vapor pressure on temperature, as shown before in Figure 9.4.

The line separating the solid and liquid phases represents the melting–freezing process as a function of temperature and pressure. This curve ends at the triple point of water and indicates the well-known fact that water freezes at 0 [°C]. But this line also indicates that as the pressure increases, the melting point of water decreases. This effect has a curious consequence: when skating on ice, the small area of the blade of the skate in touch with the ice produces a large pressure on the ice. At this pressure, the ice melts and produces a very thin film of liquid water between the blade of the skate and the ice, reducing the friction. Hence, skaters glide on the ice nearly frictionless.

The negative slope of the melting line is specific to water, and is due to the hydrogen-bonded structure of ice that occupies a larger volume than liquid water. In all other compounds, the solid phase occupies a smaller volume than the liquid phase. The increased volume of ice causes it to have a lower density than liquid water, and indicates that the volume increases, rather than decreases, when water freezes. The lower density of ice causes it to rise on top of liquid water, which causes water to "freeze from top down," rather than from the bottom. This has profound effects on aquatic life in cold temperatures.

Most solids exhibit very low vapor pressure; nevertheless, molecules do escape from solids as well and establish a solid–vapor equilibrium. An exception to the above statement is solid CO_2, or dry ice, which has a large vapor pressure of about 56.5 [atm] at 20 [°C]. Dry ice sublimes at −78 [°C], that is, it goes directly from the solid to the gaseous phase at ambient pressure, as shown by the curve marked "sublimation" in Figure 9.6. Thus, its vapor pressure is 1 [atm] at −78 [°C].

Example 9.5 Given the two vapor pressure values in the above paragraph, calculate the heat of sublimation of dry ice.

Answer: $\Delta H_{sub} = 19.2 \left[\dfrac{kJ}{mol}\right]$

9.4 Chemical Equilibrium and the Equilibrium Constant

So far, we have discussed the concepts of a dynamic equilibrium in terms of a liquid establishing its vapor pressure. We have seen that the rate of the forward reaction (evaporation) equals the rate of the reverse reaction (condensation) and that a steady-state situation is achieved. Furthermore, the equilibrium concentration in the gaseous phase, which we call the vapor pressure, is constant, as long as the temperature is kept constant. This aspect is totally new in terms of our previous (stoichiometric) view of chemistry: not only does a reaction NOT go to completion, but the point at which the reaction (the evaporation of the liquid in our case) stops is determined by the concentration of the gaseous component. We demonstrated this in Figure 9.3 when we found that, regardless of the volume of the container, evaporation occurs until the vapor pressure has reached 24 [mm Hg].

A totally analogous picture can be developed for chemical equilibria. For example, the reaction discussed in the introduction, (in the gas phase)

$$H_2 + F_2 \rightleftarrows 2HF \tag{9.1}$$

reaches a dynamic equilibrium at which the concentrations or partial pressures of H_2, F_2, and HF remain constant over time, since the forward reaction rate and the reverse reaction rate are equal at equilibrium. Thus, if these concentrations are constant, we will try to write a mathematical expression to indicate this fact. It was found experimentally that for a generalized chemical reaction

$$a\,A + b\,B \rightleftarrows c\,C \tag{9.27}$$

the quotient

$$\frac{[C]^c}{[A]^a\,[B]^b} \tag{9.28}$$

is constant. Here, A and B are the reactants, C is the product, square brackets denote equilibrium concentrations, and the stoichiometric coefficients appear in the exponent of the concentrations. Since, for any given reaction at a fixed temperature, this quotient is constant, we assign a name to it:

$$\frac{[C]^c}{[A]^a\,[B]^b} = K_{eq} \tag{9.29}$$

and call it the equilibrium constant K_{eq}. That is worth thinking about: we say that for a given chemical reaction, the ratio of the product and reaction concentrations, taken to the power of the respective stoichiometric coefficients, is constant, and **does not depend on the amounts of reactants and products**. For the reaction in Eq. 9.1, for example, the idea introduced by Eq. 9.29 implies that

$$\frac{[HF]^2}{[H_2][F_2]} = K_{eq} \tag{9.30}$$

is a constant, regardless of the amounts (or concentrations) of the reactants. Thus, K_{eq} does not depend on the concentrations of products and reactants, but the reverse is true: the concentrations of products and reactants are determined by K_{eq}.

This is a distinction that we must clearly understand. It is so important that we need to discuss it using an example you are all familiar with. Let us look at a constant that you encountered in high school algebra: the number π, which is the ratio of the circumference, C, of a circle to its diameter, d. Experimentally, it was found that the ratio, C/d, is constant:

$$C/d = \pi \tag{9.31}$$

However, the constant π does not depend on the numerical value of C or d, since it is a constant (we would not call it a constant if it were a function of C or d!!!). Rather, if d is varied, C will vary as well according to

$$C = \pi \cdot d \tag{9.32}$$

to maintain the ratio C/d.

Exactly the same is true for the chemical equilibrium constant, K_{eq}. It may be determined experimentally by measuring equilibrium concentrations of all reactants and products, but it does not functionally depend on them. Rather, the concentrations depend on K_{eq}. K_{eq} is a function of temperature only (as we shall introduce in Chapter 11 on thermodynamics), just as the vapor pressure of a liquid changes with temperature.

For the chemical reaction described by Eq. 9.1, the equilibrium constant, at a given temperature, is 115. Thus, we may write Eq. 9.30 as

$$\frac{[HF]^2}{[H_2][F_2]} = 115 \tag{9.33}$$

which holds for any starting conditions, i.e. for any concentrations of reactants and products you may choose. Thus, if you start with 1.0 [mol/L] of each of H_2 and F_2, and no HF at all, the forward chemical reaction will occur such that at equilibrium, the quotient given by Eq. 9.33 holds. Similarly, if you start with pure HF, the reverse reaction will occur until Eq. 9.33 holds.

Of course, if we had written the original equation in the opposite direction:

$$2\,HF \rightleftarrows H_2 + F_2 \tag{9.34}$$

the equilibrium constant would be given by its inverse value:

$$\frac{[H_2][F_2]}{[HF]^2} = \frac{1}{115} = 8.695 \cdot 10^{-3} \tag{9.35}$$

since it is always given by the quotient of product concentrations over reactant concentrations. The magnitude of the equilibrium constant actually tells us quite a bit about the reaction. If you look at equation 9.33, you will agree that, in order for K_{eq} to be 115, the numerator needs to be larger than the denominator. Thus, we can say that the equilibrium lies on the side of the product, HF. However, that does not imply that the reaction goes to the right, see Section 9.6. On the other hand, in Eq. 9.35, the equilibrium constant is smaller than 1; thus, the numerator is small and the denominator large, and the equilibrium lies on the reactant, HF. We conclude that if $K_{eq} > 1$, the equilibrium lies on the side of the products; if $K_{eq} < 1$, the equilibrium lies on the side of the reactants.

Example 9.6 Consider a hypothetical gas phase reaction $2A(g) + B(g) \rightleftarrows A_2B$. Write the equilibrium expression for the reactants and products in [mol/L].

Answer: $K = \dfrac{[A_2B]}{[A]^2[B]}$ in units of $[L^2/mol^2]$

9.5 Equilibrium Calculations

Since the problem was asked, but not answered, in the previous section, let us revisit it: what will be the equilibrium concentration of all three species, H_2, F_2, and HF, if we mix 1.0 [mol H_2], and 1.0 [mol F_2] in a 1.0 [L] container with no HF present. We shall set up a table of initial and final concentrations as shown below. This formalism is extremely important, and it is really necessary that you understand it, for we shall use it throughout all discussions involving equilibria.

	Initial concentrations	Equilibrium concentrations
H_2	1.0	
F_2	1.0	
HF	0	

(9.36)

That was easy. Now, how about the concentrations at equilibrium? Obviously, that is what we want to calculate. We know that the equilibrium of the reaction

$$H_2 + F_2 \rightleftarrows 2\,HF \tag{9.1}$$

lies on the right-hand side, since K_{eq} is 115. Since we start with pure hydrogen and fluorine gas, the reaction must occur to the right to produce HF. Thus, some hydrogen will react to produce HF, but we do not know how much. Thus, let us call "x" the amount of H_2 that reacts. Therefore, at equilibrium, the amount of H_2 left over will be 1.0 − x. However, from the stoichiometry of the reaction given by Eq. 9.1 we know that one mole hydrogen gas reacts exactly with one mole of fluorine gas. Thus, the fraction of fluorine gas that react with hydrogen also must be "x", and the concentration of fluorine at equilibrium is 1.0 − x as well. This is shown in the next iteration of our concentration table below:

	Initial concentrations	Equilibrium concentrations
H_2	1.0	1.0 − x
F_2	1.0	1.0 − x
HF	0	

Now, what will be the HF concentration at equilibrium? We know that "x" moles of H_2 and "x" moles of F_2 have reacted; according to Eq. 9.1, they will produce 2x moles of HF. We complete our table and obtain:

	Initial concentrations	Equilibrium concentrations
H_2	1.0	$1.0 - x$
F_2	1.0	$1.0 - x$
HF	0	$2x$

We know that the concentrations at equilibrium are related to the equilibrium constant. Thus, we may write:

$$\frac{[HF]^2}{[H_2][F_2]} = \frac{(2x)^2}{(1-x)(1-x)} = 115 \tag{9.37}$$

Taking the square root of both sides of Eq. 9.37 yields[1]

$$\frac{2x}{(1-x)} = \pm\sqrt{115} \tag{9.38}$$

$$x_+ = 0.843 \text{ or } x_- = -0.843 \tag{9.39}$$

Remember, a quadratic equation generally has two different solutions. We cannot determine *a priori* which of these actually is the solution we want. Thus, we have to reason which one of the solutions is the valid one. We started the equilibrium process with a concentration of H_2 and F_2 of 1.0 [mol/L], and we called x the amount of the starting materials that actually reacted. Thus, the solution $x_- = -0.843$ cannot be a valid solution of the quadratic equation because it would make the concentration of both H_2 and F_2 1.843 [mol/L], more than what we started with. Thus, the physically meaningful solution of the quadratic equation is $x = 0.843$. Furthermore, it is advisable to carry more significant figures than usual, so that we can carry out the test of our calculations (see below). Our equilibrium table will appear as shown above in Eq. (9.40).

	Initial concentrations	Equilibrium concentrations
H_2	1.0	$1.0 - 0.843 = 0.157$
F_2	1.0	$1.0 - 0.843 = 0.157$
HF	0	1.686

(9.40)

To test whether or not we did the math correctly, let us substitute our equilibrium concentrations into the equilibrium expression and see if these concentrations do, indeed, obey the equilibrium constant:

$$\frac{[HF]^2}{[H_2][F_2]} = \frac{(1.686)^2}{(0.157)(0.157)} = 115.3 \tag{9.41}$$

which is close enough to the equilibrium constant we started out with. Thus, we see that if we start with a certain set of conditions (the *initial* conditions), the reaction will occur until the equilibrium constant is fulfilled. In this case, the reaction proceeds forward, and produces HF until the equilibrium is reached.

In chemical reactions, the equilibrium expressions can get quite messy. Let us take, for example, the formation of ammonia from the elements:

$$N_2(g) + 3H_2(g) \rightleftarrows 2NH_3(g) \tag{9.42}$$

---

[1] This is a special case of a quadratic equation which can be solved easily by taking the square root of both sides. A more general solution of quadratic equations is presented in Section 1.5.

This is an extremely important, but difficult, chemical reaction, since nitrogen is abundant but quite inert. The equilibrium constant is very favorable at room temperature:

$$K_{298} \approx 930{,}000 \tag{9.43}$$

But at room temperature, the reaction is extremely slow, so slow, in fact, that no product is formed. This is a situation we encounter in chemistry and biochemistry quite often: the reaction should proceed in the direction the equilibrium position suggests, but the reaction is so slow that, for all practical purposes, it does not occur, or, in the parlance of Chapter 13, the rate of the chemical reaction is nearly zero. This is because the reaction has a very high "activation energy," an energy bump the reactants have to overcome to form the product. We will discuss this aspect in detail in Chapter 13 on Chemical Kinetics. Such reactions can be accelerated by increasing the temperature. Let us think about that for a second. Gasoline, which is mostly an alkane named octane with a formula of C_8H_{18}, at room temperature is perfectly "stable" in the presence of oxygen in the air, although the equilibrium for the reaction

$$C_8H_{18} + 12.5\, O_2 \rightarrow 8\, CO_2 + 9\, H_2O \tag{9.44}$$

lies far on the side of the products (we will learn in Chapter 11 how to calculate the equilibrium constant for this reaction from thermodynamic data). However, octane ignites quite violently when you drop a lit match in the container (do not do this at home): raising the temperature locally gets the reaction mixture over the energy bump.

So, for the case of the hydrogen/nitrogen mixture, we try the same approach: we increase the temperature of the reaction mixture, because the rate of reaction increases with temperature (see Section 13.5). But at elevated temperatures, for example, at 200 [°C] or 473 [K], the equilibrium constant for the reaction is less favorable:

$$K_{473} \approx 5770$$

At higher temperature, where the reaction is faster, the equilibrium shifts to the left, toward the side of the reactants. We shall discuss the temperature dependence of equilibria in Chapters 11 and 13. Thus, increasing the temperature of the reaction may be less advantageous in terms of the equilibrium position. For the industrial production of ammonia, moderate temperatures are used along with a catalyst to speed up the reaction.

Let us use the equilibrium constant at 473 [K] to discuss the equilibrium concentrations for this reaction. If you were to start with a stoichiometric mixture of the reactants, i.e. 1 [mol] of N_2 and 3 [mol] of H_2, let us figure out the form of the equilibrium expression. We start again with our table of concentrations, in analogy to Eq. 9.36:

	Initial concentrations	Equilibrium concentrations
N_2	1.0	$1.0 - x$
H_2	3.0	
NH_3	0	

and assume that a fraction "x" of the nitrogen reacts. From the stoichiometry of the reaction, Eq. 9.42, we know that for each molecule of nitrogen, three molecules of hydrogen react. Thus, the concentration of hydrogen at equilibrium is $3.0 - 3x$. Similarly, we argue that each fraction "x" of nitrogen that reacts will produce 2x moles of ammonia, since there is a factor of two between nitrogen and ammonia in Eq. 9.42. This leads to our table of concentrations as follows:

	Initial concentrations	Equilibrium concentrations
N_2	1.0	$1.0 - x$
H_2	3.0	$3.0 - 3x$
NH_3	0	$2x$

The equilibrium expression for this reaction is:

$$K = 5770 = \frac{[NH_3]^2}{[N_2][H_2]^3} = \frac{(2x)^2}{(1-x)(3-3x)^3} \tag{9.45}$$

Eq. 9.45 can be simplified to

$$5770 = \frac{(2x)^2}{(1-x)(3(1-x))^3} = \frac{(2x)^2}{27(1-x)^4} \tag{9.46}$$

In general, equations that have the unknown to third (cubic) or forth power (quartic) are difficult to solve. In the case of Eq. 9.46, however, we can take the square root of both sides:

$$2x/(1-x)^2 = \pm\sqrt{27 \cdot 5770} \tag{9.47}$$

This is just a quadratic equation, with solutions

$$2x/(1-x)^2 = \pm 394.7 \tag{9.48}$$

Using the negative sign of the right-hand side yields imaginary solutions; using the positive sign, we obtain

$$x_+ = 1.0737 \text{ or } x_- = 0.9313 \tag{9.49}$$

Again, we argue that only the second value, $x = 0.9313$ is physically possible. Thus, our table of equilibrium concentrations can be written as shown in Table 9.1:

Table 9.1 Equilibrium concentrations of hydrogen/nitrogen/ammonia reaction.

	Initial concentrations	Equilibrium concentrations
N_2	1.0	0.0687
H_2	3.0	0.2061
NH_3	0	1.8626

These concentrations fulfill the equilibrium expression within rounding error limits (5768 ≈ 5770). Notice, however, that if the concentrations of the reactants are not in the stoichiometric ratio, the equation equivalent to 9.46 may not be solved easily, since it contains terms in x^4, x^3, x^2, etc.

Example 9.7 Relate the numerical value of the equilibrium constant in Example 9.6, where the concentrations were expressed in units of [mol/L], to the value of the equilibrium constant if the concentrations are expressed in units of pressures [atm].

Answer: Let K_C be the equilibrium constant expressed in concentration units [mol/L] and K_P the equilibrium constant expressed in [atm]. Then $K_C = K_P(RT)^2$

The examples discussed so far dealt with gaseous equilibria. The gaseous concentrations can be expressed in units of [mol/L], as we have done so far, or as partial pressures, since $(n/V) = p/RT$ and (n/V) is the gaseous concentration in [mol/L]. However, the numeric value of the equilibrium constant can differ dependent on whether it is expressed as molar concentrations or partial pressures, as you have shown in Example 9.7. Furthermore, we have introduced the concept that chemical reactions generally do not go to completion, but reach a state of dynamic equilibrium. This is true for the vaporization of a liquid in a closed system to establish its vapor pressure, or for more complicated systems like the hydrogen/nitrogen/ammonia reaction discussed above. We have defined the equilibrium constant as the quotient of equilibrium concentrations of products and reactants, each taken to the appropriate stoichiometric powers. We have discussed chemical reactions that had all reactants or products in the gas phase. The equilibrium expression holds equally well for reactants and products in solution, where the concentrations are expressed in units of molarity [mol/L]. However, when a reactant or product is a pure liquid or a pure solid, its concentration does not appear in the equilibrium expression. For example, the equilibrium expression for the reaction

$$CaCO_3(s) \rightleftarrows CaO(s) + CO_2(g) \tag{9.50}$$

is simply

$$K = [CO_2] \tag{9.51}$$

because the concentration of calcium carbonate in solid calcium carbonate is constant, as is the concentration of solid CaO. These concentrations are related to the density of the solid, even if the amounts of the solids change during the reaction. We may think of the numeric value of K to incorporate the two (constant) concentrations of the solids involved.

This aspect is also important for our previous discussion of the vapor pressure. Then we deal with the evaporation process of a pure liquid:

$$H_2O(liquid) \rightleftarrows H_2O(vapor) \quad (9.2)$$

for which we can write the equilibrium expression as

$$K = [H_2O_{vapor}] \quad (9.52)$$

since, again, the concentration of liquid water does not change because liquid water has a constant concentration, of about 55.6 [mol/L]. Since the concentration of water vapor $[H_2O_{vapor}]$, in [mol/L], is related to the pressure according to $n/V = p/RT$, we see that Eq. 9.52 can be written as

$$K = p_{vap}/RT \quad (9.53)$$

Thus, we see that the vapor pressure really represents an equilibrium constant for the reaction given by Eq. 9.2.

Example 9.8 Experimentally determined equilibrium concentrations for the gas phase reaction in Example 9.6, $2A + B \rightleftarrows A_2B$ were found to be
[A] = 0.060 [mol/L] [B] = 0.040 [mol/L] [A$_2$B] = 0.72 [mol/L].
Calculate the numerical value of the equilibrium constant.

Answer: K = 5000

9.6 Direction of a Chemical Reaction and the Concentration Quotient Q

We have seen in the discussion so far that a chemical reaction proceeds until equilibrium is reached, and the equilibrium concentrations conform with the equilibrium constant, K. This constant can be reached by the reaction proceeding "to the left," or "to the right," depending on the starting conditions. Consider the reaction we discussed in Eq. 9.1:

$$H_2 + F_2 \rightleftarrows 2HF \quad (9.1)$$

$$\frac{[HF]^2}{[H_2][F_2]} = 115 \quad (9.33)$$

If we start with a mixture of pure hydrogen and fluorine, the reaction will occur to the right to reach equilibrium. On the other hand, if we start with pure HF, then the reaction will occur to the left to reach equilibrium. If we define a quotient Q of nonequilibrium concentrations (or "initial" concentrations) that has the same format as the equilibrium constant, that is (see Eq. 9.28):

$$K_{eq} = \frac{[C_{equilibrium}]^c}{[A_{equilibrium}]^a [B_{equilibrium}]^b} \quad (9.54)$$

$$Q = \frac{[C_{initial}]^c}{[A_{initial}]^a [B_{initial}]^b} \quad (9.55)$$

we see that in the case discussed above (starting with pure hydrogen and fluorine, and [HF] = 0), Q assumes the value of Q = 0. In this case, the reaction proceeds to the right. On the other hand, if we start with only HF, the denominator is zero, and Q = ∞. Thus, the reaction will proceed to the left to reach equilibrium. We summarize:

- if Q < K, the reaction proceeds to the right to reach equilibrium.
- if Q > K, the reaction proceeds to the left to reach equilibrium.
- if Q = K, no reaction occurs, and we started at equilibrium conditions.

In later discussion of equilibrium, we shall associate these cases correspond to negative, positive, and zero *free enthalpy*, respectively, where the free enthalpy is the universal sign post indicating the direction of a chemical reaction.

Example 9.9 Initial concentrations of A, B, and A$_2$B for the reaction in Example 9.8 were $[A]_i$ = 0.03 [mol/L], $[B]_i$ = 0.03 [mol/L], $[A_2B]_i$ = 0.5 [mol/L]. In which direction does the reaction proceed to equilibrium?

Answer: To the left

9.7 Numerical Determination of Equilibrium Constants from Experimental Data

So far, we have assumed that the numerical value of the equilibrium constant is given for a reaction of interest. These equilibrium constants, however, are experimentally derived quantities, and are determined by measuring the equilibrium concentrations of reactants and products at a given temperature. Once these concentrations are determined, K can be computed easily, according to Eq. 9.30. This constant is valid for any concentrations of the reactants and products, since, as the name implies, it is a constant. The only variable that can change the value of this constant is the temperature, as we shall discuss later. Thus, it is important to remember that K may be determined experimentally by measuring equilibrium concentrations, but K is not a function of these concentrations, but only a function of temperature.

Often, only one measurement of a concentration is necessary to determine the numerical value K. Consider the case of reaction 9.1 again. If we start a reaction with a hydrogen and a fluorine concentration of 1.0 [mol/L], we need to perform one measurement only to determine the numerical value of K. For example, if we find the equilibrium concentration for HF to be 1.686 [mol/L], we know that this corresponds to the term 2x in the table of equilibrium concentrations:

	Initial concentrations	Equilibrium concentrations
H_2	1.0	$1.0 - x$
F_2	1.0	$1.0 - x$
HF	0	$2x = 1.686$

from which we can calculate the equilibrium concentrations of hydrogen and fluorine, and hence, the equilibrium constant.

9.8 Perturbations of Equilibria: Le Chatelier's Principle

When a system is at equilibrium, and a stress or perturbation is applied to the system (such as changing the volume of the reaction mixture, or adding reactants or products), the system will respond in such a manner as to minimize the stress. This principle will be illustrated using the two equilibrium systems introduced earlier in the discussion of equilibria.

Let us look at a perturbation of the equilibrium situation described by Eqs. 9.33–9.40 after a perturbation is applied to the system. If you manage a chemical plant that produces HF, you may realize that one mole of hydrogen and one mole of fluorine do not give you the promised two moles of HF, but only 1.68, i.e. the yield of the process is only 84%. Can one modify the reaction equilibrium to get more products?

What will happen if we add two more moles of H_2 to the equilibrium mixture so that we have 3.0 moles of hydrogen and 1.0 mole of fluorine? We write our table of initial conditions as

	Initial concentrations	Equilibrium concentrations
H_2	3.0	$3.0 - x$
F_2	1.0	
HF	0	

Now, assume that fraction "x" of hydrogen reacts, as before. But we know from the stoichiometry of the reaction that each of the "x" H_2 molecules that react requires exactly one F_2 molecule. Thus, the fraction of F_2 left after equilibrium is established is $1.0 - x$, and the amount of HF produced is still 2x:

	Initial concentrations	Equilibrium concentrations
H_2	3.0	$3.0 - x$
F_2	1.0	$1.0 - x$
HF	0	$2x$

This leads to the equation

$$\frac{[HF]^2}{[H_2][F_2]} = \frac{(2x)^2}{(3-x)(1-x)} = 115 \tag{9.56}$$

or $4x^2 = 115(3-4x+x^2)$ or $345 - 460x + 111x^2 = 0$ (9.57)

which we solve using the formalism for quadratic equations (see Section 1.5). The solutions are $x_- = 0.983$ or $x_+ = 3.16$. As before, we have an easy time to select the solution that is physically meaningful: we cannot have 3.16 [mol] reacting if we only have 3 [mol] of reactant. Thus, we select $x = 0.983$, and see what our equilibrium concentrations come out to be:

	Initial concentrations	Equilibrium concentrations
H_2	3.0	$(3.0 - x) = 3 - 0.983 = 2.017$
F_2	1.0	$(1.0 - x) = 1 - 0.983 = 0.017$
HF	0	$(2x) = 1.966$

A check of our equilibrium concentration yields:

$$\frac{1.966^2}{(2.017)(0.017)} = 113 \text{ (close to 115)} \tag{9.58}$$

What have we accomplished? By using more hydrogen, we have shifted the equilibrium toward the side of the products. Instead of getting 1.68 moles of product, we are getting 1.966 moles of product, a 20% improvement. We see that adding more of one reactant shifts the reaction further toward the side of the products. This is Le Chatelier's principle: any stress applied to a chemical equilibrium will cause a reaction in the direction that relieves the stress. In this example, adding more of one reactant (the stress) causes the reaction to shift toward the products in order to reestablish equilibrium.

The other example involving Le Chatelier's principle to be discussed is the nitrogen/hydrogen/ammonia mixture, discussed in Eqs. 9.45–9.49, and see how the equilibrium is affected if we reduce the volume of an equilibrium mixture by a factor of two. In other words, let us squeeze the container in which the equilibrium amounts of gases are contained to half of its volume. This will affect the equilibrium, since for the reaction

$$N_2(g) + 3\,H_2(g) \rightleftarrows 2\,NH_3(g) \tag{9.43}$$

the reactants represent four moles of gas, the product, however, only two moles of gas. By perturbing the volume of the gases at equilibrium, we expect the reaction to shift to the right, in the direction of lower volume of gaseous components.

Thus, the starting conditions will use the results from the above calculations that were summarized in Table 9.1, and are here referred to as Orig.Equil.Conc. After squeezing the reaction mixture to half its volume, all concentrations go up by a factor of two, and the reaction occurs, reducing/or incrementing the concentrations as shown in the column Squeezed Conc. in Table 9.2:

Solving the resulting quartic equation can be performed using the procedure similar to the one employed in Eqs. 9.46 and 9.47, and is left here as an exercise (Example 9.10, which is quite tricky). The relevant solution of this equation is $x = 0.03923$, which gives us the new equilibrium concentrations listed in the column "Squeezed Equil." Again, the equilibrium constant is maintained for these new equilibrium conditions. The same results are obtained if we start an equilibrium with concentrations of nitrogen at 2 [mol/L], hydrogen at 6 [mol/L] and ammonia concentration of zero, that is, exactly the doubled concentrations that were used in the original calculations (Eqs. 9.46–9.50, see Example 9.10).

According to Le Chatelier's principle, one can influence the equilibrium position by changing the concentrations by decreasing the volume, by adding reactants, or by removing products. This latter approach is followed in a biochemical reaction step (within the glycolysis sequence of reactions) with unfavorable equilibrium position: the reaction product is continuously removed by a very fast secondary reaction to drive the reaction forward in spite of the unfavorable equilibrium position.

Table 9.2 Equilibrium concentrations of hydrogen/nitrogen/ammonia reaction.

	Orig. Equil. Conc.	Squeezed Conc.	Squeezed Equil.
N_2	0.0687	$2 \cdot 0.0687 - x = 0.1374 - x$	0.0982
H_2	0.2061	$2 \cdot 0.2061 - 3x = 0.4122 - 3x$	0.2945
NH_3	1.8626	$2 \cdot 1.8626 + 2x = 3.7252 + 2x$	3.8037

Example 9.10 Recalculate the $H_2/F_2/HF$ equilibrium position described by Eq. 9.41 after additional 0.1 [mol/L] hydrogen is added to the equilibrium mixture.

Answer: $[H_2] = 0.2213$ [mol/L]; $[F_2] = 0.1213$ [mol/L]; $[HF] = 1.7573$ [mol/L]

Example 9.11 Solve the quartic equilibrium equation resulting from the concentrations listed in the central column in Table 9.2. You may want to use the same procedure outlined in Eqs. 9.47 and 9.48 to reduce the quartic equation to a quadratic equation.

Answer: See answer booklet

Example 9.12 Show that the concentrations listed in the rightmost column in the table above can be obtained by the starting conditions of nitrogen at 2.0 [mol/L], hydrogen at 6.0 [mol/L], and ammonia concentration of zero, i.e. the original starting concentration doubled.

Answer: See answer booklet

9.9 Solubility and Solubility Product

A direct application of the concepts of equilibria is the solubility of sparingly soluble salts to form saturated aqueous solutions. The concept of soluble and "sparingly soluble" or insoluble salts was first mentioned in the preliminary discussion of solutions (Chapter 4). When a solid salt, say 10 [g] of AgCl, is added to one liter of water, we observed that some of the solid dissolves, but most of the solid just sinks to the bottom of the flask and just sits there, as shown in Figure 9.7. Let us first ask the question: how do we know that some silver chloride dissolved? That can be demonstrated by measuring, for example, the conductivity of the solution toward electricity since dissolved ions can migrate to the cathode or anode, thereby conducting electricity. Or, we could separate the solution from the remaining solid (by filtration, for example), and analyze the solution for the presence of ions (see below).

What we have described for AgCl so far holds for just about all salts: when salt is added to water, some of it will dissolve, and some will sink to the bottom and do nothing. It is just the amount of material that dissolves which is different for soluble and sparingly soluble salts. For a soluble salt, for example, NaCl, about 300 [g] can be added to 1 [L] of water before solid NaCl builds up at the bottom of the beaker. For other salts, only a few milligrams or even micrograms will dissolve, and most of the added salt will be sitting at the bottom of the flask.

We call a solution of a salt, where we have a solid in equilibrium with dissolved salt, a "saturated" solution. This saturation process occurs again as a dynamic equilibrium, as discussed before in the case of vapor pressure: if we add a solid salt to pure water, the process of dissolution begins. Simultaneously, however, the reverse process will start to occur: ions in the solution will collide with each other and may precipitate. Thus, the equilibrium formed between ions in solution, and the solid that will not dissolve, is a dynamic one: at equilibrium, the rate of salt dissociating and dissolving equals the rate of ions precipitating.

We shall concern ourselves in this discussion with the solubility of salts in aqueous solution. Many salts, in particular those which form minerals, are sparingly soluble in water. This makes sense: if they were soluble, they would have long been dissolved by rain. Let us take the mineral calcite, $CaCO_3$, for an example. This compound is relatively insoluble in water; however, a small amount will dissolve in water. The amount that dissolves (given by the solubility) generally increases with increasing temperature. The most important aspect of this dissolution process is that the (small) fraction of calcite that dissolves will undergo complete dissociation into ions, and solvation (hydration) of these ions occurs. In fact, the ions formed will significantly increase the electric conductivity of the solvent; therefore, we call the solution formed an "electrolyte solution."

The effect of dissolution and solvation is shown schematically in Figure 9.8. This figure points out that the process of dissolution depends on two energetic factors that are in balance: the ions in the solid are held in the lattice by electrostatic interactions, which give the crystal its stability. When ions

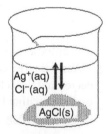

Figure 9.7 A sparingly soluble solid in dynamic equilibrium with dissolved (and dissociated) material.

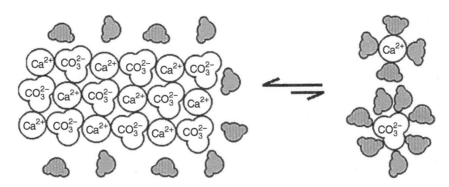

Figure 9.8 Schematic of an ionic solid, CaCO₃, dissolving and dissociating into hydrated ions.

leave the lattice and go into solution, these ions will be completely solvated, or surrounded by water molecules. The solvation process is, in general, favorable from an energetic viewpoint, since water is a polar molecule, and the ions are charged particles. Thus, whether a salt dissolves readily, or is nearly insoluble, depends primarily on the balance between solvation energy and lattice energy.

The take-home message of this section is the following: when a salt dissolves in water, complete dissociation into ions occurs for all the material that has dissolved. However, depending on the solubility of the salt, the fraction that dissolved may vary enormously, from many [g/L] for some salts, to less than 10^{-10} [g/L] for others.

9.9.1 The Solubility Product Constant, K_{sp}

The fraction of salt that dissolves will completely dissociate into ions; thus, the dynamic equilibrium we defined in the previous discussion will be between the undissolved salt and solvated ions in solution. The mineral feldspar, CaF_2, for example, in water will undergo the following process

$$CaF_2(s) \rightleftarrows Ca^{+2}(aq) + 2F^-(aq) \tag{9.59}$$

where the left-hand side represents the undissolved, ionic solid (salt) that does not seem to do anything in the process. Of course, Eq. 9.59 describes a dynamic equilibrium process where the rate of dissolution and precipitation are the same. Therefore, an equilibrium expression can be written:

$$\mathbf{K} = \frac{[Ca^{+2}][F^-]^2}{CaF_2(s)} \tag{9.60}$$

However, since $CaF_2(s)$ is a solid, its concentration is constant, as discussed at the end of Section 9.5 (Eq. 9.51). Thus, one multiplies the equilibrium constant by the constant concentration of the solid and calls this product the "solubility product" K_{sp}:

$$K_{sp} = [Ca^{+2}][F^-]^2 \tag{9.61}$$

In Eq. 9.60, the concentrations in the square brackets represent the hydrated ions. For CaF_2, the solubility product constant can be found in the literature or the internet and has the value

$$K_{sp}(CaF_2) = 4.0 \cdot 10^{-11} \tag{9.62}$$

As before in the discussion of equilibria, the magnitude of K_{sp} indicates the extent of a reaction. Thus, when K_{sp} is very small, not very much of the ionic material dissolves.

9.9.2 Solubility Calculations

From the earlier equilibrium discussion, it is clear that the information presented here is sufficient to compute the concentration of ions in solution, since we are dealing with an equilibrium process with a known equilibrium constant and predictable stoichiometry.

For the example cited above, the solution equilibrium of feldspar,

$$CaF_2(s) \rightleftarrows Ca^{+2}(aq) + 2F^-(aq) \qquad (9.59)$$

we proceed, as usual, and write a table of initial and final concentrations:

CaF$_2$	Initial amount	Final amount
Ca^{2+}	0	x
F$^-$	0	2x

Here, we carefully avoided defining a concentration for the solid CaF_2, and wisely referred to its "amount." However, we know that if a fraction x of this original amount dissolves, this fraction will completely dissociate according to Eq. 9.59, and produce a calcium ion concentration, $[Ca^{+2}]$ = x. Since the number of fluoride ions will be twice that of the calcium ions, $[F^-]$ = 2x. Thus, the expression for the solubility product will be

$$K_{sp}(CaF_2) = 4.0 \cdot 10^{-11} = [Ca^{+2}][F^-]^2 = x(2x)^2 = 4x^3 \qquad (9.63)$$

$$x^3 = 10^{-11} \text{ or } x = 2.15 \cdot 10^{-4} \qquad (9.64)$$

As pointed out before, the final answer should be rounded off to the appropriate number of significant figures; however, in order to perform the check of our calculations, we carry the extra significant figures for the time being:

$$[Ca^{+2}] = 2.15 \cdot 10^{-4} \text{ [mol/L] and } [F^-] = 4.30 \cdot 10^{-4} \text{ [mol/L]} \qquad (9.65)$$

We check the results by substituting these concentrations into the equilibrium expression:

$$\begin{aligned} K_{sp} &= [Ca^{+2}][F^-]^2 = (2.15 \cdot 10^{-4})(4.3 \cdot 10^{-4})^2 \\ &= (2.15 \cdot 10^{-4})(18.49 \cdot 10^{-8}) \\ &= 39.75 \cdot 10^{-12} = 3.975 \cdot 10^{-11} \approx 4.0 \cdot 10^{-11} \end{aligned} \qquad (9.66)$$

Thus, the question "what are the ion concentrations in a saturated solution of CaF_2?" was answered by the methods developed in earlier discussions on the equilibrium formalism. An equilibrium constant always denotes the extent of a chemical reaction: thus, K_{sp} is closely related to the actual solubility of a compound. The smaller K_{sp}, the lower the solubility.

However, the solubility of a compound does depend on two factors: the magnitude of K_{sp} and the form of the equilibrium expression. Let us compare the two sparingly soluble salts, CaF_2 with a solubility product of $4.0 \cdot 10^{-11}$ and AgCl with a solubility product of $1.6 \cdot 10^{-10}$ (all K_{sp} values are given at room temperature). One should think that CaF_2 is less soluble, since its solubility product constant is smaller than that of AgCl. However, the solubility of AgCl is given by an equation that involves the square root of K_{sp}, rather than the cube root as in the case of CaF_2.

$$AgCl(s) \rightleftarrows Ag^+(aq) + Cl^-(aq) \qquad (9.67)$$

$$K_{sp} = [Ag^+][Cl^-] = x^2 = 1.6 \cdot 10^{-10}$$

$$x = 1.14 \cdot 10^{-5} \qquad (9.68)$$

Thus, the solubility of AgCl ($1.14 \cdot 10^{-5}$ [mol/L]) is lower than that of CaF_2 ($2.15 \cdot 10^{-4}$ [mol/L]), although the solubility product is smaller for the latter.

9.9.3 Common Ion Effect

We will briefly touch upon a subject that will again be discussed in Chapter 10, the so-called common ion effect. This terminology refers to a situation where we wish to discuss the solubility of a compound, for example, AgCl, in a solution that already has a fairly high concentration of a "common ion," say the chloride ion in sea water. How does the common ion affect the solubility of AgCl in sea water? In Example 4.4 in Chapter 4, we calculated that the chloride ion concentration in sea water is approximately 0.5 [M]. We need not concern ourselves with the sodium ions that do not participate in any chemical reaction.

How can we determine how much silver chloride will dissolve in sea water? The relevant equations, from the viewpoint of the silver chloride, are still

$$AgCl\ (s) \rightleftarrows Ag^+\ (aq) + Cl^-\ (aq) \tag{9.67}$$

and

$$K_{sp} = [Ag^+][Cl^-] = 1.6 \cdot 10^{-10} \tag{9.68}$$

However, the chloride ion concentration in Eq. 9.68 is no longer determined by Eq. 9.67 alone, but also by the chloride ion concentration of the sea water. Thus, we write the total chloride ion concentration as the sum of the chloride ions from sea water, $[Cl^-]_{SW}$ and from AgCl, $[Cl^-]_{AgCl}$

$$[Cl^-]_T = [Cl^-]_{SW} + [Cl^-]_{AgCl} \tag{9.69}$$

The solubility constant of AgCl is a constant and cannot be messed with. Thus,

$$\begin{aligned} K_{sp} &= [Ag^+][Cl^-]_T = [Ag^+]\{[Cl^-]_{SW} + [Cl^-]AgCl\} = 1.6 \cdot 10^{-10} \\ &= [Ag^+]\{0.5 + [Cl^-]AgCl\} = 1.6 \cdot 10 - 10 \end{aligned} \tag{9.70}$$

We know from the discussion in the previous section that the chloride ion concentration is only $1.14 \cdot 10^{-5}$ [M] when AgCl is dissolved in pure water; this concentration is much smaller than the common ion concentration of 0.5 [M]. Thus, we are justified in setting

$$\{0.5 + [Cl^-]_{AgCl}\} \approx 0.5 \tag{9.71}$$

and obtain for the silver ion concentration

$$[Ag^+]\{0.5\} = 1.6 \cdot 10^{-10}\ \text{or}\ [Ag^+] = 3.2 \cdot 10^{-10}\ [M] \tag{9.72}$$

Thus, we see that the solubility of AgCl is reduced by five orders of magnitude in the presence of the chloride common ion. As mentioned above, similar calculations will reoccur in the discussion of acids and bases.

9.9.4 Experimental Determination of K_{sp}

As with all equilibrium constants, K_{sp} is determined experimentally. In the case of K_{sp}, this could be accomplished by measuring the solubility gravimetrically. Let us take the example of CaF_2 we discussed before. When 10.000 [g] of this compound is added to sufficient water to make 1.00 [L] of the solution, and given sufficient time to establish equilibrium, we find that 0.0168 [g], or 16.8 [mg], actually dissolve. This can be determined by separating the solution from the remaining solid by filtration, drying the solid, and weighing it. Alternatively, the solvent of the solution could be evaporated, and the remaining solid could be dried and weighed. There are, in addition, much more accurate ways to determine how much material actually dissolves; these spectral methods will be discussed in Chapter 15.

Anyway, what is more important here is the methodology to determine the solubility product from observed solubility. These are often given, for historical reasons, in [g/100 mL], [mg/100 mL] of [g/L] of solution. Let us use the value for CaF_2 given above, and refer to the solubility as S. Thus, $S(CaF_2) = 0.0168$ [g/L]. Using the molecular mass of CaF_2, 78 [g/mol], the solubility can be expressed in molar units:

$$n = m/\mathcal{M} = 0.0168/78 = 2.15 \cdot 10^{-4}\ [M] \tag{9.73}$$

Now, it is necessary to use our knowledge how this compound dissociates in aqueous solution:

$$CaF_2(s) \rightleftarrows Ca^{+2}\ (aq) + 2\ F^-(aq) \tag{9.74}$$

Since the number of moles that dissolve is $2.15 \cdot 10^{-4}$, it follows that the calcium ion concentration is $2.15 \cdot 10^{-4}$, since each mole of CaF_2 produces one mole of calcium ions. By the same token, we know that the fluoride concentration is twice that of the calcium ion concentration. Thus, K_{sp} is determined by

$$K_{sp} = [Ca^{+2}][2\ F^-]^2 = 2.15 \cdot 10^{-4}(4.3 \cdot 10^{-4})^2 = 4.0 \cdot 10^{-11} \tag{9.64}$$

In terms of the solubility S, we could have written the last equations as

$$K_{sp} = (S)(2S)^2 = 4S^3 = 4(2.15 \cdot 10^4)^3 \tag{9.75}$$

When defining the relationship between solubility product constant and solubility, one has to be extremely careful to get the stoichiometry right. This can get quite messy. Let us take the following reaction for an example:

$$Fe_2S_3 \text{ (s)} \rightleftarrows 2\,Fe^{3+} \text{ (aq)} + 3\,S^{2-}\text{(aq)} \tag{9.76}$$

If the solubility (in mol/L) of this compound is designated S, its relationship to K_{sp} would be

$$K_{sp} = (2S)^2\,(3S)^3 = 108\,S^5 \tag{9.77}$$

From this discussion, you should not be too surprised that the solubility product K_{sp} for iron (III) sulfide is very small: $K_{sp}(Fe_2S_3) = 1.4 \cdot 10^{-88}$.

Example 9.13 What is the iron ion concentration in a saturated solution of Fe_2S_3?

Answer: $[Fe^{3+}] = 2.1 \cdot 10^{-18}$ [M]

Example 9.14 What is the solubility (in [g/L]) of barium sulfate in 1.0 [L] of a 0.01 [M] solution of sodium sulfate?

Answer: $2.5 \cdot 10^{-6}$ [g]

9.9.5 Precipitation Reactions

The arguments presented here can be used to separate different cations in solution. An entire branch of Chemistry that used to be taught in General Chemistry courses dealt with the separation of mixtures of solutions based on their solubilities. The principles here is fairly straightforward. Imagine you have a solution of two metal ions, Cu^{+2} and Ni^{+2}, both at concentrations of 0.01 [M]. The solubility product constants for the corresponding metal sulfides, CuS and NiS, are

$$K_{sp}(CuS) = 8.5 \cdot 10^{-45}$$
$$K_{sp}(NiS) = 3.0 \cdot 10^{-21}$$

Given that both metal ion concentrations are originally at 10^{-2} [M], we can determine that NiS will start precipitating at a sulfide ion concentration of $3.0 \cdot 10^{-19}$ [M], whereas CuS will start precipitating at very low sulfide ion concentration of $8.5 \cdot 10^{-43}$ [M]. Thus, when adding sulfide ions to a solution containing both copper and nickel cations, we may assume that CuS will precipitate quantitatively i.e. use up all available sulfide ions before NiS can start to precipitate. The separation process can also be controlled via the pH of the solution, when H_2S is the source of the sulfide ions. H_2S is a weak, diprotic acid that dissociates according to

$$H_2S \rightleftarrows H^+ + HS^-\ K_{a1}$$
$$HS^- \rightleftarrows H^+ + S^{2-}\ K_{a2}$$

where K_{a1} and K_{a2} are the equilibrium constants (acid dissociation constants) that we will discuss in the next chapter. Thus, we see that we can control the amount of sulfide ions in the solution by controlling the pH of the solution with a buffer. As mentioned above, these reactions are important for the separation of ions by selective precipitation.

Example 9.15 In order to demonstrate the presence of fluoride ions in fluorinated drinking water, a drop (0.0500 [mL]) of 1.000 [M] calcium chloride solution was added to 1.000 [L] of water. No reaction was observed. Based on this result, what is the maximum concentration of fluoride ions in the drinking water?

Answer: $[Ca^{2+}] \leq 8.9 \cdot 10^{-4}$ [M]

This chapter is concluded with a cautionary remark. It was assumed that the equilibrium expressions presented so far hold for all concentrations. This, however, is only true for relatively low concentrations. At high concentrations of ions (for example in precipitation reactions where ions are added to drive the equilibrium to the side of products as much as possible) we find deviations from the ideal behavior discussed so far. These deviations are accounted for by introducing the concept of "activity," which is basically a factor that accounts for ion–ion interactions at high concentrations. This is along the lines of scientific discoveries and theories, in general: we start by formulating a model, and observe how well this model explains experimental results. Often, the simple model may require additional constraints: we saw before, in Chapter 8, that the ideal gas law needed to be amended to account for interactions between gaseous particles, and their small volumes, to describe the behavior of gases at high pressure and/or low temperature when condensation occurs. Similarly, the activity, rather than concentrations, are used at high concentrations of reactants or products. Also, pressures of gases used in the description of the vapor pressure of gases, or in gaseous equilibria using partial pressures, need to be corrected at extreme values of pressures by what is referred to as "fugacity," a factor that takes into account the physical interaction of gaseous species that reduces the pressures somewhat. However, the deviations due to the "activity" or "fugacity" are generally small, but need to be considered under extreme conditions.

Example 9.16 Which of the following choices will produce a precipitate if 50 [mL] of the following solutions (all at 0.01 [M] concentrations) are added to each other? This is a reminder to review Section 4.5.

(a) Silver nitrate + sodium nitrate
(b) Sodium chloride + calcium nitrate
(c) Sodium sulfate + calcium nitrate
(d) None of the above

Answer: (c) Details in answer booklet

Example 9.17 In a saturated solution of silver sulfide, Ag_2S, the Ag^+ ion concentration is $5.0 \cdot 10^{-17}$ [M]. The solubility product of silver sulfide is approximately

(a) $6.3 \cdot 10^{-50}$
(b) $2.5 \cdot 10^{-49}$
(c) $1.3 \cdot 10^{-33}$
(d) None of the above

Answer: (a) Details in answer booklet

Further Reading

Very basic reference to vapor pressure: https://www.chem.purdue.edu/gchelp/liquids/vpress.html

A more quantitative discussion of liquid/vapor equilibrium: https://chem.libretexts.org/Bookshelves/General_Chemistry/Map%3A_Chemistry_-_The_Central_Science_(Brown_et_al.)/11%3A_Liquids_and_Intermolecular_Forces/11.05%3A_Vapor_Pressure

General discussion of equilibrium: https://chem.libretexts.org/Bookshelves/Introductory_Chemistry/Chemistry_for_Allied_Health_(Soult)/08%3A_Properties_of_Solutions/8.02%3A_Chemical_Equilibrium

Examples of precipitation and solubility product: https://chem.libretexts.org/Bookshelves/General_Chemistry/Map%3A_General_Chemistry_(Petrucci_et_al.)/18%3A_Solubility_and_Complex-Ion_Equilibria/18.1%3A_Solubility_Product_Constant_Ksp

10

Acids and Bases

10.1 What Are Acids/Bases?

In earlier chapters, we have defined acids as a species that gives up a proton. This happens very easily if the H atom in a molecule is bonded to an electronegative atom or an electron withdrawing group. In this case, the bond that connects the hydrogen atom to the rest of the molecule (which we call X) is broken as shown in Eq. 10.1:

$$H-X \rightarrow H^+ + |X^- \qquad (10.1)$$

It (heterolytic bond cleavage) produces a cation and an anion. The vertical line on the X^- represents the electron pair left behind at the anion.

The other way a bond may break is referred to as homolytic bond cleavage according to

$$H-X \rightarrow H\cdot + \cdot X \qquad (10.2)$$

which produces two radicals.

The former case produces a proton, H^+, devoid of any electrons, which we refer to as the unit of acidity in aqueous solution. For example, we may consider HCl or HNO_3 (hydrochloric and nitric acids), both of which dissociate completely in aqueous solution to produce a proton and the corresponding anion. Similarly, we defined a base as a compound that gives up a hydroxide ion, OH^-, in aqueous solution. NaOH and $Ca(OH)_2$ are typical examples of bases according to our earlier definition. Water molecules themselves can undergo heterolytic bond cleavage according to:

$$H-OH \rightarrow H^+ + |OH^- \qquad (10.3)$$

where the vertical line at the anion represents the electron pair that remains with the hydroxide ion. Thus, water undergoes a process that we call autoionization or autodissociation in which water acts as both an acid and a base (more on this is discussed later).

This description of acids and bases as compounds that produce H^+ and OH^-, respectively, goes back to a Swedish scientist, S. Arrhenius, who was a major force establishing the theories of ionic compounds, their dissociation in aqueous media, and the resulting conductivity of the solutions. Although his view of acids and bases is still valid, there are two additional views of acids and bases that generalize the acid–base concepts to other systems and to nonaqueous situations.

One of them is a direct consequence of the two species responsible for acidity and basicity, H^+ and $|OH^-$. The proton is, as we know, a naked nucleus without any electrons. In the Lewis model of acids and bases, any species lacking electrons, and therefore able to accept electrons, is referred to as a Lewis acid. The hydroxide ion, on the other hand, is a relatively electron-rich species, with three nonbonding electron pairs in addition to the bond between O and H. In the Lewis model, any electron-rich molecule would be considered a Lewis base. By this token, ammonia, $|NH_3$, would be a Lewis base, whereas an electron deficient molecule like BH_3 would be considered a Lewis acid. Thus, in the Lewis model, the reaction between borane and ammonia,

$$BH_3 + |NH_3 \rightarrow H_3B-NH_3 \qquad (3.14)$$

which we introduced earlier, would be a typical acid–base reaction. This view, although somewhat unfamiliar at this point, is very useful in organic chemistry to predict chemical reactions.

A final description of acid/base behavior is the Brønsted–Lowry (or simply the Brønsted) acid/base model. In this model, the action of acids or bases is described in terms of reactions with solvent or other acid/base molecules. Let us discuss this model again in terms of an example. The hydrogen ion, H^+, is just a proton without electrons. Its low radius and $+1$ charge give it a high (positive) electric charge density, and it will seek out any negative charge density. Thus, whenever such a proton encounters an electron pair, it will bind to it:

$$H^+ + H_2O \rightarrow H_3O^+ \tag{10.4}$$

In this case, the water molecule, with its free electron pair, acts as a base, reacting with the proton. Remembering that in water the proton may be created by the autodissociation of water (Eq. 10.3), we may combine Eqs. 10.3 and 10.4 and obtain

$$H_2O + H_2O \rightarrow H_3O^+ + OH^- \tag{10.5}$$

In the Brønsted acid/base model, Eq. 10.5 would be interpreted as one water molecule, acting as an acid, gives up a proton, thereby turning into its conjugated base, OH^-. The other water molecule accepts the proton, thereby turning into its conjugated acid, H_3O^+. This model will be used extensively when we talk about the acid/base behavior of certain salts. For example, the ammonium ion, NH_4^+, reacts with water according to

$$NH_4^+ + H_2O \rightarrow H_3O^+ + NH_3 \tag{10.6}$$

Here, NH_4^+, a Brønsted acid, gives up a proton and turns into its conjugated base, ammonia. Similarly, the base water accepts a proton and turns into its conjugated acid, hydronium ion.

In different chemistry text books, one finds either the proton, H^+, or the hydronium ion, H_3O^+, as the species responsible for acidity. In a sense, neither one of these forms is completely accurate. We stated before that, due to the high charge density, the proton is unlikely to be found unattached to anything with electrons to share. From this viewpoint, H_3O^+ is certainly a better description than H^+. Yet even H_3O^+ is a charged species that will be solvated in aqueous solution; therefore, $(H_2O)_2H^+$ or even $(H_2O)_3H^+$ may be better descriptions of a solvated proton. To keep things simple, let us just refer to the quintessential unit of acidity as H^+, but remember that several water molecules may be associated with it.

Example 10.1 For the reaction $Al^{3+}(aq) + 6\ H_2O(l) \rightleftharpoons Al(H_2O)_6^{3+}(aq)$, identify the Lewis acid and base.

Answer: Al^{3+} is the Lewis acid, water is the Lewis base

Example 10.2 Write a Brønsted acid/base reaction scheme for the reaction between water and ammonia.

Answer: $H_2O(l) + NH_3(aq) \rightarrow OH^-(aq) + NH_4^+(aq)$

10.2 Strong Acids and Bases; Definition of pH and pOH

The difficulty understanding acid/base chemistry results from the fact that acids and bases come with quite different properties that we need to differentiate: they occur as strong acids and bases, weak acids and bases, and acidic and basic salts. In addition, we need to define and understand buffer solutions. These four different situations require different formalisms for answering their properties. We start the discussion here with the case of strong acids and bases. A strong acid or base is one that undergoes complete dissociation into ions.

When a strong acid is dissolved in water, it will completely dissociate into a hydrogen ion (proton) and its anion. Which acids are strong acids? Experimental results suggest that hydrochloric (HCl), hydrobromic (HBr), and hydroiodic (HI) acids, as well as nitric acid (HNO_3), sulfuric acid (H_2SO_4: the first proton only), and chloric acid ($HClO_3$) are the common strong acids we need to concern ourselves with. Let us look at the dissociation of nitric acid as an example. When this molecule is dissolved in water, the H—O bond breaks to give a proton and the nitrate anion according to

$$HNO_3 \rightarrow H^+ + NO_3^- \tag{10.7}$$

Taking the water molecules into account, Eq. 10.7 can be written as

$$HNO_3 + H_2O \rightarrow H_3O^+ + NO_3^- \tag{10.8}$$

in accordance with the Brønsted formalism. These two equations are equivalent.

Let us represent a strong acid by the formula HX, where X^- is the acid anion (Cl^-, NO_3^-, etc.). Thus, we may describe the generalized dissociation of a strong acid by

$$HX \rightarrow H^+ + X^- \tag{10.9}$$

If we prepare a solution by dissolving 0.1 [mol] of HX in sufficient water to produce a 0.1 [M] solution, then the HX concentration would be 0.1 [M] if no dissociation occurred. However, since the reaction equilibrium described by Eq. 10.9 lies completely on the right-hand side, no undissociated HX will be found, and the concentration of H^+ and X^- are both 0.1 [M] and the HX concentration is essentially zero.

Similarly, strong bases are compounds that dissociate completely into cations and hydroxide ions when dissolved in aqueous solution. The hydroxides of alkali metals are strong bases: NaOH, KOH, etc. The hydroxides of the alkaline earth metals, such as $Ca(OH)_2$, are strong bases as well. There is one major difference between polyprotic acids (acids that can lose more than one proton such as H_2SO_4 or H_3PO_4) and bases that can produce more than one hydroxide ion, such as $Ca(OH)_2$: in the case of the bases, both hydroxide ions dissociate off the cation, whereas in the polyprotic acids, the second (or third) dissociation step does not go to completion. This case will be discussed later in Sections 10.3 and 10.8.

If we represent a strong base by the symbol $M(OH)_n$, we may represent the dissociation of a strong base by

$$M(OH)_n \rightarrow M^{+n} + n\,OH^- \tag{10.10}$$

Just as in the case of strong acid, we argue that the reaction equilibrium given by Eq. 10.11 lies far on the right-hand side. Thus, when a solution of concentration c of a strong base is prepared, the resulting hydroxide ion concentration will be $n \cdot c$.

Next, we define a commonly used measure of a solution's acidity, the pH value:

$$pH = -\log[H^+] \tag{10.11}$$

Thus, for the example of a strong acid, in which

$[H^+] = 0.1\,M = 10^{-1}$ [M], the pH will be

$$pH = -\log(10^{-1}) = -(-1) = 1. \tag{10.12}$$

In analogy to the pH value, we define the pOH value as

$$pOH = -\log[OH^-] \tag{10.13}$$

Consider a solution prepared by dissolving $1.0 \cdot 10^{-5}$ [mol] of NaOH in 100 [mL] of water. Then,

$$[OH^-] = 1.0 \cdot 10^{-4}\,M \text{ and} \tag{10.14}$$

$$pOH = -\log(1.0 \cdot 10^{-4}) = 4 \tag{10.15}$$

Example 10.3 Calculate the pH of a 1.50 [M] aqueous solution of HCl.

Answer: pH = −0.18

Example 10.4 What is the pOH of 0.005 [M] aqueous solution of $Ca(OH)_2$?

Answer: pOH = 2

Example 10.5

(a) Can an acidic solution have a pH of zero?
(b) What would be the hydrogen ion concentration in such a solution?

Answer: (a) Yes (b) $[H^+] = 1$

10.3 Weak Acids/Bases

We define a weak acid as a compound where the dissociation of the acid into protons and acid anion is not complete, but proceeds to establish an equilibrium with a well-defined equilibrium constant. Thus, we have to revert to our earlier discussions of equilibria to tackle this problem. Let us look at an example. Nitrous acid, (HNO_2), is a weak acid that partially dissociates according to

$$HNO_2 + H_2O \rightleftharpoons H_3O^+ + NO_2^- \qquad (10.16)$$

or simply as

$$HNO_2 \rightleftharpoons H^+ + NO_2^- \qquad (10.17)$$

Since the dissociation step is not complete, we find three species: H^+ (or H_3O^+), NO_2^-, and undissociated HNO_2 in an aqueous solution of HNO_2. These three species are in equilibrium; therefore, their concentrations are interdependent and cannot vary arbitrarily.

For a general discussion of any weak acid, let us denote the acid as HA and its anion as A^-, and write the dissociation equation as

$$HA \rightarrow H^+ + A^- \qquad (10.18)$$

Since this reaction reaches an equilibrium, we write it with a double arrow to emphasize this aspect:

$$HA \rightleftharpoons H^+ + A^- \qquad (10.19)$$

Here, we indicated with the longer arrow that the equilibrium lies on the side of the undissociated acid. Further, we may use our knowledge from the discussion on equilibria (Chapter 9) to describe the dissociation equilibrium as follows:

$$\frac{[H^+][A^-]}{[HA]} = \text{constant} = K_a \qquad (10.20)$$

Since the expression on the left-hand side of Eq. 10.20 has a fixed numeric value, the three concentrations are interdependent. Thus, regardless of the amount of acid, HA, in a solution, the reaction indicated by Eq. 10.18 will occur until an equilibrium is established and until the concentrations of HA, H^+, and A^- fulfill Eq. 10.20. Thus, if some more HA is added to the solution, it will dissociate, thereby reducing [HA] and increasing both $[H^+]$ and $[A^-]$ such that the value of K_a is maintained. Next, we will do some simple equilibrium calculations to demonstrate this point and to develop the formalism to predict the pH of weak acids. For a strong acid, the dissociation proceeds essentially to completion; that is, there is no undissociated acid left, and the equilibrium constant, K_a, would be essentially infinity (since the denominator is zero). For weak acids, the acid dissociation constant is small, anywhere from 0.01 to 10^{-10}, or even smaller.

Let "c" be the original concentration of a weak acid HA, before any dissociation occurs. The dissociation of HA according to Eq. 10.19 gives us equal amounts of H^+ and A^-. We do not know these concentrations, so let us call them x:

$$[H^+] = [A^-] = x \qquad (10.21)$$

The amount of HA left can be expressed as $c - x$ and we can write a table of initial and final concentrations as follows:

	$[H^+]$	$[A^-]$	$[HA]$
Initial	0	0	c
Final	x	x	c − x

Substituting these values into the equilibrium expression gives

$$\frac{[H^+][A^-]}{[HA]} = K_a = \frac{x \cdot x}{c - x} \qquad (10.22)$$

$$\text{or } x^2 + x\, K_a - c\, K_a = 0 \qquad (10.23)$$

Let us do a concrete example, namely that of the dissociation of nitrous acid given by Eq. 10.17. In this case, the numeric value of the dissociation constant is

$$K_a(HNO_2) = 7.2 \cdot 10^{-4} \qquad (10.24)$$

Let us consider a solution of nitrous acid prepared by dissolving gaseous HNO_2 in sufficient water to make its concentration (before any dissociation would occur) 0.010 [M]. Thus, Eq. 10.23 reads

$$x^2 + 7.2 \cdot 10^{-4} x - 7.2 \cdot 10^{-6} = 0 \qquad (10.25)$$

Using the formula for solving quadratic equations (see Section 1.5)

$$x_{+,-} = \frac{-b \pm \sqrt{b^2 - 4ac}}{2a} = \frac{-7.2 \cdot 10^{-4} \pm \sqrt{(7.2 \cdot 10^{-4})^2 + 4 \cdot 7.2 \cdot 10^{-6}}}{2}$$

$$= \frac{-7.2 \cdot 10^{-4} \pm 5.415 \cdot 10^{-3}}{2}$$

$$x_+ = 2.34703 \cdot 10^{-3} \qquad x_- < 0$$

Thus, the concentrations of all species at equilibrium are given as follows:

	$[H^+]$	$[NO_2^-]$	$[HNO_2]$
Initial	0	0	0.01
Final	$x = 2.35 \cdot 10^{-3}$	$x = 2.35 \cdot 10^{-3}$	$c - x = 7.65 \cdot 10^{-3}$

Although you should report the result to two significant figures, it is useful to carry out the calculations using more significant figures, if you decide to verify that the resulting concentrations, indeed, agree with the equilibrium constant:

$$K_a = \frac{x \cdot x}{c - x} = \frac{(2.35 \cdot 10^{-3})^2}{7.65 \cdot 10^{-3}} = 7.22 \cdot 10^{-4} \qquad (10.26)$$

When we solved Eq. 10.25, we discarded the negative solution since the hydrogen or acid ion concentration cannot be negative. The pH of the solution is 2.6.

Example 10.6 Calculate the $[H^+]$ concentration and the pH for (a) 0.10 and (b) 0.0010 [M] aqueous solution of HNO_2.

Answer: (a) $[H^+] = 0.0085$; pH = 2.1; (b) $[H^+]$ $5.62 \cdot 10^{-4}$; pH = 3.2

The percentage of dissociation, $100[H^+]/[HA]$, depends on the concentration, as can be shown easily by repeating the above calculations for a number of different concentrations c, as shown in Example 10.7. Le Chatelier's principle applies here: by diluting the solution, both the concentrations in the numerator and denominator change. However, since the numerator has the product of two concentrations, these concentrations will change nonlinearly, whereas the denominator changes linearly. As the acid becomes more dilute, a larger fraction of it needs to dissociate to maintain K_a.

Example 10.7 Calculate the percent dissociation for 0.10, 0.01, and 0.0010 [M] aqueous solution HNO_2.

Answer: 8.5, 23, and 56% for the three concentrations

If we had written the weak acid dissociation (Eq. 10.17) using the Brønsted formalism as follows:

$$HA + H_2O \rightarrow H_3O^+ + A^- \qquad (10.27)$$

The corresponding equilibrium expression has the (bulk) water concentration in the denominator:

$$\frac{[H_3O^+][A^-]}{[HA][H_2O]} = \text{constant} = K_a \qquad (10.28)$$

Since this is a different chemical equilibrium equation, it will have a different numeric value for its equilibrium constant. We shall revisit this subject after discussing the self-dissociation of water toward the end of this section.

We now turn to the problem of weak bases that can be treated mathematically exactly like the problem of weak acids. A weak base, for example, is an aqueous solution of ammonia, NH_3. Now, you may ask "... where's the base????" In other words, where is the hydroxide ion that we know is the cause of basicity?

When we refer to ammonia as a weak base, we have to specify that we really mean "an aqueous solution of ammonia." According to the Brønsted formalism, aqueous ammonia is a base because

$$NH_3 + H_2O \rightarrow NH_4^+ + OH^- \qquad (10.29)$$

The two reactants on the left could be written as a salt-like compound, ammonium hydroxide, $NH_4^+ OH^-$; however, this compound does not exist under standard conditions. However, an aqueous solution of ammonia dissolved in water appears as if $NH_4^+ OH^-$ was partially dissociating according to Eq. 10.30:

$$NH_4^+ OH^- \rightarrow NH_4^+ + OH^- \tag{10.30}$$

For now, we refer to the undissociated weak bases simply as the adduct $NH_3 \cdot H_2O$. Then,

$$\frac{[NH_4^+][OH^-]}{[NH_3 \cdot H_2O]} = K_b \tag{10.31}$$

which we solve exactly as indicated above for the situation of weak acids by writing a table of initial and equilibrium concentrations:

	$[NH_4^+]$	$[OH^-]$	$[NH_3 \cdot H_2O]$
Initial	0	0	c
Final	x	x	c − x

This leads us to

$$\frac{[NH_4^+][OH^-]}{[NH_3 \cdot H_2O]} = K_b = \frac{x \cdot x}{c - x} \tag{10.32}$$

$$\text{or } x^2 + x K_b - c K_b = 0 \tag{10.33}$$

The numerical value of K_b of ammonia in water is $2.0 \cdot 10^{-5}$.

Example 10.8 Calculate the pOH of a 0.01 [M] aqueous solution of ammonia in water.

Answer: pOH = 3.4

10.4 The Relationship Between pH and pOH: Self-dissociation of Water

The next subject to discuss in this context is the property of water to undergo self-dissociation according to:

$$H_2O \rightarrow H^+ + OH^- \tag{10.3}$$

The equilibrium expression for Eq. 10.6 is

$$\frac{[H^+][OH^-]}{[H_2O]} = K \tag{10.34}$$

where the concentration of liquid water is $[H_2O] \approx 55.6$ [M]. (Think about where this number comes from.) The self-dissociation of water is very small; thus, we can claim that, even after some water undergoes this dissociation, the concentration of liquid water is still nearly 55.6 [M]. Therefore, we may write Eq. 10.33 as:

$$[H^+][OH^-] = K[H_2O] \tag{10.35}$$

and consider the product on the right-hand side constant, and give it a new name:

$$[H^+][OH^-] = K_w = 1.0 \cdot 10^{-14} \tag{10.36}$$

The numerical value (at room temperature) of K_w is, like all equilibrium constants, an experimentally derived value.

In order to calculate the pH value of water, we use the formalism derived above; we set

$$[H^+] = [OH^-] = x \tag{10.37}$$

since the dissociation of water produces equal amounts of protons and hydroxide ions. Consequently,

$$K_w = 1.0 \cdot 10^{-14} = x^2 \tag{10.38}$$

$$x = [H^+] = [OH^-] = 10^{-7} \tag{10.39}$$

and it follows that for pure water, pH = pOH = 7 (10.40)

and pH + pOH = 14 (10.41)

One out of every 10 million water molecules undergoes dissociation to produce hydrogen and hydroxide ions.

10.5 Common Ion Effect

In the discussion of equilibria, the term "common ion" implies that one and the same species, or ion, may be produced from two different sources. We used this term before (Section 9.9.3) when we discussed the solubility of silver chloride in a solution that contained chloride ions from another source. Let us look at another example to demonstrate this situation.

We have discussed before how to calculate the pH of a solution of a strong acid, say HCl. When one prepares a solution of 10^{-2} [M] HCl in water, the pH is 2. Similarly, for 10^{-5} [M] HCl, the pH would be 5. Let us go one step further, and calculate the pH of 10^{-8} [M] HCl. Our first guess would be that the pH is 8. But something must be seriously wrong with this guess: we add acid to water, and the solution turns basic (pH 8)? What has gone wrong here is that we have ignored the fact that we have two sources of protons that contribute nearly equally to the total hydrogen ion concentration: HCl and water. HCl, as a strong acid, completely dissociates into H^+ and Cl^-; water, on the other hand, undergoes self-dissociation according to Eq. 10.3. Thus, we have protons produced by two separate reactions that occur simultaneously. We call the H^+ in this situation the common ion. Now, let us turn our attention to getting the correct answer for the problem at hand: what is the pH of a 10^{-8} [M] solution of HCl in water? We know that HCl is a strong acid and will completely dissociate in water. Thus, we may write

$$[H^+]_{HCl} = 10^{-8} [M] \tag{10.42}$$

where the subscript HCl denotes that the proton ion concentration in Eq. 10.41 is due to the dissociation of HCl. Water also will produce a similar hydrogen ion concentration by its dissociation according to Eq. 10.3:

$$H_2 \rightarrow H^+ + OH^- \tag{10.3}$$

which is governed by the equilibrium expression

$$[H^+][OH^-] = K_w = 1.0 \cdot 10^{-14} \tag{10.35}$$

Again, this equation holds and cannot be messed with. However, what is different now is that $[H^+]$ and $[OH^-]$ will no longer be equal, as we have two sources of protons, and only one source of hydroxide ions. Thus, we rewrite Eq. 10.35 as:

$$\{[H^+]_{HCl} + [H^+]_{H_2O}\}[OH^-] = 1.0 \cdot 10^{-14} \tag{10.43}$$

The two proton concentrations in the curved parentheses are due to the "common ion." Thus, we may write

$$\{10^{-8} + [H^+]_{H_2O}\}[OH^-] = 1.0 \cdot 10^{-14} \tag{10.44}$$

However, we also know that each water molecule that dissociates will create one proton and one hydroxide ion. Thus, we call x the number of water molecules that will dissociate, and obtain

$$[H^+]_{H_2O} = [OH^-]_{H_2O} = x \tag{10.45}$$

Substituting Eq. 10.45 into Eq. 10.44 yields

$$(10^{-8} + x)(x) = 1.0 \cdot 10^{-14} \tag{10.46}$$

$$x^2 + 10^{-8}x - 1.0 \cdot 10^{-14} = 0 \tag{10.47}$$

which has a physically significant solution $x = [OH^-] = [H^+]_{H_2O} = 9.51 \cdot 10^{-8}$, and

$$[H^+]_{TOTAL} = \{[H^+]_{HCl} + [H^+]_{H_2O}\} = 1 \cdot 10^{-8} + 9.51 \cdot 10^{-8} = 1.051 \cdot 10^{-7} \tag{10.48}$$

We see that the solution is slightly acidic, and

$$[OH^-] = 9.51 \cdot 10^{-8} \tag{10.49}$$

To test our results, we calculate the product of proton and hydroxide ion concentration, and obtain

$$\begin{aligned}&[H^+]_{TOTAL} \cdot [OH^-] \\ &= 1.051 \cdot 10^{-7} \cdot 9.51 \cdot 10^{-8} = 9.99501 \cdot 10^{-15}\end{aligned} \tag{10.50}$$

which is acceptably close to K_w.

Common ion problems will be with us for most of the remainder of this chapter. Often, you may ignore the common ion effects (as in the case of 10^{-2} [M] HCl mentioned above). Here, the proton concentration due to HCl exceeds the weak proton concentration produced by the autodissociation of water, and we can safely assume that nearly all protons are due to HCl.

Another example where we need to take into account the common ion effect is the dissociation of sulfuric acid, H_2SO_4. The first dissociation step

$$H_2SO_4 \rightarrow H^+ + HSO_4^- \tag{10.51}$$

goes essentially to completion since sulfuric acid is a strong acid with $K_{a1} = \infty$. However, the second proton's dissociation

$$HSO_4^- \rightarrow H^+ + SO_4^{2-} \tag{10.52}$$

is governed by an equilibrium constant,

$$K_{a2} = 1.2 \cdot 10^{-2} \tag{10.53}$$

This dissociation constant corresponds to a fairly high degree of dissociation in the second step (medium strength acid). The dissociation given by Eq. 10.49 leads to the equilibrium expression

$$K_{a2} = \frac{[H^+][SO_4^{2-}]}{[HSO_4^-]} \tag{10.54}$$

However, the proton concentration in the numerator is not just the proton ion concentration of the second dissociation step, Eq. 10.51, but also from the first step, Eq. 10.51:

$$K_{a2} = \frac{\{[H^+]_1 + [H^+]_2\}[SO_4^{2-}]}{[HSO_4^-]} \tag{10.55}$$

where the subscripts 1 and 2 refer to the first and second dissociation step. The effect of the protons from the first step is to shift the equilibrium of the second step toward the left-hand side, i.e. reducing the degree of dissociation. In the following example, you will calculate the pH of a solution of sulfuric acid using the formalism given in Eq. 10.55.

Example 10.9 Calculate the pH of a 0.01 [M] aqueous solution of sulfuric acid.

Answer: pH = 1.83

10.6 Acidic and Basic Salts

According to the Brønsted theory of acids and bases, it comes as no surprise that certain salts may act as acids or bases. Consider the example of soap, which forms a mildly basic solution in water – a situation you may have experienced if you accidentally got soap solution in your eyes (it burns!!!). Soap is the "salt" (although a piece of hand soap may not immediately feel like a salt) of a weak fatty acid and a strong base, usually KOH. The fatty acids in soap are molecules with a long, hydrophobic tail group, and a hydrophilic head group:

$$CH_3-(CH_2)_{14}-CO_2H \qquad \text{(palmitic acid)}$$

Molecules incorporating the COOH group are referred to as organic acids. When these molecules react with KOH or NaOH, the corresponding salt is obtained:

$$CH_3-(CH_2)_{14}-CO_2K \quad \text{or} \quad CH_3-(CH_2)_{14}-CO_2Na$$

which, when dissolved in water, completely dissociates. The detergent properties of these molecules result from the fact that the fatty acid tails are able to solvate grease and nonpolar materials, whereas the polar head group ascertains that the resulting aggregate between the soap and the greasy stains is water-soluble.

The acid dissociation constant of typical organic acids is around 10^{-4} to 10^{-5}; thus, they are considered relatively weak acids. The potassium salts of these acids are, therefore, the salt of a strong base and a weak acid. Using the abbreviation $An^- = CH_3-(CH_2)_{14}-CO_2^-$, we write the reaction of the anion with water according to the Brønsted formalism as

$$An^- + H_2O \rightarrow HAn + OH^- \tag{10.56}$$

We need not concern ourselves with any reaction between the base cation, K^+, and water, since the equilibrium of the reaction

$$K^+ + H_2O \rightarrow KOH + H^+ \tag{10.57}$$

totally lies on the left-hand side. The equilibrium expression for Eq. 10.56 is given by

$$K = \frac{[HAn][OH^-]}{[An^-][H_2O]} \tag{10.58}$$

At this point, we do not know the numeric value of the equilibrium constant. A simple trick is used to relate dissociation constants of conjugated Brønsted acids and bases: we multiply numerator and denominator of Eq. 10.58 by $[H^+]$, and obtain

$$K = \frac{[HAn][OH^-][H^+]}{[H^+][An^-][H_2O]} \tag{10.59}$$

This is perfectly legal, since multiplying both the numerator and the denominator of a fraction by the same number does not change the value of the fraction. The left part of the fraction in Eq. 10.59 is just the inverse of the acid dissociation constant, whereas the right part is K_w. Since the reaction of An^- with water is sometimes referred to as a hydrolysis reaction, we call the equilibrium constant in Eq. 10.59 the hydrolysis constant and designate it by K_h. Thus,

$$K_h = K_w/K_a \tag{10.60}$$

We proceed in a similar fashion in the case of a salt of a strong acid and a weak base, for example, ammonium chloride, NH_4Cl. An aqueous solution will contain the ammonium ion, NH_4^+, and the chloride ion Cl^-. According to the Brønsted formalism,

$$NH_4^+ + H_2O \rightarrow H_3O^+ + NH_3 \tag{10.61}$$

which accounts for the acidity of the solution. Again, the reaction of the anion with water,

$$Cl^- + H_2O \rightarrow HCl + OH^- \tag{10.62}$$

can be ignored, since HCl is a strong acid. The ammonium ion, however, is a weak acid, certainly much weaker than H_3O^+. Thus, the equilibrium given by Eq. 10.61 needs to be taken into account. The equilibrium expression for Eq. 10.61 is given by

$$K = \frac{[H_3O^+][NH_3]}{[NH_4^+][H_2O]} \tag{10.63}$$

for which we need to find a numerical value. Again, we multiply the numerator and denominator of the equilibrium expression in Eq. 10.63 by $[OH^-]$, and obtain

$$K_h = K_w/K_b \tag{10.64}$$

Example 10.10 Confirm how we get from Eq. 10.61 to 10.62.

Answer: see answer booklet

Let us calculate, for a numerical example, the pH of a 0.1 [M] solution of ammonium chloride in water, assuming that K_b of ammonia is $1.8 \cdot 10^{-5}$, and that the salt ammonium chloride completely dissociates into the ions in aqueous solution. Then the reaction

$$NH_4^+ + H_2O \rightarrow H_3O^+ + NH_3$$
$$\begin{array}{cccc} c & & 0 & 0 \quad \text{(before "hydrolysis")} \\ c-x & & x & x \quad \text{(at equilibrium)} \end{array} \tag{10.61}$$

will produce an "original" ammonium ion concentration (before hydrolysis) of 0.1 [M]. After the equilibrium is reached, the ammonium ion concentration is reduced by x and is 0.1 − x. For each fraction of x that reacts with water, equal concentrations x of H_3O^+ and ammonia are produced:

$$[H_3O^+] = [NH_3] = x$$

Thus,

$$K = \frac{[H_3O^+][NH_3]}{[NH_4^+][H_2O]} = \frac{x^2}{c-x} = K_h \tag{10.63}$$

with $K_h = K_w/K_b = 1.0 \cdot 10^{-14}/1.8 \cdot 10^{-5} = 5.6 \cdot 10^{-10}$

K_h is obviously a small number; therefore, x is also a small number. Consequently, we use the common simplification that $c - x = 0.1 - x \approx 0.1$. We obtain

$$x^2 \approx 5.6 \times 10^{-11} \quad x = [H_3O^+] = 7.5 \cdot 10^{-5} \tag{10.65}$$

which corresponds to a pH of 4.1. Thus, we see that a solution of ammonium chloride (a salt of a strong acid and a weak base) in water is acidic. Similarly, we find that a solution of a strong base and a weak acid is basic.

In summary, the cation of a weak base (BOH), or the anion of a weak acid (HAn), undergoes hydrolysis according to $B^+ + H_2O \rightarrow B + H_3O^+$ and $An^- + H_2O \rightarrow HAn + OH^-$, respectively. The resulting equilibria can be described by the hydrolysis constants:

$$K_h = [B][H_3O^+]/[B^+][H_2O] \text{ and } K_h = [HAn][OH^-]/[An^-][H_2O] \tag{10.66}$$

which yield equations for the unknown $[H_3O^+]$ or $[OH^-]$ concentrations:

$$[H_3O^+]^2 = x^2 \approx c\, K_h \text{ and } [OH^-]^2 = x^2 \approx c\, K_h. \tag{10.67}$$

Example 10.11 Calculate the pH of a 0.01 [M] solution of soapy water discussed at the beginning of this section. Use K_a for the fatty acid of $1.5 \cdot 10^{-5}$.

Answer: pH = 10.5

10.7 Buffers

Buffers, or buffering solutions, are aqueous media composed of a mixture of a weak acid and its conjugated base, or a weak base and its conjugated acid, which minimize changes in pH when either H^+ or OH^- ions are added to the solution. That is a fairly hefty statement that deserved some more explanation. First of all, let us investigate the second of these arguments, the resistance to change in pH. Many chemical reactions produce acids or bases as a by-product. Take the normal metabolic oxidation of food by oxygen to carbon dioxide. In aqueous solution, carbon dioxide creates carbonic acid, which dissociates to protons and hydrogen carbonate, HCO_3^-. Thus, when metabolizing, cells would become more acidic. This is a definite no-no: many of the enzymes that catalyze metabolic reactions work only in very narrow pH ranges; furthermore, an increase or decrease in pH would raise havoc with a cell's ability to control osmolality. Thus, a cell must maintain its pH for proper functioning. Solution-phase chemical reactions, either on the laboratory or industrial scale, often require control of the pH as well, since increase or decrease of the hydrogen ion concentration could adversely affect the chemical equilibrium that governs the yield of a given reaction. Thus, it is imperative that we (or the body, in the case of biochemical reactions) maintain the pH at a constant level.

How can this be achieved? The idea of a buffer, as indicated above, is to have a solution that consists of a weak acid and its anion, or a weak base and its cation. Consider, for example, the weak acid HNO_2 and its anion NO_2^- (the nitrite ion). According to the Brønsted formalism, HNO_2 reacts with water according to

$$HNO_2 + H_2O \rightarrow H_3O^+ + NO_2^- \tag{10.16}$$

Here, HNO_2 is a weak acid, and NO_2^- its conjugated base. A buffer solution, therefore, is a solution that contains a weak acid and its conjugated base at approximately equal proportions. How can this be achieved? We know that HNO_2 is a weak acid; thus, the equilibrium described by Eq. 10.16 lies on the left side, and the nitrite ion concentration is much smaller than the nitrous acid concentration. What can we do to increase the nitrite ion concentration? Well, we use the fact that salts dissociate into ions. Thus, if we add a salt, such as $NaNO_2$ to aqueous HNO_2, we can have both the acid and its anion in solution at comparable concentrations.

The salt of the weak acid or base will always dissociate completely in solution. Thus, this dissociation will produce the conjugated base or acid, respectively. Therefore, the solution equilibrium will be governed by the appropriate acid or base equilibrium. In the case of nitrous acid/nitrite ion, for example, we know that the concentrations of nitrite and undissociated nitrous acid are ultimately related by the acid dissociation constant of HNO_2:

$$K_a = [H^+][NO_2^-]/[HNO_2] \tag{10.68}$$

If we solve Eq. 10.68 for the hydrogen ion concentration, we obtain

$$[H^+] = K_a[HNO_2]/[NO_2^-] \tag{10.69}$$

Thus, we see that in a solution that contains equal amounts of nitrous acid and nitrite anion, that is, if $[HNO_2] = [NO_2^-]$, the hydrogen ion concentration is just given by the K_a value,

$$[H^+] = K_a. \tag{10.70}$$

If we define the pK_a is the negative logarithm of K_a, we obtain

$$pH = pK_a \tag{10.71}$$

Often, Eq. 10.69 is written in logarithmic form:

$$\log[H^+] = \log K_a + \log[HNO_2] - \log[NO_2^-] \tag{10.72}$$

$$pH = pK_a + \log\{[NO_2^-]/[HNO_2]\} \tag{10.73}$$

The logarithmic forms of the buffer equations (i.e. Eqs. 10.72 and 10.73) are often referred to as the Henderson–Hasselbalch equation.

Example 10.12 Calculate the pH of a nitrous acid/nitrite ion buffer composed of 0.01 [M] nitrous acid and (a) 0.02 [M] NO_2^-, (b) 0.01 [M] NO_2^-, and (c) 0.005 [M] NO_2^-.

Answer: (a) pH = 3.45; (b) pH = 3.15; (c) pH = 2.85

Example 10.12 demonstrated that the pH of the nitrous acid/nitrite ion buffer can be adjusted around the pK_a-value by varying the ratio of acid/anion concentrations. Now, the next topic to be discussed is: How does this buffer minimize changes in the pH of a solution when H^+ or OH^- ions are added to it? The answer here lies in the fact that the components of the buffer will react with added acid or base such as to reduce their influence. Let us see what happens when we add acid or base to a nitrous acid–sodium nitrite buffer containing equal amounts of NO_2^- and HNO_2. If we add acid, i.e., H^+, to this solution, the following reaction occurs:

$$H^+ + NO_2^- \rightarrow HNO_2 \tag{10.74}$$

Since nitrous acid is a weak acid, this equilibrium lies on the right-hand side; i.e. the conjugated base, NO_2^-, has neutralized the added acid. In this process, the NO_2^- concentration has decreased, and the HNO_2 has increased. However, the hydrogen ion concentration of the solution is still governed by Eq. 10.70.

If, on the other hand, base is added to the solution of HNO_2 and NO_2^-, the weak acid, HNO_2, will react with OH^- according to

$$HNO_2 + OH^- \rightarrow H_2O + NO_2^- \tag{10.75}$$

The equilibrium of this reaction also lies on the right-hand side, since HNO_2 is a weak acid. Consequently, its conjugated base, NO_2^-, is a strong base. In the process, the weak acid, HNO_2, has neutralized the added base. Of course, the HNO_2 concentration has decreased and the NO_2^- concentration increased.

Therefore, let us calculate the pH after addition of acid or base, based on the pK_a of nitrous acid, which is 3.14 (this follows from Eq. 10.24). The pH of an equimolar sodium nitrite–nitrous acid buffer is given, according to Eq. 10.71 as

$$pH = pK_a = 3.14 \tag{10.76}$$

For simplicity's sake, let us say we have 1.0 [L] of the buffer, and that the original concentrations of HNO_2 and NO_2^- are both 0.1 [M]. Now, let us consider what happens if we add 0.01 [mol] of H^+ to this buffer solution. As discussed above (Eq. 10.74), the predominant reaction that occurs is:

$$H^+ + NO_2^- \rightarrow HNO_2 \tag{10.74}$$

The original NO_2^- concentration was 0.1 [mol/L]; of this, 0.01 [mol] will react with the added H^+ to produce HNO_2. Thus, 0.1 − 0.01 = 0.09 [mol] of nitrite will be left. On the other hand, the new amount of HNO_2 is 0.1 + 0.01 = 0.11 [mol]. Assuming that the volume of the buffer has not changed, we can calculate the new pH of the buffer, after addition of 0.01 [mol] of H^+, according to Eq. 10.73:

$$pH = pK_a + \log\{[NO_2^-]/[HNO_2]\} \tag{10.73}$$
$$= 3.14 + \log\{0.09/0.11\} = 3.14 - 0.087 = 3.05 \tag{10.77}$$

We see that after adding 0.01 [mol] of acid to the buffer, the change in pH is less than 0.1. Let us contrast this result to what would have happened if we added the same amount of acid to 1.0 [L] of pure water: if we add 0.01 [mol] of acid to 1.0 [L] of water, the new hydrogen ion concentration would be $[H^+] = 10^{-2}$ M, or pH = 2. Thus, we would have changed the pH from 7 to 2, or by 5 units. This corresponds to an increase in the hydrogen ion concentration by a whopping factor of 100 000!! Therefore, we see the buffer is doing what it is supposed to do: it minimized the pH change when acid is added to it.

Example 10.13 Repeat the calculations if 0.01 [mol] of hydroxide ions are added to 1.0 [L] of the HNO_2/NO_2^- buffer where both initial concentrations are 0.1 [M].

Answer: pH = 3.23

The same principles discussed so far for the sodium nitrite/nitrous acid buffer holds for other buffer systems as well. Examples 10.16 and 10.17 deal with another acidic buffer system based on formic acid/sodium formate. Particularly important in nature are the dihydrogen phosphate/hydrogen phosphate ($H_2PO_4^-/HPO_4^{2-}$) buffer, which accounts for the constant pH value of blood, in spite of metabolic activity that creates and consumes H^+ ions.

Basic buffers are produced, as pointed out before, by mixing a weak base and the salt of the weak base. Typical examples would be the ammonia/ammonium chloride buffer, introduced in Example 10.18. In analogy, the pOH of a basic buffer (weak base BOH and its cation, B^+) is given by

$$pOH = pK_b + \log\{[B^+]/[BOH]\} \tag{10.78}$$

The capacity of the buffer to maintain the pH under the influence of added acid or base depends on the concentration of the weak acid and the conjugate base. If, for example, both the acid (HNO_2) and base (NO_2^-) concentrations in the buffer discussed above are 0.1 [M], the buffer is able to maintain the pH even if large amounts (up to 0.05 [moles]) of acid or base are added. Whenever the added acid or base concentration is larger than the buffer's weak acid/conjugate base concentration, the buffer capacity is exceeded, and buffering action is no longer possible.

10.8 Acid–Base Titrations

Titrations are methods of analytical chemistry to determine the amount or concentration of an unknown in a sample. The idea behind acid–base titrations, for example, is to add known amounts of acid to a sample of base of unknown concentration. Once neutralization has occurred, one can back-calculate, from the amount of acid added, what the original concentration of the base was. Admittedly, the advent of modern analytical instrumentation has made the manual determination of concentrations a little obsolete; however, the principles of titrations are so important that they need to be discussed. Furthermore, an acid/base titration is still about the least expensive and most easily implemented method of analysis.

Let us get a bit more specific in the examples for acid/base titrations. Imagine that one morning, in a small lake in the Adirondack Park, all the lake trout are floating belly-up. If you know anything about fish, you will agree that upside-down trout are very unhappy. Furthermore, you may know that a frequent cause of fish dying *en masse* is high acidity of water. Thus, one would have to determine quickly and in the field what the pH of the lake water is. Titration is a viable method that requires minimal apparatus.

Thus, one would remove a sample of the lake water and add base of a very well-established concentration to the acidic water until all the acid is neutralized. From the volume of base required to reach neutrality, and from its concentration, one can figure out how many moles of base were required, and from this, how many moles of acid were in the sample of water you took from the lake. Now, using the knowledge of the previous section, one may decide what buffer to use to balance the pH of the lake, and the trout in the lake will be eternally grateful.

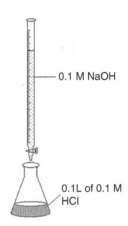

Figure 10.1 Example of an acid–base titration.

In order for this titration to work, we need to consider a number of prerequisites. First, we need a way to determine when neutrality is reached during a titration. Neutrality can be established in a number of ways; the cheapest and easiest is using an indicator, which is a compound that changes color between acidic and basic solutions. Grape juice, for example, is dark blue in basic solution and dark red in acidic solution. We shall discuss the color changes in indicators in some more detail later (see Section 10.8.3). Second, we need to calibrate, or standardize, the base (or in general terms, the titer) that we use in the titration. Third, we need to be able to measure the volume of the titer added to the titrant quite accurately. For this, we use volumetric burettes that you most likely have seen in the laboratory. In order to minimize error, we will use titers of low concentration such that we can determine the number of moles of titer accurately (see Figure 10.1). Finally, we need to understand how the pH changes in such an acid base titration.

10.8.1 Titration of a Strong Acid with a Strong Base

Next, we shall calculate the changes in pH of an acidic solution as we add small aliquots of base to it. Let us start with a flask that contains 0.10 [L] of 0.10 [M] HCl. In a real titration, one would add a drop of indicator that tells us when we have reached neutrality, and a magnetic stirring bar. We fill a burette with an aqueous solution of NaOH, for example, 0.10 [M]. We know that the pH of the original acid is 1 (since $[H^+] = 10^{-1}$ [M]). Now, we start by adding 5.0 [mL] (0.005 [L]) of 0.1 [M] NaOH to the acid. We need to figure out what the pH of the acid is after we added the first aliquot of base. We do this by calculating the total number of moles of $[H^+]$ in the original solution:

$$n = C_M V = 0.10 \cdot 0.10 = 0.010 \, [L \cdot mol/L = mol] \tag{4.9}$$

and the number of moles of base added in the first aliquot

$$n = 0.1 \cdot 0.005 = 0.00050 \quad [L \cdot mol/L = mol]$$

The reaction that occurs when we add the base is $H^+ + OH^- \rightarrow H_2O$. Since this reaction essentially goes to completion, we see that we produced 0.00050 [mol] of water, and have $0.01 - 0.0005 = 0.0095$ [mol] of acid left. In addition, we have added 0.005 [L] to the 0.1 [L] of acid. Thus, the total volume of the solution is 0.105 [L].

This gives us a new hydrogen ion concentration,

$$[H^+] = 0.0095/0.105 = 0.0905 \, [M], \tag{10.79}$$

and a new pH value of 1.04. We repeat these calculations for each new aliquot of added base, and obtain pH values as a function of added base as given in Table 10.1. This table contains in the left-most column the volume of 0.1 [M] NaOH added (expressed in [L]), the total volume after each addition of base, the number of moles of acid left, the acid concentration, and the pH of the solution after each addition of base. Figure 10.2 shows a plot of the pH as a function of the volume of added base. There are several important facts about the shape of this titration curve. First of all, we see that the pH changes rapidly at the point where the number of moles of base added equals to the number of moles of acid in solution:

$$n_{H^+} = n_{OH^-} \tag{10.80}$$

Table 10.1 pH, as a function of added base, for a strong acid/strong base titration.

Vol. of base added [L]	Total vol. [L]	Acid left [mol]	Acid conc. [mol/L]	pH
0.0050	0.1050	0.00950	0.09048	1.04
0.0100	0.1100	0.00900	0.08182	1.09
0.0150	0.1150	0.00850	0.07391	1.13
0.0250	0.1250	0.00750	0.06000	1.22
0.0350	0.1350	0.00650	0.04815	1.32
0.0450	0.1450	0.00550	0.03793	1.42
0.0550	0.1550	0.00450	0.02903	1.54
0.0650	0.1650	0.00350	0.02121	1.67
0.0750	0.1750	0.00250	0.01429	1.85
0.0800	0.1800	0.00200	0.01111	1.95
0.0850	0.1850	0.00150	0.00811	2.09
0.0900	0.1900	0.00100	0.00526	2.28
0.0920	0.1920	0.00080	0.00417	2.38
0.0940	0.1940	0.00060	0.00309	2.51
0.0950	0.1950	0.00050	0.00256	2.59
0.0960	0.1960	0.00040	0.00204	2.69
0.0970	0.1970	0.00030	0.00152	2.82
0.0980	0.1980	0.00020	0.00101	3.00
0.0990	0.1990	0.00010	0.00050	3.30
0.0995	0.1995	0.00005	0.00025	3.60
0.1000	0.2000	0.00000	0.00000	7.00
0.1005	0.2005	0.00005	0.00025	10.40
0.1010	0.2010	0.00010	0.00050	10.70
0.1015	0.2015	0.00015	0.00074	10.87
0.1025	0.2025	0.00025	0.00123	11.09
0.1035	0.2035	0.00035	0.00172	11.24
0.1050	0.2050	0.00050	0.00244	11.39
0.1060	0.2060	0.00060	0.00291	11.46
0.1070	0.2070	0.00070	0.00338	11.53
0.1080	0.2080	0.00080	0.00385	11.59
0.2000	0.3000	0.01000	0.03333	12.52

We call this point the equivalence point, because the number of equivalents of acid and base are equal. This rapid change in pH makes the idea of titrations practical: since literally one drop of base changes the pH from acidic to basic, we can determine the equivalence point very accurately by using indicators that change their colors over a narrow pH range. More about this later.

For a strong acid/strong base titration, we see that at the equivalence point, the number of moles of acid and base is equal (see Eq. 10.80). Therefore, the number of moles of counterion is equal as well, $n_{Na^+} = n_{Cl^-}$. Therefore, all the acid is neutralized by base, and the number of moles of sodium ions equals the number of moles of chloride ions. We basically have a neutral solution of sodium chloride in water. Addition of base after the equivalence point, therefore, is the same as adding base to pure water, and the shape of the titration curve beyond the equivalence point reflects this fact.

The other important observation about the shape of the curve in Figure 10.2 is the slow change of pH at the beginning of the titration. Although we are adding base to the acid, the pH changes very sluggishly in the beginning, requiring over 80 [mL] of base before the pH changes by one unit!

10.8.2 Titration of a Weak Acid with a Strong Base

As we shall see, this is a situation quite different from what we have encountered previously in Section 10.8.1. First of all, before we even add base, the pH is determined by the weak acid dissociation equilibrium. Let us look at a specific example such as a titration of acetic acid with NaOH. For simplicity, we abbreviate the acetate anion as Ac^-. In order to keep the differences to the last problem as small as possible, let us use the same concentrations and volumes as starting condition.

Thus, we start with 0.10 [L] of 0.10 [M] acetic acid solution, which has a $K_a = 1.8 \cdot 10^{-5}$. This is a weak acid; thus, its pH at the starting point of the titration, before any base is added, is given by Eq. 10.81:

$$\frac{[H^+][Ac^-]}{[HAc]} = 1.8 \cdot 10^{-5} \tag{10.81}$$

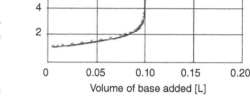

Figure 10.2 Titration curve of a strong acid with a strong base.

This can be solved using the formalism developed in Section 10.3, Eqs. 10.22–10.25, to give a pH of the original acid of pH = 2.87.

Now, what is happening at the equivalence point? Before, in the strong acid/strong base titration, we have seen that at the equivalence point, the number of moles of acid and base was equal:

$$n_{H^+} = n_{OH^-} \tag{10.80}$$

Therefore, the number of moles of counterion, $n_{Na^+} = n_{Ac^-}$. Thus, we have a solution of NaAc in water. However, we know that a solution of a salt of a strong base and weak acid is not neutral, because the acid anion, Ac^-, undergoes hydrolysis. We use Eqs. 10.57–10.60:

$$[OH^-]^2 = x^2 \approx c\, K_h \tag{10.82}$$

where K_h is the hydrolysis constant, $K_w/K_a = 10^{-14}/1.8 \cdot 10^{-5} = 5.56 \cdot 10^{-10}$

We solve for x and obtain $x = [OH^-] = 7.45 \cdot 10^{-6}$

$$pH = 8.87 \tag{10.83}$$

As expected, the pH at the equivalence point is basic, since we are dealing with the hydrolysis of a salt of a weak acid and a strong base (see Point 2 in Figure 10.3).

Finally, we shall deal with the part of the titration where the amount of base added is larger than zero, but less than the amount required to reach the equivalence point. Let us look at a specific instance, when we have added 10 [mL] or 0.01 [L] of 0.10 [M] NaOH to the acetic acid. The original number of moles of acetic acid is

$$n_{HAc} = 0.10\,[L] \cdot 0.10\,[mol/L] = 0.010\,[mol] \tag{10.84}$$

and the amount of base added at this point is

$$n_{OH^-} = 0.01\,[L] \cdot 0.10\,[mol/L] = 0.001\,[mol] \tag{10.85}$$

The actual reaction in the solution is:

$$HAc + OH^- \rightarrow H_2O + Ac^- \tag{10.86}$$

Since we added 0.001 [mol] of base, we have created 0.001 [mol] of acetate, and the remaining concentration of the acid is

$$[HAc] = 0.010 - 0.001 = 0.009 \tag{10.87}$$

Therefore, now we have a situation where we have a free acetic acid, and its anion, acetate in solution. This is, of course, a buffer (Point 1 in Figure 10.3). In order to solve for the pH after the addition of 10 [mL] of base, we use Eq. 10.70:

$$pH = pK_a + \log\{[A^-]/[HA]\} \tag{10.71}$$

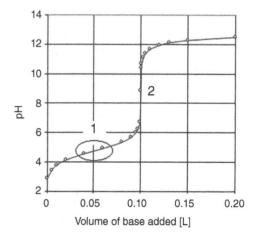

Figure 10.3 Titration curve of a weak acid with a strong base. Point 1: buffer region. Point 2: hydrolysis point.

Table 10.2 pH, as a function of added base, for a weak acid/strong base titration.

Vol. of base added [L]	Total vol. [L]	Acid left [mol]	Acid conc. [mol/L]	Log(acetate/acid)	pH
0.0000	0.100	0.0100	0.00000		**2.87**
0.0050	0.105	0.0095	0.00050	−1.279	3.46
0.0100	0.110	0.0090	0.00100	−0.954	3.79
0.0200	0.120	0.0080	0.00200	−0.602	4.14
0.0400	0.140	0.0060	0.00400	−0.176	4.56
0.0600	0.160	0.0040	0.00600	0.176	4.92
0.0800	0.180	0.0020	0.00800	0.602	5.34
0.0900	0.190	0.0010	0.00900	0.954	5.69
0.0950	0.195	0.0005	0.00950	1.279	6.02
0.0970	0.197	0.0003	0.00970	1.510	6.25
0.0990	0.199	0.0001	0.00990	1.996	6.74
0.1000					**8.87**
0.1005					*10.4*
0.1010					*10.7*
0.1025					*11.09*
0.1050					*11.39*
0.1100					*11.68*
0.1200					*11.96*
0.1300					*12.12*
0.1500					*12.3*
0.2000					*12.52*

and get

$$pH = 4.74 + \log\{0.001/0.009\} = 3.79 \quad (10.88)$$

The pH values of a weak acid/strong base titration are shown in Table 10.2. Note that in this table, the two values in bold print, i.e. for 0 and 100 [mL] of base added, are the ones calculated previously (Eqs. 10.81 and 10.83). Notice also that the values in italic, i.e. for added base volumes above 100 [mL], were taken from Table 10.1. This is because past the equivalence point, the two titrations become identical in terms of the pH response. This occurs because all the weak acid has been used up, and in the presence of excess base, the hydrolysis reaction

$$Ac^- + H_2O \rightarrow HAc + OH^- \quad (10.86)$$

is pushed toward the side of the reactants. In other words, hydrolysis is no longer important, and the titration proceed like the strong acid/strong base situation. A graphic representation of the titration of a weak acid with a strong base is shown in Figure 10.3.

A titration curve such as the one shown in Figure 10.3 permits an accurate determination of the pK_a value of a

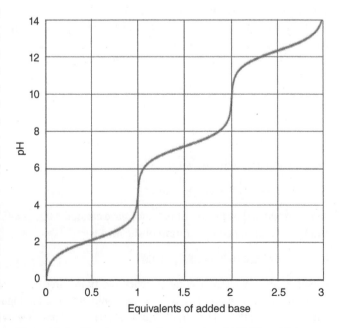

Figure 10.4 Titration curve of a triprotic acid, H_3PO_4, with base.

Table 10.3 Commonly used indicators.

Indicator name	Color (acidic)	Color (basic)	pH range	pK_{In}
Methyl orange	Red	Yellow	3.2–4.4	3.7
Bromothymol blue	Yellow	Blue	6.0–7.6	7
Phenol red	Yellow	Red	6.8–8.4	7.9
Phenolphthalein	Colorless	Magenta	8.2–10.0	9.4

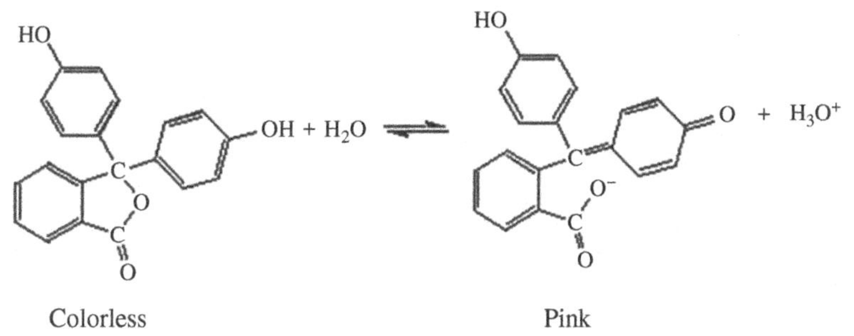

Figure 10.5 Structural changes in the phenolphthalein molecule upon deprotonation. Remember that in organic chemistry and biochemistry, the symbol "C" for carbon atoms is omitted and indicated by the intersection of covalent bonds.

Colorless   Pink

weak acid or a weak base: when half the volume required to reach the end point has been added, we know that at this equivalence point, pH = pK_a. In the titration of (see Figure 10.4) a triprotic acid such as H_3PO_4, two equivalence points are reached when one or two equivalents of base have been added, and buffer solutions of different pH values are established after 0.5, 1.5, and 2.5 equivalents of base have been added (the $H_3PO_4/H_2PO_4^-$, $H_2PO_4^-/HPO_4^{2-}$, and HPO_4^{2-}/PO_4^{3-} buffers with pH 2.1, pH 7.2, and pH 12.4, respectively).

When a strong base or a weak base is titrated with acids, titration curves are obtained that look very much like Figures 10.2 and 10.3, respectively, but flipped around a horizontal axis. These curves can be constructed following the logic explained for the construction of Tables 10.1 and 10.2.

10.8.3 Acid–Base Indicators

Finally, let us discuss briefly the method used commonly in acid–base titrations to determine the endpoint of the titration. Depending on the titration we are performing, we wish to employ an indicator that changes color in the neutral range (for a strong acid/base titration), in the basic range (for a weak acid/strong base titration), or in the acidic range (for a strong acid/weak base titration). Examples of such indicators, and the pH range of their color change, are given in Table 10.3.

The color changes observed in an indicator is due to the fact that indicators themselves are weak acids or bases with pK_{In} values much smaller than those of the analytes in a titration. Here, pK_{In} indicates the pK_a or pK_b value of the indicator. The color changes are due to the fact that protonation or deprotonation of a chromophore, that is, the grouping of the molecule that actually has a visible spectrum, can change the molecular bonding scheme, particularly the π-electrons, sufficiently to cause a color change. This is shown in Figure 10.5 for the indicator phenolphthalein in both the deprotonated and the protonated forms.

Example 10.14 The [OH^-] of a 0.1 [M] solution of a weak base with $K_b = 10^{-4}$ in water is approximately:

(a) $3 \cdot 10^{-3}$ [M]
(b) $1 \cdot 10^{-3}$ [M]
(c) $3 \cdot 10^{-2}$ [M]
(d) $1 \cdot 10^{-2}$ [M]

Answer: (a) Details in answer booklet

Example 10.15 The pH of a dilute solution of ammonium chloride in water is

(a) slightly acidic
(b) slightly basic
(c) strongly basic
(d) none of the above

Answer: (a) Details in answer booklet

Example 10.16 The pH of an equimolar sodium formate/formic acid buffer is

(a) basic
(b) neutral
(c) acidic
(d) depends on the concentrations used

Answer: (c) Details in answer booklet

Example 10.17 After addition of 0.001 [mol] of H^+ ions to 100 [mL] of sodium formate/formic acid buffer with [HCOOH] = 0.1 [M] and [HCOO$^-$] = 0.05 [M], the new the new pH is

(a) 3.2
(b) 3.3
(c) 3.4

Answer: (b) Details in answer booklet

Further Reading

https://chem.libretexts.org/Bookshelves/Physical_and_Theoretical_Chemistry_Textbook_Maps/Supplemental_Modules_(Physical_and_Theoretical_Chemistry)/Acids_and_Bases/Acid/Overview_of_Acids_and_Bases

11

Thermodynamics: Energy, Energy Conversions, and Spontaneity

11.1 Energetics of Chemical Reactions

In this chapter, we shall investigate one of the main reasons why humans got involved with chemistry: chemical reactions, in addition to making new compounds, can be used to produce heat and do work for us. The development of our technological society (whether you like it or not!) was driven originally by the invention of the steam engine and the combustion engine (which do work for us) rather than by carrying out chemical reactions for the sake of chemistry. So, chemistry should be viewed as a major contributor to technology and the industrialization process.

The production of heat during chemical processes is relatively obvious: many reactions, such as explosions or combustions, occur with a spectacular release of heat. For such a process, the reaction products are found to be hotter than the reactants, since the heat of the reaction heats up the products. Other reactions require heat to occur. This is manifested by the products being colder than the reactants. Both these kinds of reactions do occur in nature, and a number of examples will be discussed later in this chapter that will release or absorb heat.

Most chemical reactions you may have performed in the laboratory are carried out in such a way that they do not produce useful work. Yet, the car battery you may have used to start a car is a typical device that produces work (turning over the engine in your car) from a chemical reaction. A battery stores chemical energy that is released on demand to perform work. The combustion of gasoline in a car engine is, of course, another chemical process we use to get work. We will return to both of these processes at a later time. In our discussion of thermodynamics, we shall attempt to predict and calculate the energetic changes (i.e. the release or absorption of heat), and how some of this energy can be converted to work.

In this thermodynamics chapter, we shall also investigate, on a macroscopic level, why chemical reactions occur. On a microscopic, or molecular level, we have already addressed this point, when we discussed the formation of bonds, the rearrangements of electrons etc., However, in this chapter, we shall address this question from a viewpoint of energetics. The point discussed above, namely that reactions occur which require or release energy, poses a dilemma: it appears that chemical reactions may go "uphill," against the expected release of energy. We shall deal with this aspect in later parts of this chapter.

Some textbooks present thermodynamics, the study of the "motion of heat," in two different chapters. The first one is on thermochemistry (the heat change of chemical reactions), the definition and measurements of heat, and the first law of thermodynamics. This corresponds to Sections 11.1–11.7 in this presentation. A follow-up chapter deals with the entropy, free energy, and equilibrium considerations of thermodynamics, presented here in Sections 11.8–11.10.

11.2 Thermochemistry

11.2.1 Definition of Energy, Work, and Heat

Before we can start a discussion of the energetics of chemical reactions, we need to review the definitions of energy, heat, and work. Let us start with work, w. Mechanical work is defined as the product of force F and distance, for example, by lifting a weight against the force of gravity:

$$w = F \cdot x \tag{11.1}$$

We assume that the weight is moved in a direction opposite to the action of the force (up, in the case of work against gravity); otherwise, an angle-dependent term has to be included in Eq. 11.1. Since force equals mass · acceleration, the mechanical work performed when lifting a mass against the gravitational acceleration g (9.81 [m/s^2]) by the earth by a distance x is

$$w = m \cdot g \cdot x \qquad (11.2)$$

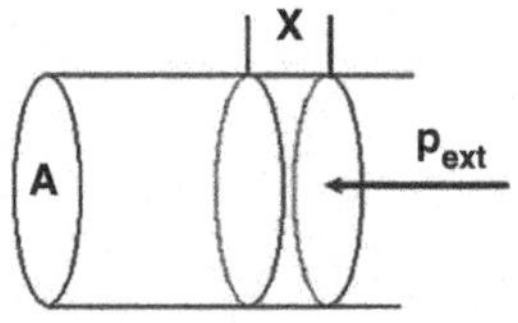

Figure 11.1 Expansion of a gas against constant external pressure.

Thus, in order to lift a weight against the gravitational force, we have to perform work.

Moving a weight perpendicularly to the direction of the force (for example, parallel to and on a horizontal surface) requires no work against the gravitational force, but still might have to overcome friction, air resistance, and the like. Therefore, even such a process will require work to be expended. Since the force F is defined as mass · acceleration, it has units of [kg m/s^2 = N (Newton)], and mechanical work or energy has units of [kg m^2/s^2 = Nm = J (Joule)].

One form of mechanical work is particularly important in science and engineering, namely the work performed when a gas, confined in a cylinder with a moveable piston, expands against a constant external pressure (see Figure 11.1). This is, of course, the principle behind car engines or steam engines. This form of work is often referred to as volume work, and is the guiding principle behind the conversion of heat (see below) of a chemical reaction into work. Let us assume that the area of the piston that is moved is A, and that the piston travels a distance x during the expansion. Using Eq. 11.1, w = F · x, and the definition of pressure as force divided by area

$$p = F/A \text{ or } F = pA \qquad (11.3)$$

we can write the volume work w_{vol} as

$$w_{vol} = p \cdot A \cdot x \qquad (11.4)$$

for a motion of the piston by a distance x. Here, the product A · x represents the volume change ΔV (area · distance) for the expansion process. Thus, we may write

$$w_{vol} = p \cdot \Delta V \qquad (11.5)$$

Volume work has units of [L atm], as suggested by Eq. 11.5. The unit [L atm] represents work or energy; thus, any work expressed in [L atm] can be converted to [J]. You may recall from our discussion of the gas laws and the kinetic models of gases that the gas constant R was given as 0.082 [L atm K^{-1} mol^{-1}] or 8.3 [J K^{-1} mol^{-1}]. From this, we see that 1 [L · atm] ≈ 101 [J].

Example 11.1 What is the mechanical work required for a 60 [kg] athlete to run up the Empire State Building in New York City during the annual Empire State Building climb? The height covered by this race is 320 [m].

Answer: 188 [kJ]

Example 11.2 What volume work is performed by blowing up a balloon to a volume of 15 [L] against atmospheric pressure?

Answer: 15 [L atm] = 1.5 [kJ]

Another form of work or energy is electrical work, for example, work performed when the battery in a car is used to start the car, or in a Tesla, to move the car around. Electric work is defined by the product voltage, current, and time

$$w_{el} = \varepsilon \cdot i \cdot t \qquad (11.6)$$

where the voltage ε is measured in volt (v), the current i in ampere (A), and the time t in seconds (s). By definition,

$$1 [v \cdot A \cdot s] = 1 [W \cdot s] \text{(Watt second)} \qquad (11.7)$$

where the watt is the unit of electric power and is the product of voltage times current. Furthermore,

$$1 [v \cdot A \cdot s] = 1 [J]. \qquad (11.8)$$

More on electrical work, and on how to produce an electrical voltage from chemical reactions, will be presented in Chapter 12. The definition of electrical and mechanical energy makes it easy to estimate the cost of the energy we use up in everyday life. Electrical energy is purchased by the kWh (kilowatt hour), and a little reflection shows that 1 [kWh] = 3 600 000 [W·s] = 3 600 000 [J] = 3600 [kJ]. One [kWh] of electricity costs about 25 cents, depending on where you live. In order to perform the required 188 [kJ] of work to run up the Empire State building, one expends just 1.3 cent worth of electricity. Therefore, electricity seems to be a cheap way to get some work done, and it may explain why we use an elevator instead of running up the building.

Next, let us concentrate on heat. The heat content, or heat, of any matter is related to the temperature of the matter: the higher the temperature of the material (the hotter it is), the more heat it contains. Therefore, we may say that temperature is a property of matter, which can be measured with a thermometer. Heat is a form of energy, which may change the temperature of matter. This above statement, cast into a mathematical format, can be written as

$$q \propto \Delta T \tag{11.9}$$

which says exactly what we wrote above: the change in temperature, indicated by ΔT, is proportional to the heat q flowing into or out of the matter under consideration. Here, the symbol Δ refers to change; so ΔT is the change in temperature, which we define as $\Delta T = T_f - T_i$, where f and i denote final and initial conditions.

How can we make the expression in Eq. 11.9 a true equation? We all know from simple experience that more matter requires more heat to bring about a temperature change. Think about making a cup of tea in the morning: if you put more water on the stove or hot plate, more heat (longer time) is required to bring it to a boil. This is shown schematically in Figure 11.2. In this figure, the fat arrows indicate heat added to the water in the beaker (ignore the effect of the beaker itself for the time being). Thus, it appears that the mass m of matter to which heat is added affects the temperature change:

$$q \propto m \, \Delta T \tag{11.10}$$

Still, this is not a proper equation, because equal masses of different materials will respond differently to a given amount of heat. This is addressed by defining a material-specific constant known as the "heat capacity" or "specific heat" of matter, and obtain:

$$q = m \, c_p \, \Delta T \tag{11.11}$$

We are using the symbol c_p for the heat capacity in this chapter, where the subscript "p" indicates that the transfer of heat occurs at constant pressure. This aspect is important mainly when discussing gases, where the transfer of heat can proceed at constant pressure or at constant volume.

Liquid water has a specific heat of 4.18 [J/g °C], indicating that 4.18 [J] of heat will cause a temperature change of 1 [°C] when added to 1 [g] of water.[1] Metallic iron, for comparison, has a specific heat of 0.45 [J/g °C]. Table 11.1 lists the heat capacities, expressed in [J/g °C] and [J/mol °C] for three representative materials. We will discuss the magnitudes of the heat capacities listed in Table 11.1 later in this section.

Equation 11.11 is the basis of measurements known as calorimetry. Notice that in thermodynamics, T is expressed in Kelvin (K), since the centigrade scale is an arbitrary scale. However, in calorimetry we are dealing with temperature changes,

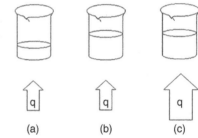

Figure 11.2 Schematic of a thought experiment in which heat q (fat arrows) is added to different amounts of water in a beaker. A little reflection shows that the same amount of heat added to twice the amount of water (Figures 11.2a and 11.2b) causes a smaller different temperature increase in Figure 11.2b. A large amount of heat added to equal amounts of water (Figures 11.2b and 11.2c, indicated by the fatter arrow) causes a larger temperature increase in Figure 11.2c.

Table 11.1 Heat capacities of representative materials.

	c_p [J/g °C]	c_p [J/mol °C]
He (g)	5.19	20.8
H$_2$O (l)	4.18	75.2
Fe (s)	0.45	25.2

---

1 Formerly, the calorie (cal) was used as a unit of heat, with 1 cal = 4.18 [J]. The calorie was defined as the amount of heat required to raise the temperature of 1 [g] of water by 1 [°C]. In this unit, the specific heat of water is 1 [cal/g °C].

and a change in temperature, ΔT, of 1 [K] is the same as 1 [°C], since the centigrade scale differs from the Kelvin scale by an added constant, namely 273. Thus, in the discussion of calorimetry, ΔT, may be expressed in units of [°C].

Next, we need to ask the question: what, really, is heat? We have seen in the discussion of gases that the temperature of a gas is a measure of the kinetic energy of the gaseous atoms or molecules. Similarly, we can describe the temperature of any matter – solid, liquid, or gaseous – to be related to the motion of the atoms or molecules in the matter. In contrast to gases, where the gaseous particles move freely (but bounce into each other and the walls of the container), the atoms or molecules in liquid and solids are constrained by the forces acting between these particles. However, the atoms in solids and liquids are in constant vibrational motion: in hot matter, the amplitude of vibration of the particles is large, and there are many different ways the atoms and molecules can vibrate. The specific heat, c_p, which relates temperature to heat, is a measure of the number of degrees of freedom of vibration the atoms or molecules have. The specific heat c_p has units of [J/mol °C] or [J/g °C] and is one of the fundamental properties of matter, although we have used it strictly as a proportionality constant relating heat and temperature.

Returning to the example of adding heat to water mentioned above, let us estimate what amount of heat we need to make our early morning, eye-opening cup of coffee or tea. Assume the tap water has a temperature $T_i = 10$ [°C]. Brewing tea or coffee works best when the water boils, therefore $T_f = 100$ [°C] and $\Delta T = 90$ [°C]. A cup holds about 100 [g] of water; furthermore, we assume that no heat is lost to the container and the surroundings. Thus, the heat required to reach the boiling point of the water is approximately (using Eq. 11.11)

$$q = 100 \cdot 4.18 \cdot 90 \, [g \cdot (J/g \, °C) \cdot (°C) = J]$$
$$\approx 3.6 \cdot 10^4 \, [J] = 36 \, [kJ] \tag{11.12}$$

The amount of electricity needed to heat the water is 36 000 [J]/3 600 000 [J/kWh] $\approx$ 0.01 [kWh], or only about 0.2 cents worth of electricity.

This heat may be derived from electric energy, and we see here as well as in earlier parts of this discussion that we can convert between forms of energy: we can convert energy to heat, and energy to work. This is the take-home message of this discussion: heat and work are forms of energy; energy may be viewed as the potential or ability to perform work or create heat. The electric energy you purchase may be derived from heat (nuclear reactions or the burning of fossil fuel), or from the conversion of potential energy into electricity (hydroelectric power) or from solar energy.

11.2.2 Calorimetry: Measurement of Heat Flow

In this section, we will introduce the methods of measuring the heats of chemical reactions, and chances are that you will have to carry out some calorimetric measurements in the chemistry laboratory. Therefore, we need to investigate here the principles of calorimetry. Since chemistry is an empirical science, that is, a science based on the observation of experimental results, this section also gives us an excellent view how science and scientific principles can be developed from experimental observations.

Imagine that you are to perform a simple experiment, namely the mixing of 50.0 [g] of water at 80 [°C] (we call this the hot temperature, T_h) with 50.0 [g] of water at 20 [°C] (T_c). If we assume that the specific heat of water does not depend on temperature, and that the container in which we mix the two water samples does not absorb any heat at all, we come to the conclusion that the final temperature of the mixed water should be the average, or 50 [°C]. What happens, however, if we take only 10.0 [g] of hot (80 [°C]) water, and add it to 90.0 [g] of cold (20 [°C]) water? We have to keep track of the heat lost by the hot water, and the heat gained by the cold water, using Eq. 11.11. Not knowing what the final temperature of the mixture will be, we call this temperature T_x. The heat lost by the hot water is given by

$$q_h = 10.0 \cdot 4.18 \cdot (T_x - T_h) \, [g(J/g°C)(°C) = J] \tag{11.13}$$

and the heat gained by the cold water is

$$q_c = 90.0 \cdot 4.18 \cdot (T_x - T_c) \, [g(J/g°C)(°C) = J] \tag{11.14}$$

where we have defined the temperature difference, as always, as the final condition minus the initial conditions. Since we disallow heat loss to the surroundings, all the heat lost by the hot water sample ends up in the cold water. Thus, it appears that $q_c = q_h$.

We now introduce a bookkeeping scheme, which we shall use throughout the remainder of the discussions in this book. From our outside viewpoint, one of the samples loses heat, while the other gains the same amount. We define a sign scheme similar to the way one would balance a checkbook: loss of heat, for the hot water, is entered in the books as a negative heat, just as a check you write gets subtracted from the balance. The heat gained by the cold water is entered with a positive sign and corresponds to a deposit into your account. Therefore, the magnitudes of $q_c = q_h$ are the same (no heat is lost in the total process), but the signs of the two components must be different, since one represents a gain, the other a loss. Thus, we write

$$q_c = -q_h \tag{11.15}$$

Thus, we may write

$$10.0 \cdot 4.18 \cdot (T_x - 80) = -90.0 \cdot 4.18 \cdot (T_x - 20) = 90.0 \cdot 4.18 \cdot (20 - T_x) \tag{11.16}$$

Multiplying out the products, we arrive at

$$41.8(T_x - 80) = 376.2(20 - T_x)$$
$$(41.8 + 376.2)T_x = 418\,T_x = 7524 + 3344 = 10868 \tag{11.17}$$

$$T_x = 26\,[°C] \tag{11.18}$$

This result makes intuitively sense, since we have more cold water than hot water. Thus, the resulting temperature should be closer to the cold-water temperature. Let us go a step further. Obviously, you need a container in which to mix the two samples of water. Assume you have the cold water in the cheap kind of calorimeter the chemistry lab instructor hands you under the name of "benchtop calorimeter," but which you recognize as a 1-cent Styrofoam coffee cup. These Styrofoam cups are routinely used as teaching calorimeters because they are good insulators – they do not transmit much heat to the surroundings (remember, their original task is to keep your coffee hot). Furthermore, a Styrofoam cup has low heat capacity itself. Why is this important? If the cold water is in the Styrofoam cup, the cup assumes the temperature of the cold water, 20 [°C] in this case. Now we add the hot water, and the temperature of the water mixture should reach 26 [°C], as we calculated before. However, even if the heat capacity of the Styrofoam cup is low, some heat will be diverted and heat the cup instead of the cold water. What now? We could determine the mass of the Styrofoam cup, look up the specific heat of Styrofoam, and figure out the heat absorbed by the calorimeter using Eq. 11.11, according to

$$q_{cc} = m_{cc} c_{cc} \Delta T \tag{11.19}$$

where the subscript cc refers to the coffee cup calorimeter. Instead of trying to determine the heat lost to the calorimeter by its mass and heat capacity, one usually resorts to a different method, namely to determine it experimentally. To this end, we lump the mass and specific heat of the calorimeter into one number K_{cc}, which we call the calorimeter constant, and write the heat absorbed by the calorimeter instead as

$$q_{cc} = m_{cc} c_{cc} \Delta T = K_{cc} \Delta T \tag{11.20}$$

Then, we carry out a calibration experiment as follows. We perform the exact same mixing experiment of 90.0 [g] of water at 20 [°C] and 10.0 [g] of water at 80 [°C]. We find that the observed final temperature of the mixed water sample is 25.5 [°C] rather than 26 [°C] we calculated above (Eq. 11.38). We can determine the calorimeter constant for this situation as follows. Assume that the heat lost by the hot water is still

$$q_h = 10.0 \cdot 4.18 \cdot (T_f - T_h) = 10.0 \cdot 4.18 \cdot (25.5 - 80) = -2278.1\,[J] \tag{11.21}$$

since the final temperature, T_f, is known from measurement (25.5 [°C]). This heat flows to both the cold water and the calorimeter:

$$\begin{aligned} q_c &= 90.0 \cdot 4.18 \cdot (T_f - T_c) + K_{cc}(T_f - T_c) = 90.0 \cdot 4.18 \cdot (25.5 - 20) + K_{cc}(25.5 - 20) \\ &= 2069.1 + 5.5\,K_{cc} \end{aligned} \tag{11.22}$$

Using $q_c = -q_h$, we can solve for the calorimeter constant K_{cc}:

$$K_{cc} = 38\,[J/°C] \tag{11.23}$$

Thus, we see that the calorimeter itself, the Styrofoam cup, has a heat capacity of 38 [J/°C] which implies that it took 209 [J] of heat when the hot water was mixed with the cold water, and the calorimeter's temperature increased by 5.5 [°C].

When different materials are allowed to come into thermal contact, for example, if we drop a hot piece of metal into water, the different heat capacities of both materials will have to be taken into account. However, the problem can be solved exactly the same way we did it above for two samples of water, using Eqs. 11.13–11.15. In fact, calorimetric measurements were performed to experimentally determine the heat capacities of various materials; the specific heats in Table 11.1 and in your main textbook were derived exactly from calorimetric observations. The example below presents such an experiment.

Example 11.3 A 10.0 [g] piece of copper at 90.0 [°C] was dropped into 90.0 [g] of water at 20.0 [°C] in a calorimeter that has a calorimeter constant of 83.6 [J/°C] (the water and the calorimeter are assumed to have the same initial temperature). The final temperature of the water, copper, and calorimeter was found to be 20.58 [°C]. Calculate the heat capacity of copper.

Answer: c_p (Cu) = 0.384 [J/g °C]

11.3 The First Law of Thermodynamics

So far in the discussion in this chapter, we have touched upon some definitions of work (volume work, electrical work) and heat, and we have discussed the measurement of heat flow in detail in Section 11.2.2. The first law of thermodynamics states that both heat and work are forms of energy that can be interconverted, such that the change in energy, ΔE, of a process can be defined as the sum of heat and work:

$$\Delta E = q + w \tag{11.24}$$

where the change in energy is defined as the difference between the energy of the final and initial states,

$$\Delta E = E_f - E_i \tag{11.25}$$

for a closed system, that is, a system where no matter is exchanged with the surroundings. For a chemical reaction we refer to this energy as the "internal energy." This terminology implies that a chemical "system," for example the reactants and the products in a chemical reaction, have an inherent energy content, as shown in Figure 11.3. Here, we consider the initial state to be the reactants, and the final state the reaction products. ΔE is the energy difference between the initial and final states (i.e. between reactants and products) that can be released as heat, work, or a combination of both. We will come back to this diagram shortly. The intermediate energy bump between the original and the final states is known as the activation energy that may be required to get the reaction going. More on this in Section 11.4.

Since the energy of the system can be converted from one form into another, the first law of thermodynamics implies that the total energy of a system is constant, and/or that energy can neither be created nor destroyed. In summary, the *first law of thermodynamics* states that

- The energy of an isolated system is constant but can be interconverted between various forms (potential, kinetic energy), and between heat and work.
- The universe, as such, can be considered an isolated system; thus, the energy of the universe is constant.
- In chemical reaction, the change in energy can be released in a combination of heat and work. All three quantities, energy, heat, and work, have the same units. The work may be volume work (see below), electrical work, or any other form of work, such as work provided by the contraction of a muscle.

Many chemical reactions involve the production and/or consumption of gases. Therefore, many reactions involve a change in the volume, ΔV, between reactants and products. An example would be the combustion of gasoline. Let us assume, for simplicity, that gasoline is a hydrocarbon with the formula C_8H_{18} (octane):

$$C_8H_{18}(l) + 12.5\,O_2(g) \rightarrow 8\,CO_2(g) + 9\,H_2O(g) \tag{11.26}$$

This reaction involves 12.5 moles of gas on the reactant side and 17 moles of gas on the product side. Notice that we have ignored the volume of the liquid octane, since one mole of octane, 114 [g] at a density of octane 0.7 [g/mL], only occupies a volume of ca. 160 [mL]. This amount is negligible compared to the change in gaseous volumes, which is 4.5 moles of gas, or about $4.5 \cdot 24$ [L] ≈ 108 [L] at 25 [°C]. At a constant external pressure of 1 [atm], the volume work here is

$$w_{vol} = p \cdot \Delta V \tag{11.5}$$
$$= 108\,[\text{L atm}] \cdot 101\,[\text{kJ/(L atm)}] \approx 10.9\,[\text{kJ}] \tag{11.27}$$

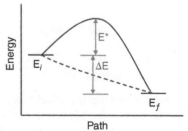

Figure 11.3 Energy diagram for a reaction where E_i is higher than E_f. Therefore $\Delta E = E_f - E_i < 0$.

and much higher at the temperature at which the products are found. Here, we used the energy conversion given in Example 11.2.

We use the sign convention that this work, performed by the chemical reaction given in Eq. 11.26, is lost by the system, and therefore write

$$\Delta E = q - p \cdot \Delta V \tag{11.28}$$

The heat q produced by this reaction is significant: burning gasoline produces a lot of heat (see Example 11.4). ΔE, the change in internal energy for this reaction, is the sum of work and heat and must be constant according to the first law of thermodynamics; however, we may vary the relative amounts of heat and work, as a simple consideration shows: if one gallon of gas is just burned by dropping a match into a container of gasoline (do not try this at home!!!) we get a lot of heat, and a big, ugly fire, but the only work produced is the volume work described by Eq. 11.27. However, if we burn the gallon of gas in the engine of a car, we can extract work, and produce less heat. In general, we want to minimize the heat loss of the reaction to maximize the work output. However, we can never extract all the energy change, ΔE, as work. This is the subject of the second law of thermodynamics, discussed in Section 11.8.

11.4 State Functions

Energy is a fundamental quantity in nature that can neither be created nor destroyed. Any change in energy during a process therefore depends only on the difference between the final and the original energy states, and not the path required to get from one state to the other. This is shown in Figure 11.3, and can be visualized as follows. Imaging you are hiking down a mountain, starting from E_i and ending at E_f. You can take a direct path indicated by the dotted line in Figure 11.3, and your change in (potential) energy will just be $\Delta E = E_f - E_i < 0$. Alternatively, you can take a path that first ascends to the summit marked with the asterisk. This requires an energy E^*, but on the way down your potential energy change is now $\Delta E + E^*$. Thus, your total change in potential energy is still ΔE. Quantities that depend only on the final and initial state of a system are known as state functions. The change of energy, ΔE, is such a state function. The energy diagram in Figure 11.3 shows that, even in the presence of an intermediate state of higher energy, the energy difference between the final state f and the initial state i is ΔE. The "activation energy" E^* mentioned earlier may be required to get the reaction going, and will be discussed in more detail in Chapter 13.

Whereas the internal energy is a state function, neither heat nor work are. In other words, since the sum of heat and work equals the energy change, the relative amount of heat and work may change, depending along what path the reaction is carried out, as pointed out above for the combustion of gasoline. Heat and work, therefore, are referred to as path functions.

This implies that although energy cannot be gained or lost, work can be lost to heat. For example, water in a reservoir on top of a mountain may be used to drive a turbine and generator, thereby creating the ability to perform electric work. However, if one just drained the reservoir down a river without a generator, the entire potential energy would be wasted: the friction of the water impinging on the ground produces heat, and all its potential energy will be wasted.

Another example for carrying out a chemical reaction along different pathways is the different ways to discharge a car battery. A fully charged battery stores a significant amount of energy (on the order of 4 [MJ], see Chapter 12, Eq. 12.67). This energy can be used to create useful work: it can run an electric motor (the starter) to turn over the engine, or even to move your car. This energy also could be wasted as heat: if you accidentally leave the light on overnight, the battery will be discharged by morning, and all you have to show for your money is warm light bulbs. Even more spectacular, but not recommended as an experiment, is the fast discharge of a battery by shorting the + and − poles with a jumper cable. The discharge produces sufficient heat to melt the cables and may cause significant other damage to the battery or you without producing work.

11.5 Definition of Enthalpy

As pointed out in Section 11.1, one aim of thermodynamics (the movement of heat) is the prediction of whether a chemical reaction will produce heat, and if so, how much heat. In other words, we aim to predict whether a reaction is exothermic (produces heat) or endothermic (requires heat). But how can we predict the positive or negative heat of a reaction if heat itself is a path function? In other words, it depends on how the reaction is being carried out?

Prediction of heat flow during a chemical reaction can be achieved by defining a heat-related quantity while specifying a particular path of a reaction, and thereby, avoiding the ambiguity of dividing the energy difference between heat and work. This is achieved by defining the "enthalpy" of a reaction as

$$H = E + p \cdot V \tag{11.29}$$

If a chemical reaction produces or consumes gases, the product of $p \cdot V$ represents a work term. Without going into the details of the derivation, which involve some differential calculus, and combining the first law of thermodynamics

$$\Delta E = q - p\Delta V \tag{11.28}$$

with the above definition in Eq. 11.29, one can show that

$$\Delta H = q + V\Delta p \tag{11.30}$$

Eqs. 11.28 and 11.30 contain all the pertinent information we need to know about q, E, and H. When we carry out a reaction under constant pressure conditions, say in an open flask under atmospheric pressure, the external pressure does not change ($\Delta p = 0$); therefore

$$q = \Delta H \text{ (at constant pressure)} \tag{11.31}$$

If we carry out a reaction in a closed container, such as in a bomb calorimeter, the volume does not change ($\Delta V = 0$), and

$$q = \Delta E \text{ (at constant volume)} \tag{11.32}$$

What we have achieved by specifying the path, and therefore, standardizing the work term, is to make the heat flow of a reaction a path-independent quantity. Since most chemical reactions are carried out in containers open to the atmosphere (i.e. at constant pressure), the $V \Delta p$ term in Eq. 11.30 is zero, and the enthalpy becomes a surrogate for the heat flow.

The discussion of the last section was devoted to the justification of defining a new function, ΔH, the enthalpy, which is a state function and can be measured in a calorimeter just as q was, if we keep the pressure constant. Similarly, when the volume is kept constant, as in a bomb calorimeter, the energy change of a reaction, ΔE, can be measured directly as the heat.

11.6 Hess' Law and Reaction Enthalpies

Since the enthalpy is a state function and does not depend on the reaction path, we can break down a chemical reaction into partial steps, and add the reaction enthalpies for each step to arrive at the enthalpy of the total reaction. Let us look at a simple example. The combustion of solid carbon, such as coal in a blast furnace where the amount of oxygen is limited, proceeds in two steps:

$$C(s) + \tfrac{1}{2}O_2(g) \rightarrow CO(g) \quad \Delta H_1 \tag{11.33}$$

$$CO(g) + \tfrac{1}{2}O_2(g) \rightarrow CO_2(g) \quad \Delta H_2 \tag{11.34}$$

In the presence of unlimited oxygen, the reaction may proceed directly in one step:

$$C(s) + O_2(g) \rightarrow CO_2(g) \quad \Delta H_T \tag{11.35}$$

and we see that the sum of Eqs. 11.33 and 11.34

$$C(s) + \tfrac{1}{2}O_2(g) \rightarrow CO(g)$$
$$CO(g) + \tfrac{1}{2}O_2(g) \rightarrow CO_2(g)$$
$$\overline{C(s) + \tfrac{1}{2}O_2(g) + \cancel{CO(g)} + \tfrac{1}{2}O_2(g) \rightarrow \cancel{CO(g)} + CO_2(g)}$$
$$= \quad C(s) + O_2(g) \rightarrow CO_2(g)$$

equals Eq. 11.35. Thus, the overall reaction was broken down into two partial reaction steps, and the total change in enthalpy, ΔH_T, is the sum of the two partial enthalpies,

$$\Delta H_T = \Delta H_1 + \Delta H_2 \tag{11.36}$$

The enthalpy change, ΔH, for step 1 above (Eq. 11.33) can be measured in a calorimeter and was found to be

$$C(s) + \tfrac{1}{2}O_2(g) \rightarrow CO(g) \quad \Delta H_1 = -110.5 \, [\text{kJ/mol}] \tag{11.33}$$

The observed enthalpy for this reaction is written with a negative sign, indicating that the reaction is exoenthalpic ("exoenthalpic" often is referred to as "exothermic"). Since this reaction releases enthalpy (heat), this heat has the effect of heating the product. In other words, hot, gaseous CO is produced in the reaction. When we write the reaction enthalpy, we define what is known as the standard conditions and assume that the temperature of reactant and products remain constant, at 25 [°C]. We accomplish this by removing the reaction enthalpy to an external reservoir of heat (for example, the calorimeter) to restore the products to the original temperature. Furthermore, we define the standard conditions for one mole of reaction such that reactants and products are in the state they naturally occur at 25 [°C] and 1 [atm] pressure. Finally, since we are interested in the *change* in enthalpy during a chemical reaction (rather than the absolute values of enthalpy) we define that elements in their standard state have standard enthalpies of zero (more on this later).

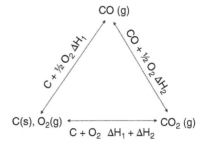

Figure 11.4 Principle of additivity of partial enthalpies (Hess' law).

The second step of the two-step combustion scheme is the further oxidation of CO to CO_2:

$$CO(g) + \tfrac{1}{2} O_2(g) \rightarrow CO_2(g) \quad \Delta H_2 = -283.0 \, [\text{kJ/mol}] \tag{11.34}$$

Since we defined ΔH to be a state function, the enthalpy changes of the combined reaction, ΔH_T, is

$$C(s) + O_2(g) \rightarrow CO_2(g) \quad \Delta H_T = \Delta H_1 + \Delta H_2 = -393.5 \, [\text{kJ/mol}] \tag{11.35}$$

must be the same as the sum of the two partial steps. Schematically, we may represent this result in terms of the graph in Figure 11.4 from which we see the connection between the quantities involved.

Notice that Eq. 11.35 represents the equation for the formation of CO_2 from the elements at their standard state. Thus, the enthalpy value listed, -393.5 kJ/mol, also represents the standard *enthalpy of formation* of carbon dioxide:

$$\Delta H_f^\circ(CO_2) = -393.5 \, [\text{kJ/mol}] \tag{11.37}$$

What we developed here for a simple two-step reaction can be extended to more complex reactions, and we may use the procedure of breaking down a reaction into partial steps to help us predict the enthalpies of reactions from tabulated values of standard enthalpies of formation. For example, to determine the heat given off (the enthalpies) when typical fuels are burned, let us look at a combustion reaction of a common fuel, liquefied butane, C_4H_{10}, available in any cigarette lighter or camping stove cartridge. What is the enthalpy of combustion, i.e. the enthalpy released when one mole of butane reacts with the oxygen in air?

According to the formalism developed above, we start with a balanced chemical equation. Notice that the following calculations are carried out for gaseous butane, rather than liquefied butane. This aspect will be the subject of Example 11.8:

$$C_4H_{10}(g) + 6.5 \, O_2(g) \rightarrow 4 \, CO_2(g) + 5 \, H_2O(g) \tag{11.38}$$

In our stepwise approach, we break down this reaction into partial steps that are depicted in Figure 11.5. First, we break butane down into the elements (in their standard state):

$$C_4H_{10}(g) \rightarrow 4C(s) + 5H_2(g) \quad -\Delta H_f^\circ(C_4H_{10}) \tag{11.39}$$

This step requires the negative enthalpy of formation of butane. Next, we take the four moles of carbon resulting from Eq. 11.39 and react them with four moles of oxygen according to

$$4C(s) + 4O_2(g) \rightarrow 4 \, CO_2(g) \quad 4 \Delta H_f^\circ(CO_2) \tag{11.40}$$

Finally, the hydrogen created in Eq. 11.39 reacts with oxygen according to

$$5H_2(g) + 2.5 \, O_2(g) \rightarrow 5 \, H_2O(g) \quad 5 \Delta H_f^\circ(H_2O) \tag{11.41}$$

You can convince yourself that the sum of Eqs. 11.39–11.41, indeed, equals Eq. 11.38. Therefore, the enthalpy of the combustion reaction, given by Eq. 11.38, is

$$\Delta H_{\text{combustion}}^\circ = 4 \Delta H_f^\circ(CO_2) + 5 \Delta H_f^\circ(H_2O) - \Delta H_f^\circ(C_4H_{10}) \tag{11.42}$$

$$= 4(-394) + 5(-242) - (-125) = -2661 \, [\text{kJ/mol}] \tag{11.43}$$

The enthalpies of formation of water, carbon dioxide, and butane needed for the numerical calculation of ΔH° in Eq. 11.42 can be found in tables in your main text or online. Let as visualize, once more, how this process works (cf. Figure 11.5). The first two terms in Eq. 11.42 represent the enthalpy released when the products, water and carbon dioxide, are created from the elements. This is true because of the definition of $\Delta H°_f(CO_2)$ and $\Delta H°_f(H_2O)$, which implies that the compounds are formed by the elements in their standard state.

The third term in Eq. 11.42, $\Delta H°_f(C_4H_{10})$ is the enthalpy of the formation of butane from the elements, carbon and hydrogen. Since its sign is negative, it means, in fact the reverse process, namely

$$C_4H_{10} \rightarrow 4C + 5H_2 - \Delta H°_f(C_4H_{10}) \qquad (11.44)$$

So, we have broken the overall reaction, given by Eq. 11.38 into the following intermediate steps:

$$C_4H_{10}(l) \rightarrow 4C(s) + 5H_2(g) \quad -\Delta H°_f(C_4H_{10})$$
$$4C(s) + 4O_2(g) \rightarrow CO_2(g) \quad 4 \cdot \Delta H°_f(CO_2)$$
$$5H_2(g) + 5/2\ O_2(g) \rightarrow 5H_2O(g) \quad 5 \cdot \Delta H°_f(H_2O)$$

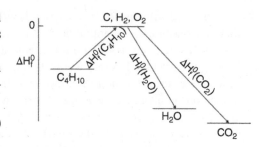

Figure 11.5 Stepwise reactions using standard enthalpies of formation to compute the enthalpy of combustion of butane.

Thus, we were justified using Eq. 11.42 to describe the enthalpy released for one mole of butane being burned. We call the enthalpy released by one mole of butane its molar enthalpy of combustion. Of course, if a different amount "n" is used, the enthalpy needs to be multiplied by this factor. The enthalpies of the formation of many compounds have been determined, in a reverse process, from measured enthalpies of combustion.

The process outlined in Eqs. 11.39–11.44 can be summarized by Eq. 11.45 for a general chemical reaction at standard conditions as follows:

$$\Delta H°_{reaction} = \Sigma_i \{\Delta H°_f(products)\}i - \Sigma_j \{\Delta H°_f(reactants)\}j \qquad (11.45)$$

where the summation is over all reactant and products. All enthalpies of formation must be multiplied by the corresponding stoichiometric coefficients.

Example 11.4 The standard enthalpy of combustion of gasoline (assumed to be octane, C_8H_{18}) according to Eq. 11.26, $C_8H_{18}(l) + 12.5\ O_2(g) \rightarrow 8\ CO_2(g) + 9\ H_2O(g)$, is -5080 [kJ/mol]. Using the standard enthalpies of formation of water and carbon dioxide given in the chapter, calculate the enthalpy of formation of octane.

Answer: $\Delta H°_f(C_8H_{18}) = -250$ [kJ/mol]

Example 11.5 The enthalpy for the reaction

$$C(diamond) + O_2(g) \rightarrow CO_2(g)$$

is -397 kJ mole. Given the enthalpy of formation of CO_2 and Hess' law to calculate the enthalpy for the reaction

$$C(graphite) \rightarrow C(diamond)$$

Show all logical steps.

Answer: $\Delta H°_{reaction} = 3$ [kJ/mol]

Example 11.6 Fermentation of glucose to alcohol and carbon dioxide occurs according to the equation

$$C_6H_{12}O_6(aq) \rightarrow 2C_2H_5OH(aq) + 2CO_2(g)$$

Calculate the enthalpy released for 0.10 [mol] of glucose fermenting.

Answer: -53 [kJ/mol]

11.6.1 Enthalpy of Crystal Formation: Lattice Energy of MgO

How far the concept of the additivity of enthalpies can be extended is demonstrated in Figure 11.6, which shows the determination of the lattice energy of a compound. The lattice energy is defined for a compound such as magnesium oxide, also known as magnesia, MgO, as the energy for the process

$$Mg^{2+}(g) + O^{2-}(g) \rightarrow MgO(s) \qquad (11.46)$$

which is the enthalpy for the gaseous ions to form an ionic lattice. This energy can be written as the sum of the following stepwise reactions. We start with the elements, magnesium and oxygen, in their standard states (point "a" in Figure 11.6). To go from point "a" to the product, MgO, involves the standard enthalpy of formation, $\Delta H_f^\circ(MgO)$, indicated by point "f."

The other path to get to point "f" that involves the desired lattice energy is presented as follows:

First, solid magnesium is converted to gaseous magnesium, indicated at point "b":

$$Mg(s) \rightarrow Mg(g) \quad \Delta H_{sub} \tag{11.47}$$

Next, gaseous Mg atoms are ionized to arrive at point "c" (see Eqs. 5.27 and 5.30):

$$Mg(g) \rightarrow Mg^{2+}(g) + 2e^-$$
$$IE = IE_1 + IE_2 \tag{11.48}$$

Next, the O_2 molecule is broken into the oxygen atoms, overcoming the bond energy of diatomic oxygen and bringing us to point "d":

$$\tfrac{1}{2} O_2(g) \rightarrow O(g) \quad E_{bond} \tag{11.49}$$

Next, the oxygen atom is converted to the oxygen ion,

$$O(g) + 2e^- \rightarrow O^{2-}(g) \quad E_a \tag{11.50}$$

where E_a is the electron affinity introduced in Eq. 5.31. Notice that all the energies in Eqs. 11.47 through 11.50 can be found in tables online. This brings us to point "e" in Figure 11.6, and the next step in our thought experiment is just to add up all the energies of the stepwise scheme to arrive at the desired lattice energy.

Example 11.7 Determine the numeric value for the lattice energy of MgO using online data and the scheme represented in Figure 11.6.

Answer: 3788 [kJ/mol]

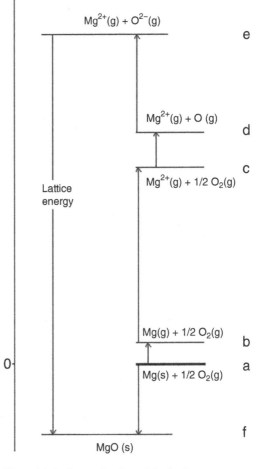

Figure 11.6 Determination of the lattice energy using Hess' law.

11.7 Enthalpy of Phase Transitions

Earlier in this chapter, we discussed that the increase in temperature of a substance as a function of added heat is given by

$$q = m\, c_p \Delta T \tag{11.11}$$

This equation is true only if the compound being heated does not undergo a phase transition. During a phase transition, the temperature does not change, even if heat is added at a constant rate. This is demonstrated in Figure 11.7, which shows the temperature of 500 [g] of water when heat is added at a constant rate 1 [kJ/min], starting from ice at -10 [°C] and ending at water vapor at 110 [°C]. In order to construct this graph, the specific heat of solid, liquid, and gaseous water {2.09 [J/(g °C)], 4.18 [J/(g °C)], and 1.99 [J/(g °C)], respectively}, and the heat (enthalpy) of melting and vaporization (333 [J/g] and 2260 [J/g], respectively) were used. Using Eq. 11.11, it is easy to see that the heat required to heat the ice from -10 to 0 [°C] (marked as region "a" in Figure 11.7) is

$$q_A = 500 \cdot 2.09 \cdot (T_f - T_i) = 500 \cdot 2.09 \cdot 10 = 10.45\,[\text{kJ}] \tag{11.51}$$

Similarly, to heat the water sample from 0 to 100 [°C] (marked as region "b") is

$$q_B = 500 \cdot 4.18 \cdot 100 = 209\,[\text{kJ}] \tag{11.52}$$

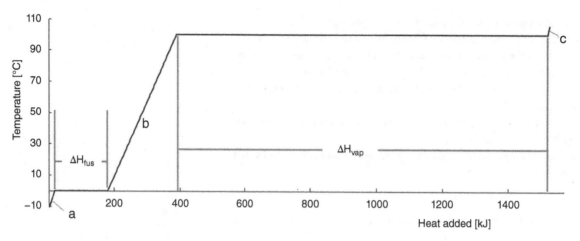

Figure 11.7 Heating curve of water with phase transitions.

and from 100 to 110 [°C] (marked as region "c")

$$q_C = 500 \cdot 1.99 \cdot 10 = 9.95\,[\text{kJ}] \tag{11.53}$$

The heat required for the melting phase transitions is

$$\Delta H_{fus} = 500 \cdot 333 = 166.5\,[\text{kJ}], \tag{11.54}$$

and for the vaporization

$$\Delta H_{vap} = 500 \cdot 2260 = 1130\,[\text{kJ}] \tag{11.55}$$

Inspection of the graph in Figure 11.7 reveals that the temperature stays constant while there is an equilibrium between two phases: when ice and water coexist, the temperature does not increase, even though heat is added. At this point, the heat capacity of the ice/water system is infinity, and all the energy added is used to break the solid into the liquid. This is, of course, a well-known fact in everyday life: as long as you have ice cubes in your ice coffee or soft drink, the temperature of the drink stays constant, at 0 [°C], although heat from the surroundings continuously adds energy to your drink. By the way, all the ice added by the barista does not make your ice coffee any colder, because, as long as any liquid (assumed to be mostly water) is in equilibrium with ice, the temperature of the system is 0 [°C]. There is a large amount of ice in your soft drink because ice is cheaper than the drink.

A similar situation exists at the boiling point. Consider a pot of water on a hot plate, such as a kitchen stove. When the water is boiling, we have an equilibrium of liquid water and water vapor, and its temperature stays constant, at 100 [°C], although we continuously add heat. The heat is used to overcome the strong intermolecular forces – the hydrogen bonds – that exist in liquid water, and whatever you have in the boiling water (for example, the egg you want to hard-boil) is exposed to a constant temperature of 100 [°C], without overheating. Only if all the water has evaporated can the temperature of the water vapor increase, as shown by the region marked "c" in Figure 11.7.

Example 11.8 (a) A 20 [g] ice cube at −5 [°C] is added to 250 [mL] of soda (assumed to be pure water at this point, with a density of 1.00) at room temperature. Ignore the heat capacity of the container. What is the final temperature of the soda?

Answer: $T_x = 17.1$ [°C]

11.8 Entropy

In the previous sections, we discussed the energetics of chemical reactions that are governed by the first law of thermodynamics. According to this law, energy can be converted between work and heat, and between different entities that participate in a process such as chemical reactions. So, let us look at a process with which we are all familiar: a hot cup of coffee sitting on an office desk. If nobody drinks the coffee, we all know what is going to happen: heat flows from the hot cup of coffee (the system) to the surroundings (the desk, the air in the office). Thus, the room and the desk will become somewhat warmer, but imperceptibly so, and the coffee assumes the temperature of the surroundings. This is a common process, but the reverse process, perfectly legitimate from the viewpoint of the first law, never happens: in this never-occurring process, the surroundings become a little cooler, and transfer their heat to the coffee, and spontaneously bring the coffee back to its original hot temperature. Sorry, but no go.

Let us look at another example of a process that could happen, but has never been observed to happen. Assume you are the dealer of cards in a casino. You take two unused, fresh decks of cards, 52 cards in each deck. One has blue, the other deck has red backs of the cards. You shuffle the two decks together, and with your magic hands, cut the resulting stack of 104 cards exactly in the middle, and keep shuffling them. After a while, red and blue cards will be pretty equally distributed among the two decks after you cut them in the middle. You keep doing this for a while, and nothing really changes in the distribution of red and blue cards in the two decks. However, it just could happen that, after you cut the stack of 104 cards in the middle, you have reestablished by chance that one deck consists of red, the other of blue cards only.

Let us take this concept a little further. Air, as you recall, is a mixture of about 80% N_2 molecules and 20% O_2 molecules that whiz around at high velocities with frequent collisions between them, see Chapter 8. It is not forbidden by the first law of thermodynamics, nor any other law discussed so far, that all the oxygen molecules move to the front 20% of the room you are sitting in, and the nitrogen molecules all move to the back 80% of the room. If you are sitting in the front, you would be oxidized quickly, while in the back, you suffocate due to the lack of oxygen. Not a pretty picture, but unlikely to happen. So, we see that, although perfectly possible, there seems to be another law we have not yet discussed, a law that introduces a quantity which is related to the likelihood, or probability, of a process occurring spontaneously. This quantity is known as the *entropy*.

Before going any further in this thought process, we need to define exactly what we mean by saying a process is spontaneous. Let us discuss this by using a mechanical analog, shown in Figure 11.8. A ball sitting on an inclined ramp, as shown in Figure 11.8a, will "spontaneously" roll down the ramp, whereas in the situation depicted in Figure 11.8b, the ball would rest in the little well forever, unless it receives a little nudge to get over the energy bump. Many chemical reactions are actually stuck in a similar situation as the ball in Figure 11.8b: they would proceed toward lower energy if they receive an "activation energy" as shown in Figure 11.3. Think about gasoline in the presence of oxygen (air): it appears perfectly stable until a spark from a spark plug in a car engine or a match provide the activation energy to get the reaction started which then proceeds to lower energy. So, we call a spontaneous process one that proceeds "downhill," or toward lower energy even if it requires an activation. By the way, the term "energy" was used here without going into any detail about what form this energy is, and should have been called "free energy." More on the concepts of free energy later (see Section 11.9). The concepts of activation energy, and its influence on the rate of a chemical reaction (i.e. how fast a reaction occurs) will be taken up again in Chapter 13. The gist of this last paragraph is that a spontaneous reaction proceeds toward lower "free energy," although it may need a little spark (or something of this order) to get going.

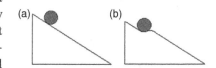

Figure 11.8 Process ("reaction") that proceeds (a) without and (b) with activation.

In the last paragraph, we have stated that a spontaneous process is one in which the free energy of the system decreases or goes downhill. However, we can also define spontaneity in terms of the entropy change in a process. What is this quantity known as entropy, and designated the symbol "S"? The rest of Section 11.8 will be devoted to shed light on this question. First of all, entropy is a state function, which we defined in Section 11.4. Second, the entropy may be viewed as a quantity somehow related to disorder or probability. Much more on this below. But from this simple statement, we can conclude that more ordered states have lower entropy than disordered states. Thus, we may qualitatively state that the entropy of solids is less than that of liquids, and less than that of gases:

$$S_{solid} < S_{liquids} < S_{gases} \tag{11.56}$$

$$S_{pure\ substances} < S_{mixtures} \tag{11.57}$$

$$S_{low\ T} < S_{high\ T} \text{ for the same substance} \tag{11.58}$$

Eq. 11.58 may be visualized to indicate that at higher temperature, the motion of atoms or molecules occurs with larger amplitude, and hence, with more randomness.

Next, it is necessary to discuss an expression for the entropy change, ΔS, of a chemical reaction or other process. Let us go back to the example of the cup of coffee nobody drank. In this example, we call the "coffee" the "system." The mug, as well as the air surrounding the mug, the entire room, the walls of the room, and everything else that is affected, is referred to as the "surroundings." The system and the surroundings make up the "universe." Thus, we say that any changes in the system will affect the surroundings somehow, and we define the

$$\text{universe} = \text{system} + \text{surroundings} \tag{11.59}$$

An exoenthalpic reaction always will release enthalpy (heat) into the surroundings. Thus, as the system loses enthalpy, the surrounding picks up an equal amount. Similarly, in an endoenthalpic reaction, the enthalpy of the system increases for the reaction. In this case, the enthalpy "content" of the surroundings decreases. But remember, according to the first law of thermodynamics, the change in energy (or enthalpy, within the constraints we discussed in Section 11.5) of the universe must be zero, i.e.

$$\Delta H_{universe} = \Delta H_{system} + \Delta H_{surroundings} = 0 \text{ or} \tag{11.60}$$

$$\Delta H_{system} = -\Delta H_{surroundings} \tag{11.61}$$

In Eqs. 11.56–11.58, we have discussed the entropy content associated with certain conditions like physical state or temperature. When we deal with a chemical reaction or process, the entropy of the system may increase or decrease. An example would be a sample of ice above 0 [°C] that spontaneously melts. A liquid is always more random, less ordered than a solid. Thus, for the system (the ice sample)

$$\Delta S_{melting} > 0 \tag{11.62}$$

using Eq. 11.56. At temperatures below 0 [°C], water will freeze spontaneously with an increase in order, or a decrease in randomness, of the system. So, for the freezing reaction we write the entropy change of the system:

$$\Delta S_{freezing} < 0 \tag{11.63}$$

Thus, we see that entropy *of the system* is not a signpost for spontaneity either since for a spontaneous process, the change in entropy of the system may be positive or negative. However, from this example we gain another insight: whether or not a process or reaction is spontaneous depends on the surroundings and the system, not just the system alone. This becomes clearer if we consider the enthalpy changes accompanying the melting/freezing processes. When ice melts, heat has to be added to the system (see Section 11.7 and Figure 11.7). The heat of fusion (see Eq. 11.54)

$$\Delta H_{fusion} = 5.99 \, [\text{kJ/mol}] \tag{11.54}$$

must come from the surroundings that cool down in the process. Conversely, when liquid water freezes, it releases this enthalpy, and the surroundings get warmer. Thus, in both cases, the entropy of the surroundings also changes, and we write the entropy changes of the universe,

$$\Delta S_{universe} = \Delta S_{system} + \Delta S_{surroundings} \tag{11.64}$$

is the sum of the entropy changes of the system plus that of the surroundings. In contrast to Eq. 11.60, where the sum of the enthalpies of the system plus that of the surroundings added to zero, the entropy change of the universe is always positive for a spontaneous process:

$$\Delta S_{universe} = \Delta S_{system} + \Delta S_{surroundings} > 0 \tag{11.65}$$

This is due to the fact that for any spontaneous process, energy is degraded (exhausted) from a more useful to a less useful form. For the freezing of the water sample, the enthalpy of freezing is released into the surroundings which warms up a little, but this heat is dispersed into the surroundings from a concentrated to a more dispersed and less useful form. For the melting process, heat has to be added to the system. This heat comes from the surroundings which cools a little, or from the combustion of fuel, or from an electric heater or the sun. The process that created the heat required for the melting of water increased the entropy of the surroundings such that Eq. 11.65 holds. So, we conclude that entropy is a quantity that is related to the distribution of energy into more random, less ordered states. In the next section, we shall demonstrate that disordered states are more probable than ordered states, and therefore, the concentrated, useful forms of energy are less likely than disordered, less useful distributions.

11.8.1 Entropy and Probability

In this section, we shall demonstrate that for many systems, disorder is more probable than order, and thus, will occur more frequently. Let us demonstrate this concept by using a few simple examples. At the start of Section 11.6, we discussed a system, two decks of cards that are shuffled together, and saw that it is improbable to spontaneously attain a state of order after repeated shuffling. In order to get a more quantitative view, let us look at an example of coin flips. Imagine you flip a quarter coin and we know that there are two possible outcomes, heads (H) or tail (T). You have a 50:50 chance for heads or tails, and the chance of getting either answer is one in two, or 50%.

Coin A	Coin B
H	H
H	T
T	H
T	T

(a)

Coin A	Coin B	Coin C
H	H	H
H	H	T
H	T	H
H	T	T
T	H	T
T	H	H
T	T	H
T	T	T

(b)

Figure 11.9 Outcome of (a) two or (b) three coin toss experiments.

If you perform the same experiment with two quarters, Coin A and Coin B, what are the chances that both show heads or tails? In order to answer that, one needs to establish the total number of possible outcomes, which are listed in Figure 11.9a. Thus, the chance that both coins simultaneously show H or T is still two in four, or 50%. For three coins, the possible outcomes are shown in Figure 11.9b. Figure 11.9 reveals two observations: first, if N coins are tossed, and there are two possible outcomes for each coin (H or T), then there will be 2^N possible outcomes. Thus, there are four different arrangements (outcomes) in the case of two coins, and eight in the case of three coins.

In addition, we see from Figure 11.9b that the chance of hitting the "ordered" states, H or T is 2/8, or 25%. A plot of the number of outcomes is shown in Figure 11.10 for $N = 2$, $N = 3$, and $N = 10$. We see from this graph that the "ordered" outcomes are less probable than any of the "disordered" outcomes ("ordered" and "disordered" here are used in the sense of all heads or tails vs. any mixed outcomes). For $N = 10$, there are $2^N = 1024$ total outcomes ("states") and only two ordered states.

If we extend this principle to larger numbers, say 50, we see that it is very unlikely (nearly impossible) to reach the ordered state, since

$$2^{50} = 1{,}125{,}899{,}906{,}842{,}624$$
$$\approx 1.125 \cdot 10^{15}$$

Thus, the chance of all coins falling heads up is 1 in 1125899906842624 which equals the chance for all coins fall tails up. So, we see that the chance of hitting the "completely ordered" states, all H or all T, is 2/1125899906842624, or 1 in 562949953421312. Let us estimate how long it may take to reach one of these states: let us assume that you can toss the coins at a rate of once a second. In order to toss the coins $5.63 \cdot 10^{14}$ times, you would toss for $1.56 \cdot 10^{11}$ hours, or $6.52 \cdot 10^9$ days, or $1.78 \cdot 10^7$ years. That is nearly 18 million years, longer than the existence of *Homo sapiens*. Of course, all "heads up" could happen at the first toss. The chance of that happening is small, but possible.

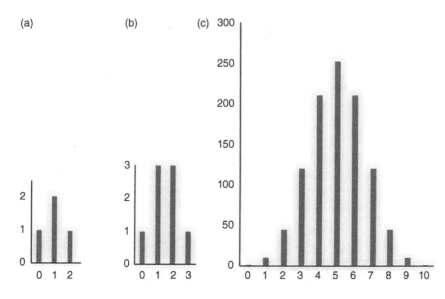

Figure 11.10 Probabilities of ordered vs. disordered coin toss outcomes for (a) 2, (b) 3, and (c) 10 coins. Figures 11.10a and 11.10b correspond to the results depicted in Figure 11.9.

The aim of the previous discussion was to demonstrate that there are many more disordered states than ordered states – many more, in fact. If we extend our thought process to chemical systems where the number of atoms or molecules is on the order of Avogadro's number, we find that there are many more disordered microstates possible, and entropy describes ways to distribute energy into microstates. Think there would be how many disordered states if we carried out the coin toss experiment discussed above with one mole of coins, rather than 50 coins! Thus, the disordered arrangements far outnumber the ordered arrangements. There "arrangements" are referred to as "microstates" Ω in statistical thermodynamics, and the entropy of a system is expressed by Eq. 11.66:

$$S = (R/N_A) \ln \Omega \quad [J/K] \tag{11.66}$$

where R is, as usually, the gas constant, N_A is Avogadro's number. The ratio $R/N_A = k$, the Boltzmann constant, was defined before (Eq. 8.43) and is a very small number ($1.38 \cdot 10^{-23}$ [J/K]). On the other hand, the number of microstates is huge in chemical systems, since the number of atoms or molecules is very large. The distribution of energy into these microstates depends on temperature, since at higher temperatures, more energy is available that can be distributed into the microstates.

Example 11.9 Calculate the number of microstates and the corresponding entropy for a system which has two outcomes (like the coin toss discussed above) for N = 20000 "particles."

Answer: $\Omega = 3.98 \cdot 10^{6020}$; $S = 1.9 \cdot 10^{-19}$ [J/K]

Notice that the definition of entropy given in Eq. 11.66 is an absolute value, not a change in entropy. Thus, when you look up thermodynamic quantities in tables online, you will notice that enthalpies and free enthalpies (see Section 11.9) are given as relative values ($\Delta H°$ or $\Delta G°$), respectively, whereas the entropy $S°$ is given as an absolute value at a given temperature and pressure. Therefore, the entropy change of a chemical reaction is given by

$$\Delta S°_{reaction} = \Sigma_i \{S°_f(\text{products})\}_i - \Sigma_j \{S°_f(\text{reactants})\}_j \tag{11.67}$$

whereas the enthalpy change was given by Eq. 11.45 as

$$\Delta H°_{reaction} = \Sigma_i \{\Delta H°_f(\text{products})\}_i - \Sigma_j \{\Delta H°_f(\text{reactants})\}_j \tag{11.45}$$

In Eq. 11.67, the entropies of elements in their standard state are nonzero!

11.8.2 Entropy and Heat Flow

In the previous section, a relationship between the absolute entropy of a system (ensemble) was obtained that was based on the distribution of energy into available microstates and we saw that the number of microstates is huge for chemical systems. Next, we shall discuss the entropy changes of real and spontaneous processes. Consider a system of two identical pieces of metal that originally are at different temperatures and a process in which they are brought into contact. In this process, heat will flow spontaneously from the hotter to the colder piece, tending to equalize the temperatures. Heat will always flow spontaneously from high to low temperature, never in the opposite direction. Let us assume the two pieces of metal are completely isolated from the surroundings; thus, we know that the change of the entropy of the surroundings

$$\Delta s_{surroundings} = 0 \tag{11.68}$$

Since the heat transfer occurred spontaneously, we know from Eq. 11.65 that the entropy of the universe increased. Therefore, the entropy of the system – the two pieces of metal in thermal contact – must have increased:

$$\Delta s_{system} > 0 \tag{11.69}$$

since the entropy of the surroundings did not change. From our discussion of calorimetry in Section 11.2.2, we know that the amount of heat transferred between the hot and cold pieces of metal is larger when the temperature difference between the two pieces of metal is larger. Thus, we assume that the entropy change for a larger temperature difference is larger, in other words, the process will be "more spontaneous." Thus, we may conclude that the entropy change is proportional to the amount of heat transferred in the process.

$$\Delta S \propto q \tag{11.70}$$

This can be justified by our understanding of the heat (content) in terms of the random thermal motion of the atoms in the pieces of metal: high temperature implies more randomness; low temperature implies less randomness. Thus, the entropy change of a process in which heat is transferred from one part of the system to another is related to the amount of heat transferred, as written in Eq. 11.70.

What is the proportionality in Eq. 11.70? Eq. 11.66 suggests that the units of entropy are energy/temperature. Thus, q/T would have units of entropy, and, indeed, this is the thermodynamic definition of entropy:

$$\Delta S = q/T \tag{11.71}$$

The proof of this definition is beyond the scope of this discussion, but it may be paraphrased as follows. If we accept Eq. 11.71, namely that the change in entropy is proportional to the amount of heat transferred, then we have to ask ourselves: does the temperature at which the heat is transferred, influence the entropy change? The answer here is "yes," and the temperature actually appears in the denominator. Thus, if equal amounts of heat are transferred, the change in entropy is larger at low temperature: the same amount of heat will cause a smaller entropy change at high temperature, than at lower temperature. If things are already disordered (messy), a little heat makes things just a little messier. When the temperature is low, and things are in a much higher state of order, the same amount of heat will create a much larger disorder!

Next, we shall argue how entropy is related to spontaneity. In order to do this, we try to establish just how much entropy is created in the example of the two blocks of metal above. We write the change in entropy of the hot block of metal according to Eq. 11.71 as

$$\Delta S_h = -q/T_h \tag{11.72}$$

while the entropy change for the cold block is

$$\Delta S_c = q/T_c \tag{11.73}$$

As we know from the discussion of calorimetry that the heat, q, in Eqs. 11.72 and 11.73 is equal, albeit with opposite signs. However, the changes in entropy for the two blocks are different, since their original temperatures were different. The total change in entropy, ΔS_T for the process is the sum of Eqs. 11.71 and 11.72:

$$\Delta S_T = -q/T_h + q/T_c = q(T_h - T_c)/(T_h T_c) \tag{11.74}$$

Since $\Delta s_{surroundings} = 0$ (Eq. 11.67), we see that for the spontaneous process – the temperature equilibration between two blocks of metal – the entropy change of the universe is positive. This may be considered the mathematical formulation of the second law of thermodynamics:

$$\Delta S_{universe} > 0 \text{ for spontaneous processes} \tag{11.75}$$

We can generalize this result, and state that *for any spontaneous process, the change in entropy of the universe is positive*. The first law of thermodynamics states that for any process, the energy is constant. The second law of thermodynamics states that for any spontaneous process, heat is dissipated into entropy thereby increasing the entropy of the universe. In other words, although the total energy of a process is constant, some of this energy is lost to disorder and is no longer useful energy.

Thus, for any spontaneous process, the entropy – disorder of the universe – increases. This statement is true for your own metabolism: you expend between 5000 and 8000 [kJ] per day, known as your base metabolic rate. This energy is mostly released as heat into the environment. If you exercise, some of the energy expenditure may be changed into work. This large increase of entropy of the surroundings allows you – the system – to maintain entropy-lowering processes inside the body. As soon as life expires, and metabolic activity ceases, the entropy of the surroundings is no longer increased, and entropy takes hold of the remains of life: decomposition into a high entropy state starts.

At equilibrium, no changes occur in the concentration of reactants, and any heat flow is considered "reversible" (implying that an infinitesimally small temperature change in either direction will cause a reversible change in the direction of the heat flow).

$$\Delta S_{universe} = 0 \text{ at equilibrium (no changes)} \tag{11.76}$$

11.8.3 Entropy as an Indicator of Energy Exhaustion

The rather thought-provoking statement in the last paragraphs leads us to a related definition of entropy, as an exhaustion of energy into less useful forms. In more advanced courses on thermodynamics, students are exposed to an ideal construct known as the "Carnot cycle" that describes a process how heat can be reversibly converted to work (or vice versa) by the expansion of an ideal gas in alternating thermal contact with a hot and cold heat reservoir, and the efficiency of such an idealized process can be related to the temperature difference between the hot and cold heat reservoir. In this ideal, reversible process, there is no change in entropy. However, in real processes – like the ones used to convert heat to work, as in a powerplant that uses fuel to create the heat to be converted – we find that the temperature of the hot reservoir decreases and the temperature of the cold reservoir increases during the work-producing cycle. Hence, powerplants constantly need to burn fuel to keep the hot reservoir hot, and use huge cooling towers to keep the low-temperature reservoir cold. Without these entropy-creating processes, the temperature of the hot and cold reservoir would equilibrate, and the system would have approached a state of energy exhaustion where no useful work could be extracted. Thus, one may consider the entropy of a system as an indicator of whether or not useful work can be extracted from it.

11.9 Free Enthalpy

In the past section, we have devised a criterion for the spontaneity of a chemical reaction, or, in fact, any process, that occurs spontaneously. This criterion was based on the entropy change of the entire universe. This criterion is always correct, so it is truly a signpost for spontaneity. However, it is somewhat of a pain to discuss a chemical reaction in terms of the entropy change of the universe, rather than defining a criterion that holds for the system. We can achieve this by defining the free enthalpy, and we go about it as follows. We start with

$$\Delta S_{universe} = \Delta S_{system} + \Delta S_{surroundings} \tag{11.77}$$

and multiply Eq. 11.77 by $(-T)$:

$$-T\Delta S_{universe} = -T\Delta S_{system} - T\Delta S_{surroundings} \tag{11.78}$$

Let us consider an ex-enthalpic reaction (although the argument works equally well for an endoenthalpic reaction). The reaction enthalpy, ΔH_{system}, is released into the surroundings. Thus, we can write the entropy gain of the surroundings as

$$T\Delta S_{surroundings} = -\Delta H_{system} \tag{11.79}$$

Substituting Eq. 11.79 into Eq. 11.78, we get

$$-T\Delta S_{universe} = \Delta H_{system} - T\Delta S_{system} \tag{11.80}$$

The right-hand side of Eq. 11.80 only contains quantities related to the system. We now define

$$\Delta G_{system} = -T\Delta S_{universe}, \text{ and obtain} \tag{11.81}$$

$$\Delta G_{systwm} = \Delta H_{system} - T\Delta S_{system} \tag{11.82}$$

This is the definition of the free enthalpy, which was introduced qualitatively earlier in this chapter. Since we are dealing here with quantities that refer to the system, it is customary to drop the subscript "system" and present Eq. 11.82 as

$$\Delta G = \Delta H - T\Delta S \tag{11.83}$$

Since the free enthalpy was defined as

$$\Delta G_{system} = -T\Delta S_{universe} \tag{11.80}$$

It follows that

$\Delta G < 0$ corresponds to a spontaneous process
$\Delta G = 0$ to an equilibrium situation, and (11.84)
$\Delta G > 0$ to a process that occurs spontaneously in the opposite direction

This free enthalpy is generally referred to as "Gibbs free enthalpy." There also exists Helmholtz *free energy* (or simply referred to as the free energy) and defined as

$$\Delta A = \Delta E - T\Delta S \tag{11.85}$$

that holds for processes carried out at constant volume and therefore is based on the internal energy, ΔE, rather than the enthalpy (see discussion in Section 11.5). Most of the following discussion centers around the *free enthalpy* since many processes in chemistry and biochemistry are taking place at constant pressure, but this discussion can easily be adopted for the *free energy* as well.

There are several important considerations about the free enthalpy. First of all, ΔG truly is the system's criterion for a spontaneous reaction: a negative ΔG always implies a spontaneous reaction. A negative ΔG can be caused by a strongly negative enthalpic term, or a strongly positive entropic term. Thus, reactions will be spontaneous if they are ex-enthalpic and proceed with a large positive entropy change. The frequently quoted statement that a reaction "goes toward lowest energy and highest entropy," however, is incorrect since it would not allow a reaction to proceed that is endoenthalpic, or has negative entropy change. The statement should read "reactions proceed toward lower free enthalpy. Lower free enthalpy can be caused by lowered enthalpy **or** increased entropy."

The last aspect of ΔG (and ΔA) actually has to do with the name, "free enthalpy" (or "free energy"). The term "free" was utilized to indicate that ΔG actually specifies the part of an energy change of a reaction or process that can be harnessed to create work. Here, the term "energy" is a measure of a process' ability to cause change. If you are lifting a weight, energy in the form of potential energy is obtained by metabolism in your muscles. This energy conversion proceeded by converting some of the metabolic enthalpy into work, while some of it was lost in the form of heat, as defined in Eq. 11.70 as $T\Delta S$. Thus, ΔG is the difference between the change in metabolic enthalpy and the heat lost during the process to the surroundings and determines the maximum work (change in useful energy) a process can provide.

11.10 Free Enthalpy and Equilibrium

For any given reaction, the change in ΔS and ΔG at standard conditions, known as $\Delta S°$ and $\Delta G°$, can be calculated from tabulated data just as the standard enthalpies can be calculated from enthalpies of formation. Notice however, that standard entropies for elements are not zero at standard conditions, since they represent absolute values that depend on the temperature. Thus, entropy of formation values of elements in their standard state are nonzero.

Next, we wish to establish one of the most fundamental aspects of thermodynamics, a relationship between ΔG, $\Delta G°$, and the equilibrium constant of a chemical reaction. This will be accomplished using an intuitive thought process that is not fully rigorous to relate the standard free enthalpy to chemical equilibrium, and requires a leap-of-faith at one point.

We showed in Eq. 11.66 that entropy can be expressed as

$$S = (R/N_A) \ln \Omega, \tag{11.66}$$

where Ω is the number of ways a system can be arranged. Thus, for an expansion of an ideal gas from V_1 to V_2, (where $V_2 > V_1$), it is reasonable to accept that

$$\Delta S = R \ln V_2/V_1 \tag{11.86}$$

since a larger volume for the same amount of gas offers more microstates, i.e. the gas can arrange itself in more different ways. Since
$V_2/V_1 > 1$, it follows that

$$\Delta S = R \ln V_2/V_1 > 0 \tag{11.87}$$

Since $p = nRT/V$, we may write

$$\Delta S = R \ln p_1/p_2 \tag{11.88}$$

The expansion of an ideal gas proceeds spontaneously ($\Delta S_{univ} > 0$) without change in energy E nor heat q; therefore, we find that

$$\Delta S_{univ} = \Delta S_{sys}$$
$$\Delta G_{sys} = -T\Delta S_{univ} = -RT \ln p_1/p_2 = RT \ln p_1/p_2 < 0 \tag{11.89}$$

We define ΔG, like ΔH, with respect to a standard state, such as a gas at 1 [atm]:

$$\Delta G = RT \ln p/p_o \tag{11.90}$$

Here comes a leap of faith: if we can define ΔG for the expansion of an ideal gas in terms of an initial state, p, and a standard state, p_o, we may similarly define ΔG for chemical reaction as the quotient of concentrations at initial conditions, Q, and a quotient of concentrations at equilibrium, K, and get in analogy to Eq. 11.89

$$\Delta G = RT \ln Q/K \text{ or } \Delta G = RT \ln Q - RT \ln K \tag{11.91}$$

where the quotient at initial conditions was previously defined (Eq. 9.56).

For a reaction A + B → C, Q, and K are

$$Q = [C]_i/[A]_i[B]_i \text{ and} \tag{9.56}$$
$$K = [C]_e/[A]_e[B]_e \tag{9.55}$$

where the subscripts "i" and "e" stand for initial and equilibrium concentrations, respectively. The standard state, corresponding to $\Delta G°$, is defined with all concentrations or pressures to be 1 [M] or 1 [atm]; therefore, [A] = [B] = [C] = 1. Thus, Q = 1 and ln Q = 0.

$$\Delta G° = -RT \ln K, \text{ or } K = e^{-\frac{\Delta G°}{RT}} \tag{11.92}$$

Suddenly, it appears that $\Delta G°$ is nothing but a glorified equilibrium constant: if $\Delta G°$ is a positive number, say 10 [kJ/mol], K will be small (0.0175), indicating that the equilibrium lies on the side of the reactants.

Example 11.10 For the reaction A + B → C, verify the value of the equilibrium constant at 25 [°C] for $\Delta G° = 10$ [kJ/mol] and discuss the equilibrium concentrations.

Answer: K = 0.0175 [C] is small, [A] and [B] are large for K to be small

For a chemical reaction, $\Delta G°$ can be calculated from tabulated data according to

$$\Delta G°_{reaction} = \Sigma_i \{\Delta G°_f(\text{product})\}_i - \Sigma_j \{\Delta G°_f(\text{reactant})\}_j \tag{11.93}$$

in analogy to the way we calculated standard enthalpies from enthalpies of formation:

$$\Delta H°_{reaction} = \Sigma_i \{\Delta H°_f(\text{products})\}_i - \Sigma_j \{\Delta H°_f(\text{reactants})\}_j \tag{11.45}$$

This is extremely useful: for any given reaction, we can calculate what the concentrations are at equilibrium, from a tabulation of thermodynamic values of reactants and products. Furthermore, the standard state $\Delta G°$ can be calculated from Eq. 11.94

$$\Delta G° = \Delta H° - T \Delta S° \tag{11.94}$$

We can write Eq. 11.91 as

$$\begin{aligned}\Delta G = RT \ln Q/K &= RT \ln Q - RT \ln K \\ &= RT \ln Q + \Delta G°,\end{aligned} \tag{11.95}$$

$$\boldsymbol{\Delta G - \Delta G° = RT \ln Q} \tag{11.96}$$

This equation is the Holy Grail of thermodynamics, because it tells us, for any reaction and any initial concentrations, in which direction the reaction will proceed to reach equilibrium. Let us try to represent what we have done in the form of a picture (see Figure 11.11). Consider the chemical reaction:

$$A_2 + B_2 \rightarrow 2AB$$

(for example, the hydrogen–fluorine–hydrogen fluoride reaction discussed in Chapter 9 on equilibria). We have seen that, regardless of the starting conditions, the reaction will reach an equilibrium. At the equilibrium, the ratio of $[AB]^2/[A_2][B_2]$ is a constant (the "equilibrium constant") that only depends on T, and not on the starting conditions. This equilibrium condition is described by K or $\Delta G°$. This is shown in Figure 11.11. Here, we plotted the composition of all participants of the reaction on the abscissa: on the left, we indicate a mixture of $A_2 + B_2$, with no AB present. On the right, we indicate pure AB, with no A_2 or B_2 present. In between these two limits is a range where we have mixtures of all three species. Thus, the x-axis represents all possible concentration ratios of products and reactants. Along the ordinate, we display the free enthalpy of the system. The equilibrium between all three species is represented by the minimum free enthalpy, $\Delta G°$. Whether we approach the minimum free enthalpy from the left or the right depends on the starting conditions: if we start with only $A_2 + B_2$, AB needs to be formed to reach

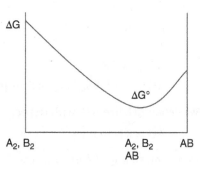

Figure 11.11 Free enthalpy representation of the reaction $A_2 + B_2 \rightarrow 2$ AB.

equilibrium, that is the reaction proceeds forward (to the right) to reach equilibrium. If we start with pure AB, the reaction occurs backward, because ΔG is positive. Remember, $\Delta G > 0$ implies the reaction occurs in opposite direction as written. Thus, AB will decompose to form $A_2 + B_2$. Thus, we see that ΔG may switch sign, depending on the starting condition, and indicates in which direction the reaction will occur to reach equilibrium.

At this point, it is very important to clarify a point we will encounter in the next chapter when we discuss standard cell potentials, $\Delta\varepsilon°$, or when browsing through some less rigorously defined discussion on energetics and spontaneity. Often, one may find a statement that a large negative $\Delta G°$-value is a sign of spontaneity. This is particularly so in some older books on bioenergetics, in which the $\Delta G°$-values for biochemical reactions, such as those in the tricarboxylic acid cycle, are used to indicate the spontaneity of a reaction. This is incorrect because of the fact that the direction of spontaneity depends on the starting conditions. However, if the starting conditions are the standard state i.e. all concentrations are 1 [mol/L] or 1 [atm] at 25 [°C], then

$$\Delta G - \Delta G° = RT \ln Q = RT \ln 1 = 0 \text{ and} \qquad (11.97)$$
$$\Delta G = \Delta G° \qquad (11.98)$$

In other words, if we start the reaction with all concentrations at 1 [mol/L] or 1 [atm], the reaction will proceed to the position indicated in Figure 11.11, and the equilibrium will lie far on the right side for large, negative values of $\Delta G°$.

Example 11.11 Now that we have established that ΔG, and not $\Delta G°$, is the sign post for spontaneity of a reaction, let us backpedal and establish under which conditions $\Delta G°$ may be used as an indicator of spontaneity.

Answer: See answer booklet

Since the discussion above may sound a bit confusing, let us go through an example of a chemical reaction, at different temperatures and with different initial concentrations, and see where we end up. This example, in all its gory detail, really should help understand the principles discussed so far. As usual, detailed answers to this problem are given in the answer booklet, but even here in the text, more details are provided to aid in the understanding of the principles involved.

Example 11.12 Consider the decomposition reaction of phosphorous pentachloride into phosphorous trichloride and chlorine according to

$$PCl_5(g) \rightarrow PCl_3(g) + Cl_2(g)$$

The thermodynamic data for all species are given in Table 11.2.

(a) From the thermodynamic data given, calculate $\Delta G°$ and K at 25 [°C].
(b) Calculate $\Delta G°$ and the equilibrium constant at 250 [°C]. Assume that $\Delta H°_f$ and $\Delta S°$ are temperature-independent.
(c) Does the equilibrium lie on the side of the products or reactants at room temperature? Can you comment on the spontaneity of the reaction?
(d) Do the values of $\Delta G°$ (at the different temperatures) agree with your expectation? Explain.
(e) Calculate ΔG for the following starting conditions (given in partial pressures) at 250 [°]C: $p(PCl_5) = 0.35$ [atm]; $p(PCl_3) = p(Cl_2) = 0.1$ [atm]. In which direction will the reaction proceed to reach equilibrium?
(f) What will be the partial pressures at equilibrium with the starting conditions given in (e)?

Table 11.2 Thermodynamic data of PCl_5, PCl_3, and Cl_2.

	PCl_5	PCl_3	Cl_2	
$\Delta H°_f$	−374	−287	0	[kJ/mol]
$S°$	365	312	223	[J/mol K]

Answers:

(a) $\Delta H°_{reaction} = \Delta H°_f(PCl_3) + \Delta H°_f(Cl_2) - \Delta H°_f(PCl_5)$
$= 87 \,[kJ/mol]$. The reaction is endoenthalpic.
$\Delta S°_{reaction} = S°(PCl_3) + S°(Cl_2) - S°(PCl_5)$
$= 170 \,[J/K \,mol]$. The reaction proceeds with an increase in entropy
$\Delta G° = \Delta H° - T\Delta S°$
$= 87 - 298 \cdot 0.170 \,[\{kJ - K(kJ/K)\}/mol] = 36.3 \,[kJ/mol]$
$K = 4.24 \cdot 10^{-7}$. The positive value of $\Delta G°$ and the small value of K, indicate that the equilibrium lies on the side of the reactant PCl_5

(b) $\Delta G°_{250} = \Delta H° - T\Delta S° = 87 - 523 \cdot 0.170 = -1.9 \,[kJ/mol]$. At 250 [°C] the value of $\Delta G°$ is negative.
$K = 1.55$. At 250 [°C], the equilibrium constant is just a bit above 1. Thus, we have more PCl_3 and Cl_2 than PCl_5.

(c) At room temperature, $\Delta G°$ is positive, therefore, K is very small, $4.24 \cdot 10^{-7}$. Thus, the equilibrium lies far on the side of the reactant, PCl_5.
At elevated temperature, $\Delta G°$ becomes negative, since the entropy term dominates. The equilibrium lies on the side of the products, PCl_3 and Cl_2.
Nothing can be said about the spontaneous direction of the reaction which depends on the starting conditions!!!

(d) The reaction produces two moles of gaseous products from one mole of reactant. So, we expect the reaction to proceed with a positive entropy change. At 250 °C, $\Delta G°$ is negative since the $-T\Delta S$ term becomes dominant. $\Delta G°$ behaves just as expected!

(e) $p(PCl_5) = 0.35$ atm; $p(PCl_3) = p(Cl_2) = 0.1$ atm. In which direction will the reaction proceed to reach equilibrium?

$\Delta G - \Delta G° = RT \ln Q$

$\Delta G_{250} = \Delta G°_{250} + RT \ln [p(PCl_3) \,p(Cl_2)/p(PCl_5)]$
$= -1900 + 8.3 \cdot 523 \,\ln[0.1 \cdot 0.1/0.35] = -17.3 \,kJ/mol$

The reaction occurs spontaneously as written, namely

$PCl_5(g) \rightarrow PCl_3(g) + Cl_2(g)$

(f) We set up our table of initial and equilibrium concentrations, as follows:

	PCl_5	PCl_3	Cl_2
Initial	0.35	0.1	0.1
Equilibrium	0.35 − x	0.1 + x	0.1 + x

$K = 1.55 = (0.1 + x)^2/(0.35 - x); \,x = 0.264$

	PCl_5	PCl_3	Cl_2
Initial	0.35	0.1	0.1
equilibrium	0.086	0.364	0.364

Check whether or not these concentrations fulfill the equilibrium constant:

$0.364^2/0.086 = 1.54 \approx 1.55$

Further Reading

Thermochemistry (Sections 11.1–11.7) is nicely summarized in https://chem.libretexts.org/Bookshelves/General_Chemistry/Map%3A_Chemistry_-_The_Central_Science_(Brown_et_al.)/05%3A_Thermochemistry and its subsections.
For sections 10.8 through 10.10, see
https://en.wikipedia.org/wiki/Entropy
https://en.wikipedia.org/wiki/Thermodynamic_free_energy

12

Reduction–Oxidation (Redox) Reactions and Electrochemistry

In the previous chapters, we have discussed processes that can produce work. After all, it is nice to press the button in an elevator, rather than work up four floors to your destination. The work done by the elevator motor, electric work, can be calculated as discussed before in Chapter 11. The electric energy required for the elevator to run comes from the combustion of fossil fuel, from hydroelectric and wind energy, or from nuclear reactions. In the first of these, the combustion of fossil fuels, chemical energy is converted to heat, as discussed in the chapter on thermodynamics. This heat is used to produce superheated steam, which drives a turbine, to convert heat into kinetic energy. The kinetic energy, in turn, is used to produce electric energy.

According to the first law of thermodynamics, it is possible to convert energy from one form to another. However, the detour of using a thermal process – i.e. the production of steam and motion – is quite inefficient, because only a relatively small amount of the energy is converted, and a large fraction of the energy is lost to random energy or entropy. Thus, it would be better to convert the chemical energy to electric energy without the detour of heat, steam, and kinetic energy, and keep the temperature as constant as possible in the course of the energy conversion.

This can be accomplished by using the principles of electrochemistry, which permit the direct conversion of chemical energy to electric energy. This conversion proceeds with high efficiency: when you use an AA battery, electric energy is produced from a chemical reaction. A small amount of heat is produced, which represents the entropy term necessary for the conditions of spontaneity we discussed before, but the majority of the chemical energy is converted to electric energy.

This conversion of chemical energy to electric energy is intimately connected with the transfer of electrons, a process that is referred to as oxidation and reduction. In the next section, we shall define these processes, i.e. oxidation and reduction, in more detail. For the time being, let it suffice to state that the gain of electrons by an atom or ion is a reduction, and the loss of electrons by an atom or ion is an oxidation. These processes will be demonstrated using a simple electrochemical reaction, which takes place when elementary aluminum (a small piece of aluminum foil) is dropped into an aqueous solution of copper (II) chloride, $CuCl_2$.

In this case, a spontaneous reaction occurs in which the aluminum foil dissolves (disappears), and reddish-brown copper is deposited on the remaining aluminum foil, or as a spongy material from the solution. The overall chemical reaction can be written as

$$2Al(s) + 3CuCl_2(aq) \rightarrow 2AlCl_3(aq) + 3Cu(s) \tag{12.1}$$

Since the chloride ions are just spectator ions that do not undergo oxidation or reduction, we can write a "net ionic equation"

$$2Al(s) + 3Cu^{2+}(aq) \rightarrow 2Al^{3+}(aq) + 3Cu(s) \tag{12.2}$$

Notice that in the net ionic equation, the atoms as well as the charges are balanced.

From our definition of oxidation and reduction above, we may state that for the reaction in Eq. 12.1, aluminum is oxidized, since it loses electrons, whereas Cu^{2+} is reduced to Cu by gaining electrons. To show the gain and loss of electrons more clearly, we can break down reaction 12.2 into "half-reactions" that explicitly account for the electrons:

$$2Al(s) \rightarrow 2Al^{3+}(aq) + 6e^- \quad \text{(oxidation)} \tag{12.3}$$

$$3Cu^{2+}(aq) + 6e^- \rightarrow 3Cu(s) \quad \text{(reduction)} \tag{12.4}$$

Thus, we may refine our definition of oxidation and reduction as follows: an oxidation half-reaction produces electrons, whereas a reduction reaction consumes them. Notice that the individual half-reactions could be written as

Understanding Essential Chemistry, First Edition. Max Diem.
© 2025 John Wiley & Sons, Inc. Published 2025 by John Wiley & Sons, Inc.

$$\text{Al(s)} \rightarrow \text{Al}^{3+}(\text{aq}) + 3\,e^- \quad \text{(oxidation)} \tag{12.5}$$

$$\text{Cu}^{2+}(\text{aq}) + 2\,e^- \rightarrow \text{Cu(s)} \quad \text{(reduction)} \tag{12.6}$$

but that the stoichiometric coefficients in Eqs. 12.3 and 12.4 are necessary to balance charges, atoms, and electron flow in the overall equation.

This simple example teaches a lot about the principles of electrochemistry, but two questions pop up immediately: First, was not our goal to produce electrical energy from the reaction? If so, where is the electrical energy? By reacting aluminum directly with copper (II) chloride, we do not seem to get anything you may call an electrochemical cell, or "battery (dry cell)." This question will be answered in Section 12.2.

The second question is aimed at chemical reactions where the electron flow cannot be established as easily as in the aluminum/copper (II) chloride reaction. Let us take the chemical reaction that is similar to that occurring in the AA battery (see Section 12.3) you buy in any store. This reaction can be represented by the *(unbalanced)* chemical process

$$\text{Zn} + \text{KMnO}_4 \rightarrow \text{ZnO} + \text{MnO}_2 \tag{12.7}$$

Which elements are oxidized, and how do we define the electron flow? To answer this question, we need to define an "electron book-keeping" procedure based on oxidation numbers.

12.1 Oxidation State and Oxidation Numbers: Balancing Redox Equations

Consider a simple chemical reaction we discussed on a number of occasions: the reaction of solid, metallic sodium with elementary chlorine to form ionic sodium chloride:

$$\text{Na(s)} + \tfrac{1}{2}\text{Cl}_2(\text{g}) \rightarrow \text{NaCl(s)} \quad [= \text{Na}^+ + \text{Cl}^-] \tag{12.8}$$

Obviously, the sodium in sodium chloride is a cation, and the chloride is an anion. Thus, the reaction in Eq. 12.8 can be broken down into two "half-reactions," which are useful to track the transfer of electrons:

$$\text{Na(s)} \rightarrow \text{Na}^+ + e^- \tag{12.9}$$

$$\tfrac{1}{2}\text{Cl}_2(\text{g}) + e^- \rightarrow \text{Cl}^- \tag{12.10}$$

Notice that the sum of Eqs. 12.9 and 12.10 yields Eq. 12.8, since the electron occurs on both sides and cancels. We call the "half-reaction" (Eq. 12.9), the production of an electron, an oxidation, and the "half-reaction" in Eq. 12.10, the consumption of an electron, a reduction reaction.

The concept introduced here, the gain and loss of electrons, is relatively clear when the product is an ionic compound, which consists of cations and anions. What about the situation when the product is a molecular compound, such as carbon dioxide in the following reaction, the combustion of coal to carbon dioxide:

$$\text{C(s)} + \text{O}_2 \rightarrow \text{CO}_2(\text{g}) \tag{12.11}$$

In order to keep track of the electron balance, we introduce the concept of oxidation numbers. Please note that oxidation numbers are an artificial concept not associated with a real, measurable quantity such as electron affinity or ionization potential. Rather, oxidation numbers are assigned artificially using a set of rules that allow us to track the flow of electrons, and therefore, define the species that are oxidized and reduced.

The rules for assigning oxidation numbers are presented here in abbreviated form, and your main textbook has more details. The essence of these rules is as follows:

- Elements in their elementary form have oxidation numbers of zero.
- In a monatomic ion, the oxidation number is the same as the ionic charge.
- Oxygen in a compound has an oxidation number of (-2). There are a few exceptions to this rule, such as in OF_2 (oxygen has an oxidation number of $+2$, since F is more electronegative than oxygen) and in the peroxide ion, O_2^{2-}, where its oxidation number is -1. For the element hydrogen, the oxidation number is ($+1$), with a few exceptions. These are the alkali metal hydrides such as LiH, where H has an oxidation number of -1. Similarly, in BeH_2, H has an oxidation number of -1.
- The sum of all oxidation numbers in a compound is zero.
- The sum of all oxidation numbers in a polyatomic ion equals the ionic charge.

Thus, using these rules, we see that in Eq. 12.8, the oxidation numbers are as follows:

$$Na(s) + \tfrac{1}{2} Cl_2 \rightarrow NaCl(s) \tag{12.9}$$
Ox#   0   0   (+1) (−1)   (12.10)

and in Eq. 12.11

$$C(s) + O_2 \rightarrow CO_2(g) \tag{12.10}$$
Ox# 0   0   (+4) 2(−2)   (12.11)

In principle, we may look at this reaction as an oxidation–reduction ("redox") pair, where

$$C(s) \rightarrow C^{4+} + 4e^- \tag{12.12}$$

represents the oxidation half-reaction, because electrons are produced on the right-hand side of the equation. Similarly, the half-reaction

$$4e^- + O_2 \rightarrow 2O^{2-} \tag{12.13}$$

represents the reduction, since electrons appear on the side of the reactants and are consumed. Thus, we define a process where the oxidation number increases toward positive numbers an oxidation and a process where the oxidation number increases toward negative numbers a reduction.

Example 12.1 Determine the oxidation numbers of all elements in the chromate ion, CrO_4^{2-}, in formaldehyde, H_2CO, methanol, CH_3OH, hydrogen peroxide, H_2O_2, lithium cobalt oxide, $LiCoO_2$, lead dioxide, PbO_2, nitrate ion, NO_3^-, potassium dihydrogen phosphate, KH_2PO_4, lithium niobate, $LiNbO_3$, (the last two compounds being important optical materials), sodium silicate, Na_2SiO_3 (a major component of glass).

Answer: See answer booklet

We now turn to the subject of balancing redox equations, using the method of half-reactions. Redox equations are usually balanced as net ionic equations, omitting spectator ions that do not participate in oxidation or reduction. Using again the dry cell battery chemistry as an example, we need to balance the reaction given before:

$$Zn + KMnO_4 \rightarrow ZnO + MnO_2 \tag{12.7}$$

we write it as a (still unbalanced) ionic equation as follows:

$$Zn(s) + MnO_4^- \rightarrow Zn^{2+} + MnO_2 \tag{12.14}$$

Now, it is a bit easier to see who is getting oxidized and reduced. Next, we break the overall process into two separate half-reactions for the oxidation and reduction.

We see, by the rules of oxidation numbers and redox processes defined above, that Zn is being oxidized to Zn^{2+} ion. In terms of a balanced half-reaction, we write this as

$$Zn(s) \rightarrow Zn^{2+} + 2e^- \tag{12.15}$$

There are a number of important aspects about any redox half-reactions. They must be written such that

- The stoichiometry is balanced (i.e. all atoms occurring on the reactant side must occur on the product side).
- The overall charge is balanced.
- The change in oxidation number is accounted for by electron flow.

Eq. 12.15 fulfills these criteria: the increase in oxidation number of Zn to Zn^{2+} is accompanied by the release of two electrons, the charges are balanced (both sides of Eq. 12.15 present charge neutrality), and the number and nature of atoms are equal on both sides.

The reduction half-reaction is more complicated. Looking at the oxidation numbers of the species involved, we see that Mn changes from +7 in the permanganate ion to +4 in MnO_2:

$$MnO_4^- \rightarrow MnO_2 \tag{12.16}$$
(+7)   (+4)

with the oxygen maintaining its oxidation number of −2. The next step, after identifying the changes in oxidation number, is to balance the flow of electrons: for the manganese atom to go from +7 to +4, three electrons need to be added:

$$3\,e^- + MnO_4^- \rightarrow MnO_2 \tag{12.17}$$

The next step involves balancing the charges on both sides. We see that on the left-hand side of Eq. 12.17 we have four negative charges; on the right-hand side, we have neutrality. In addition, we see that the oxygen atoms are not balanced. Thus, we balance charges and atoms in one step, by adding H^+ or OH^- such as to balance the charges. If we do this step correctly, the overall stoichiometry is balanced as well. In order to reach charge balance in Eq. 12.17, we may add four positive charges in the form of $4\,H^+$ on the left-hand side:

$$3\,e^- + MnO_4^- + 4\,H^+ \rightarrow MnO_2 \tag{12.18}$$

We see that the charges are balanced, but the overall stoichiometry is not. The overall stoichiometry is balanced by assuming that the four protons react with two of the oxygen atoms from the MnO_4^- ion to form two water molecules:

$$3\,e^- + MnO_4^- + 4\,H^+ \rightarrow MnO_2 + 2\,H_2O \tag{12.19}$$

In Eq. 12.19, the electron flow, oxidation numbers, charges, and stoichiometry are properly balanced, and this equation represents a correctly balanced half-reaction for the reduction of the permanganate ion to manganese dioxide.

The process of balancing the charges with H^+ or OH^- ions holds only for chemical reactions that take place in aqueous solution. If chemistry is carried out in the solid phase, or in solvents other than water, different means for balancing redox equations apply. However, 99% of all electrochemical reactions we are concerned with take place in aqueous solution; thus, we need not concern ourselves with reactions other than those taking place in water. Whether to balance the charges of a half-reaction with H^+ or OH^- ions is, in most cases, unimportant. For example, we could balance Eq. 12.18

$$3\,e^- + MnO_4^- \rightarrow MnO_2 \tag{12.18}$$

by adding 4 OH^- ions on the right-hand side:

$$3\,e^- + MnO_4^- \rightarrow MnO_2 + 4\,OH^- \tag{12.20}$$

and balancing the stoichiometry by adding two water molecules on the left-hand side:

$$2\,H_2O + 3\,e^- + MnO_4^- \rightarrow MnO_2 + 4\,OH^- \tag{12.21}$$

Both Eqs. 12.19 and 12.21 represent totally legitimate ways to balance the same half-reaction. In practice, electrochemical reactions often are carried out in acidic or basic solutions. When balancing a half-reaction, the conditions (acidic or basic) are generally specified. When acidic conditions are specified, one would balance the half-reaction using protons; in basic conditions, one would use hydroxide ions.

Finally, we assemble the two balanced half-equations

$$Zn(s) \rightarrow Zn^{2+} + 2\,e^- \text{ and} \tag{12.15}$$

$$3\,e^- + MnO_4^- + 4\,H^+ \rightarrow MnO_2 + 2\,H_2O \tag{12.19}$$

into one overall, balanced redox equation. In doing so, we need to ascertain that the number of electrons produced and consumed are the same. We do so by finding the smallest multiple (6) for electrons consumed and produced:

$$3\,Zn(s) \rightarrow 3\,Zn^{2+} + 6\,e^- \tag{12.22}$$

$$6\,e^- + 2\,MnO_4^- + 8\,H^+ \rightarrow 2\,MnO_2 + 4\,H_2O \tag{12.23}$$

and adding up the half-equations:

$$3\,Zn(s) + 2\,MnO_4^- + 8\,H^+ \rightarrow 3\,Zn^{2+} + 2\,MnO_2 + 4\,H_2O \tag{12.24}$$

Eq. 12.24 would be considered a properly balanced, net ionic equation. In terms of counterions, the equation could be written, for example, as

$$3\,Zn(s) + 2\,HMnO_4 + 6\,HCl \rightarrow 3\,ZnCl_2 + 2\,MnO_2 + 4\,H_2O \tag{12.25}$$

However, for nearly all practical purposes, the form presented in Eq. 12.24 is the preferred description of the actual reaction that occurs; furthermore, it is the form required for the calculation of the cell voltage, to be discussed in the next section.

An interesting situation occurs when the same element is being oxidized and reduced in a chemical reaction. Consider, for example, the chemical reaction that occurs when the element chlorine, Cl_2, is bubbled through water at a slightly basic pH:

$$Cl_2 + H_2O \rightarrow Cl^- + ClO^- \tag{12.26}$$

In this example, we proceed as above by identifying the reduction and oxidation half-reaction, and balancing them separately:

$$2e^- + Cl_2 \rightarrow 2Cl^- \tag{12.27}$$

This reduction reaction is balanced as written.

The oxidation reaction is from elemental chlorine to hypochlorite, ClO^-, in which chlorine has an oxidation number of +1:

$$Cl_2 \rightarrow 2ClO^- + 2e^- \tag{12.28}$$

We start balancing Eq. 12.28 by adding 4 OH^- on the left-hand side, since the reaction occurs in basic solution

$$4OH^- + Cl_2 \rightarrow 2ClO^- + 2e^- \tag{12.29}$$

Balancing the stoichiometry, we find

$$4OH^- + Cl_2 \rightarrow 2ClO^- + 2e^- + 2H_2O \tag{12.30}$$

Combining Eqs. 12.27 and 12.30 yields

$$4OH^- + 2Cl_2 \rightarrow 2ClO^- + 2Cl^- + 2H_2O \tag{12.31}$$

which can be simplified to

$$2OH^- + Cl_2 \rightarrow ClO^- + Cl^- + H_2O \tag{12.32}$$

Reactions, in which the same element is oxidized and reduced, are known as disproportionation reactions. Here, chlorine goes from an oxidation state of zero to +1 and −1, i.e. it is both oxidized and reduced.

Example 12.2 Balance the following redox equations:

(a) $Cl_2 \rightarrow Cl^- + ClO_3^-$ (basic solution)
(b) $Zn(s) + NO_3^- \rightarrow NH_4^+ + Zn^{2+}$ (acidic solution)
(c) $Cr_2O_7^{2-} + Fe^{2+} \rightarrow Fe^{3+} + Cr^{3+}$ (acidic solution)
(d) $Cu(s) + HNO_3 \rightarrow Cu^{2+} + NO(g) + H_2O$
(e) $Fe(OH)_2 + O_2(g) + H_2O \rightarrow Fe(OH)_3$

Answer: See answer booklet

12.2 Galvanic Cells, Electric Work, and Electromotive Force

What we have achieved so far in our discussion – the description of oxidation and reduction processes, and the bookkeeping of electron flow – is only one aspect of what we set out to do: the conversion of chemical energy into electrical energy. Up to now, in our examples of the Al/Cu reaction, we did not generate useful work in the form of an electric voltage or current, but described a spontaneous reaction that changed the oxidation numbers of a few elements.

What do we need to do to have a chemical reaction do work, such as powering your calculator? First, we must answer the question: what is electrical work or electric energy? Let us go back to the chapter on thermochemistry (Eq. 11.6):

$$w_{el} = \varepsilon \cdot i \cdot t \quad \text{(voltage} \cdot \text{current} \cdot \text{time)}, \tag{11.6}$$

where ε is the voltage (measured in volt [v]), i is the current (measured in ampere [A]), and t is the time in seconds [s].

1 [v·A·s] (volt ampere second) is also referred to as 1 [W·s] (watt second), where a watt is the unit of electric power and is the product of voltage times current. Furthermore, 1 [v·A·s] = 1 [J]. Bulk electricity is purchased by the kWh (kilowatt hour), with 1 [kWh] = 3 600 000 [W·s] = 3 600 000 [J] = 3600 [kJ].

The term "current" implies that something is flowing, and indeed, current can be described as the flow of charge. In fact, the current i is defined as:

$$i = Q/t \tag{12.33}$$

where Q is the symbol for electric charge. In this book, a script Q is used for electric charge, a regular, uppercase Q for the concentration quotient (see Chapter 9), and a regular, lowercase q for heat. Just as one would describe the current of water in a pipe in terms of gallons of water per second, we describe electric current as a flow of electrons; thus, electric current has units of electric charge per second. We encountered electric charge before, in our discussion of elementary particles, when we defined the charge of an electron to be $1.6022 \cdot 10^{-19}$ Coulomb, abbreviated [C] (see Table 2.1 in Chapter 2).

This is a small number; however, chemists deal in molar quantities. Thus, we designate the charge of one mole of electrons as 1 Faraday [F].

$$1F = N_A \cdot 1.6022 \cdot 10^{-19} = 6.023 \cdot 10^{23} \cdot 1.6022 \cdot 10^{-19} = 96{,}485 \, [C/mol] \tag{12.34}$$

The current of 1 ampere corresponds to a flow of charge of one mole of electrons through a conductor, per second. Thus, we write the electric work as

$$w_{el} = \varepsilon \cdot i \cdot t = \varepsilon (Q/t)t = \varepsilon Q \, [vC = J] \tag{12.35}$$

The purpose of the discussion, so far, was three-fold. First, we demonstrated that electric current is, indeed, measured in the flow of electrons (1 mole of electrons/second), and thus, can be related to chemically relevant aspects (how many moles of electrons can be pushed through a wire). Second, the discussion so far shows that, in order to get electrical work out of a chemical reaction, one needs to catch the electrons and pass them through a wire in order to get a flow of current. This aspect was, obviously, absent when we dropped aluminum foil into a solution of copper (II) chloride. Third, if we catch the electrons given up by the oxidation half-reaction, and pass them through a wire to the reduction half-reaction (where they are used up), we need to "push" them through the wire. The force, or potential, responsible for the flow of electrons through a wire is the voltage ε, which is also referred to as "electromotive" force (EMF).

In order to understand the concept of voltage and current, we may resort to an analog situation of water flow in a household system. The pressure inside a water pipe in typical household plumbing is about 30 [psi]. This corresponds to the voltage of the "hot" wire in an electric outlet (115 volt). This pressure or voltage exists even if no current is flowing. Water can be made to flow by opening a faucet, and a current of water, measured in gallons/second, is obtained. This current is larger the higher the pressure inside the pipe, and the larger the opening in the pipe. Similarly, an electric current can flow if we provide a path for it to flow to a lower potential. Such a path would be a wire. The larger the diameter of a wire, and the higher the voltage, the more current will flow. The diameter of the wire is a quantity related to the "resistance" of the wire.

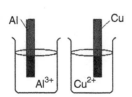

Figure 12.1 Spatial separation of oxidation and reduction reaction.

Thus, we need to establish the voltage ε created by an electrochemical reaction, such as

$$2Al(s) + 3Cu^{2+}(aq) \rightarrow 2Al^{3+}(aq) + 3Cu(s) \tag{12.1}$$

As indicated, we need to separate the electron-producing from the electron-consuming half-reactions. This can be achieved by using two containers, in which the oxidation and reduction half-reactions occur separately. For the example of the Cu/Al reaction, one of the containers is filled with a solution of a soluble aluminum salt such as $AlCl_3$. Into this solution, we immerse a solid electrode of aluminum. The other container is filled with a solution of copper (II) chloride, with an electrode of solid copper immersed in it. This is shown schematically in Figure 12.1. In this way, we have physically separated the electron-producing from the electron-consuming half-reactions.

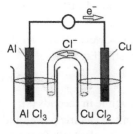

Figure 12.2 Schematic of a Cu/Al galvanic cell.

The solid aluminum electrode is oxidized, giving up electrons. Copper ions are reduced to produce solid copper, and require electrons. Thus, we connect the two metal electrodes by a wire, enabling the electrons to flow. This is shown in Figure 12.2. However, as the chemical reaction occurs, solid Al gets oxidized to Al^{3+}, which dissolves. Thus, positive charges are produced in the $Al/AlCl_3$ beaker. In order to avoid the left beaker to become positively charged, some negative charges need to flow into it. Similarly, the right-hand side, the $Cu/CuCl_2$ beaker, would become more negatively charged, since electrons flow toward the copper, and copper ions are reduced to copper, leaving Cl^- ions behind. Thus, to avoid the right-hand beaker to become negatively charged, the excess negatively charged chloride ions have to flow out of it. We achieve this by devising a "salt bridge" that allows the flow of chloride ions from the copper side to the aluminum side without mixing the salt solutions. This salt bridge allows ions to migrate without mixing the aluminum chloride and copper chloride, and could be a tube filled with an aqueous solution of chloride ion.

The experimental arrangement shown in Figure 12.2 is often referred to as a galvanic cell. We call the Al electrode at which electrons are produced and at which oxidation occurs the anode and give it a negative sign. The copper electrode, at which reduction occurs and where electrons are consumed, is the cathode and is designated by a positive sign. Notice that an electric "load" with a resistance (shown as the circle) was drawn into the wire connecting the cathode and the anode; this load could be a resistor, a light bulb, or an electric motor. Without such a load, the electric discharge would occur infinitesimally fast, and the reaction would be over in a very short time.

We have achieved what we set out to do: we have devised a system that takes the same chemical reaction that occurs spontaneously when we drop aluminum foil into copper (II) chloride solution, and modified it such that useful electric work is obtained. In order to quantify the work, we need to determine the voltage produced by the cell. This can be done experimentally by connecting a volt meter between the cathode and the anode, and observe a cell voltage of 1.10 [v]. This voltage is referred to as $\Delta\varepsilon°$ if the reaction is carried out at standard conditions (all solution concentrations at 1 molar, and room temperature and atmospheric pressure).

Standard cell potentials were measured for a large number of electrochemical cells consisting of different oxidation/reduction pairs. From these observations, a table was constructed that related the tendency of a given half-reaction to interact with other half-reactions. In other words, from the observed *standard cell potential*, a table of *potentials for half-reactions* was constructed such that a standard cell potential could be predicted by combining appropriate half-reactions.

These half-reactions are referenced to the hydrogen ion/hydrogen gas reduction half-reaction, which is arbitrarily assigned the standard potential of 0.00 [v] at 25 [°C]:

$$H^+ (1.0 [M]) + e^- \rightarrow \tfrac{1}{2} H_2 (g, 1 [atm]) \tag{12.36}$$

Just as we referenced standard enthalpies of formation to elements in their standard state, we reference electrochemical potentials of all half-reactions to the standard hydrogen electrode. This electrode can be reproduced experimentally reasonably easily: Hydrogen, at 1 [atm] pressure, is bubbled through an aqueous solution that is 1.0 [M] in H^+. A platinum mesh serves both as a catalyst, and to conduct the electron that is liberated in the process. Thus, the potentials of half-reactions reported in Table 12.1 were obtained by measuring each metal's oxidation or reduction potential against the hydrogen electrode and reporting the observed voltages. By custom, all these potentials are listed for a reduction process, and are therefore referred to as "standard reduction potentials" (SRPs) and given the symbol $\varepsilon°$. If the half-reaction listed occurs as an oxidation in a given redox equation, its potential has to be reversed and the direction of the reaction is reversed as well.

Given this table, it is easy to determine the cell potentials for the reaction given by Eq. 12.1. We proceed as follows, starting with the half-reactions 12.37 and 12.38:

Table 12.1 Standard reduction potentials for a few common half-reactions[1]

Reaction	SRP (v)
$O_2 + 4 H^+ + 4 e^- \rightarrow 2 H_2O (l)$	1.23
$Cu^{2+}(aq) + 2 e^- \rightarrow Cu(s)$	0.34
$H^+ (aq) + e^- \rightarrow \tfrac{1}{2} H_2 (g)$	0.00
$Fe^{2+}(aq) \; 2 e^- \rightarrow Fe(s)$	−0.45
$Zn^{2+}(aq) + 2 e^- \rightarrow Zn(s)$	−0.76
$Al^{3+}(aq) + 3 e^- \rightarrow Al(s)$	−1.66
$Li^+(aq) + e^- \rightarrow Li(s)$	−3.04

1 The list of SRPs presented in Table 12.1 is, of course, a very abbreviated one, and you will find a detailed and comprehensive listing of these potentials in the chapter on electrochemistry in your main text, or online.

$$Cu^{2+}(aq) + 2e^- \rightarrow Cu(s) \qquad \varepsilon^\circ = 0.34 \tag{12.37}$$

$$Al^{3+}(aq) + 3e^- \rightarrow Al(s) \qquad -1.66 \tag{12.38}$$

Since we know that the reaction occurs in the direction that dissolves solid Al into Al^{3+}(aq) ions, we reverse Eq. 12.38 and the SRP:

$$Cu^{2+}(aq) + 2e^- \rightarrow Cu(s) \qquad 0.34 \tag{12.39}$$

$$Al(s) \rightarrow Al^{3+}(aq) + 3e^- \qquad 1.66 \tag{12.40}$$

In order to balance the overall redox equation, we multiply the half-reactions (but not the potentials) by 3 and 2, respectively. The potentials are not multiplied since the number of electrons transferred in the process are accounted for by the number "n" in the equation for the electric work, $w_{el} = n \cdot \Delta\varepsilon^\circ \cdot F$.

$$3\,Cu^{2+}(aq) + 6e^- \rightarrow 3Cu(s) \qquad 0.34 \tag{12.41}$$

$$2Al(s) \rightarrow 2\,Al^{3+}(aq) + 6e^- \qquad 1.66 \tag{12.42}$$

and add the half-reactions equations:

$$3\,Cu^{2+}(aq) + 2Al(s) \rightarrow 3Cu(s) + 2\,Al^{3+}(aq) \qquad \Delta\varepsilon^\circ = 2.00\,[v] \tag{12.43}$$

Another example of a galvanic cell is the Cu/Zn reaction:

$$Zn(s) + Cu^{2+}(aq) \rightarrow Zn^{2+}(aq) + Cu(s) \qquad \Delta\varepsilon^\circ = 1.10\,[v] \tag{12.44}$$

for which the cell potential was obtained by reversing the half-reaction:

$$Zn(s) \rightarrow Zn^{2+}(aq) + 2e^- \qquad 0.76 \tag{12.45}$$

and add it to the copper reduction half-reaction:

$$Cu^{2+}(aq) + 2e^- \rightarrow Cu(s) \qquad 0.34 \tag{12.37}$$

Here, we have reversed the reduction reaction of zinc,

$$Zn^{2+}(aq) + 2e^- \rightarrow Zn(s) \qquad -0.76 \tag{12.46}$$

and ended up with a positive value of $\Delta\varepsilon^\circ$ (Eq. 12.44). However, what would happen if we picked the Cu half-reaction to reverse?

$$Cu(s) \rightarrow Cu^{2+}(aq) + 2e^- \qquad -0.34 \tag{12.47}$$

$$Zn^{2+}(aq) + 2e^- \rightarrow Zn(s) \qquad -0.76 \tag{12.46}$$

Adding the two half-reactions, we obtain for the reverse reaction

$$Zn^{2+}(aq) + Cu(s) \rightarrow Zn(s) + Cu^{2+}(aq) \qquad \Delta\varepsilon^\circ = -1.10\,[v] \tag{12.48}$$

We define (with a caveat) that a positive value of $\Delta\varepsilon^\circ$ indicates the spontaneous direction. The caveat of this statement is analogous to the discussion in Section 11.10, where we were concerned with the direction of spontaneity as related to ΔG and ΔG°. $\Delta\varepsilon$ and $\Delta\varepsilon^\circ$ are related to the free energies ΔG and ΔG° by the equations

$$\Delta G = -n \cdot F \cdot \Delta\varepsilon \tag{12.49}$$

and

$$\Delta G^\circ = -n \cdot F \cdot \Delta\varepsilon^\circ \tag{12.50}$$

These relationships will be discussed in more detail in Section 12.4 below. At this point, however, we can say that a positive value of $\Delta\varepsilon$ is an indication of the spontaneity of a reaction, since it implies a negative value of ΔG according to Eq. 12.49 that we have associated with the spontaneous direction of a reaction (see Eq. 11.84). Similarly, $\Delta\varepsilon^\circ$ is an indicator of the direction of a reaction when we start at standard conditions, as will be elaborated upon in Section 12.4. Thus, when determining the EMF of a galvanic cell and obtaining a negative value, as in Eq. 12.48, the reaction occurs in the opposite direction at standard conditions.

We now may pick up again the discussion of the electric work we have elaborated in Eq. 12.35 above. At standard conditions, the electric work performed by the Al/Cu reaction discussed above is

$$w_{el} = \Delta\varepsilon° \cdot i \cdot t = \varepsilon(Q/t)t = \Delta\varepsilon° \qquad Q[vC = J] \tag{12.35}$$

If one mole of electrons is transferred in the reaction, the charge Q would be just one Farad. If "n" moles of electrons are transferred, we write the charge transported as

$$Q = n \cdot F \tag{12.51}$$

and the electric work as

$$w_{el} = n \cdot F \cdot \Delta\varepsilon° \tag{12.52}$$

Notice that the electric work is numerically equal to the change in standard free enthalpy, $\Delta G°$.

Since the reaction

$$2Al(s) + 3Cu^{2+}(aq) \rightarrow 2Al^{3+}(aq) + 3Cu(s) \tag{12.2}$$

involves the transfer of six electrons, the work is (using the definition of the Faraday, Eq. 12.34)

$$w_{el} = \Delta\varepsilon° \cdot n \cdot Q = 6 \cdot 2.0 \cdot 96{,}485 = 1.158 \cdot 10^6 [J] \approx 1.16 [MJ] \tag{12.53}$$

which is, as chemical reactions go, quite a bit of energy.

Example 12.3 Using standard reduction potentials found in Table 12.1, determine the cell potentials and the spontaneous direction (at STP) for the following reactions:

(a) $Fe(s) + Zn^{2+} \rightarrow Zn(s) + Fe^{2+}$
(b) The hydrogen–oxygen fuel cell: $2H_2(g) + O_2(g) \rightarrow 2H_2O(l)$ (acidic solution).

Answer: (a) $\Delta\varepsilon° = -0.31$ [v] for the reaction as written. At STP, the reaction will occur spontaneously in the opposite direction.
(b) $\Delta\varepsilon° = 1.23$ [v]

Example 12.4 A current of 0.965 [A] is passed for 1000 [s] through an electrolysis cell to electrodeposit copper according to $Cu^{2+} + 2e^- \rightarrow Cu(s)$. What mass of copper will be deposited?

Answer: m = 0.32 [g]

12.3 Batteries

Hundreds of everyday processes are powered by electricity, and those processes that occur while not connected to an electric outlet require storage of electric power in batteries. Batteries generally fall into two categories: one-way, such as the AA and AAA batteries (dry cells) you buy in a hardware or drug store, and rechargeable batteries you find in your cell phone, or in hybrid or electric cars. In this section, three battery types will be discussed in some detail: the Zn/MnO_2 dry cell as an example of a one-way battery, and the lead acid and lithium-ion rechargeable batteries found in gas-powered cars (to start the engine) and in electric or hybrid vehicles, respectively.

12.3.1 Alkaline Dry Cell (AA Battery)

The overall reaction in an alkaline dry cell is

$$Zn(s) + 2MnO_2 \rightarrow ZnO + Mn_2O_3 \tag{12.54}$$

with the half-reactions

$$\begin{aligned} Zn(s) + 2OH^- &\rightarrow ZnO + H_2O + 2e^- &\varepsilon° &= 1.28\,[v] \\ 2MnO_2 + H_2O + 2e^- &\rightarrow Mn_2O_3 + 2OH &\varepsilon° &\approx 0.31\,[v] \\ & &\Delta\varepsilon° &= 1.59\,[v] \end{aligned} \tag{12.55}$$

Notice: In acidic solution, the SRP of MnO_2 to Mn^{+3} is quite a bit higher at 0.95 [v]. A typical AA-size alkaline dry cell consists of an outer container to prevent leakage, a metallic zinc anode, a paste of manganese dioxide surrounding a carbon cathode that conducts the electron current at the cathode, and a separator between the cathode and anode that prevents short-circuiting the cell but allows the hydroxide ions to migrate through it.

A typical AA cell might have a capacity of about 1 [Ah, often quoted as 1000 milliampere hours], implying that a current of one ampere [A] at ca. 1.6 [v] can be drawn from it for one hour. This information allows the determination of the electric energy[2] E_{el} contained in an AA cell:

$$E_{el} = \Delta\varepsilon^\circ \cdot i \cdot t$$
$$= 1.6 \cdot 1 \cdot 3600\,[\text{v A s}] = 5.8\,[\text{kJ}] \quad (12.35)$$

It also allows us to estimate the mass of Zn converted to ZnO, since we know that the total charge Q transported is given by

$$Q = i \cdot t = 1 \cdot 3600\,[\text{As} = \text{C}]$$

Thus, the number of moles of electrons is (see Eq. 12.49)

$$n = Q/\boldsymbol{F} = 3600/96485\,[\text{C/C mol}^{-1}] = 0.037\,[\text{mol of electrons}] \quad (12.56)$$

Since each mole of Zn produces 2 moles of electrons, the number of moles of Zn oxidized in the discharge of an AA cell is about 18 [mmol] or about 1.2 [g] of Zn. This explains the size (and mass) of an AA cell. The energy content of a dry cell calculated above (Eq. 12.56), corresponds to a molar energy of about 300 [kJ/mol].

12.3.2 Lead–Acid Battery

Next, another example of an electrochemical reaction will be presented in detail, namely the lead–acid battery found in most cars. This lead–acid battery system is not only a very important economic factor, it also may serve as a model to discuss reversibility and electrochemical equilibrium. The neat thing about the lead acid battery is that it can be charged and discharged, (i.e. run in the "spontaneous" and "nonspontaneous" directions) many thousand times, where as a normal dry cell which you buy in the store cannot be recharged, although rechargeable Ni-Cd cells are available.

A "charged" lead–acid battery is shown schematically in Figure 12.3. It consists of a lead dioxide (PbO_2) and a lead (Pb) electrode immersed in sulfuric acid. Since PbO_2 is a powdery solid, and a poor conductor of electricity, the lead oxide is actually coated onto a metallic lead electrode. In a typical car battery, the lead electrodes are separated by a porous spun class fiber mat that permits ions to migrate through it.

When the battery is discharged, the following half-reactions occur:

$$PbO_2 + 2e^- + 4H^+ \rightarrow Pb^{2+}(aq) + 2H_2O \quad \text{(reduction/cathode)} \quad (12.57)$$

This half-reaction was charge-balanced using hydrogen ions, since the reaction occurs in acidic solution.

$$Pb \rightarrow Pb^{2+}(aq) + 2e^- \quad \text{(oxidation/anode)} \quad (12.58)$$

The overall reaction is given by

$$PbO_2 + Pb + 4H^+ \rightarrow 2Pb^{2+}(aq) + 2H_2O \quad (12.59)$$

or, taking into account the counterion, by

$$PbO_2 + Pb + 2H_2SO_4 \rightarrow 2PbSO_4 + 2H_2O \quad (12.60)$$

As we shall see later, there are more than just cosmetic differences between Eqs. 12.59 and 12.60, which substantially affect the cell potential.

The discharge process is represented by Figure 12.3a. Thus, in the discharge process, Pb^{2+} in the form of $PbSO_4$ is produced from lead atoms with oxidation numbers of +4 and 0. This process is the opposite of a dis-proportionation mentioned earlier and is known as con-proportionation. The lead sulfate coats the lead electrodes, and when the battery is completely discharged, both electrodes are equal, and the EMF of the cell is zero.

---

2 We use the term electric work and electric energy interchangeably. As we shall see in Section 12.4, the maximum electric work that can be obtained from an electrochemical reaction is the free enthalpy, and corresponds to the electric energy.

When the battery is recharged, which occurs whenever the alternator (generator) is producing electricity, the process depicted in Figure 12.3b occurs. Instead of the load shown in Figure 12.3a, an external source of DC electricity (from the alternator) now drives the electron flow in the opposite direction:

$$2\,Pb^{2+}(aq) + 2\,H_2O \rightarrow PbO_2 + Pb + 4\,H^+$$

The voltage of this external source must be slightly larger, and have the opposite sign, than the cell potential. Now, dis-proportionation of the Pb^{2+} to +4 and 0 oxidation states occurs, and this process is driven by the external potential.

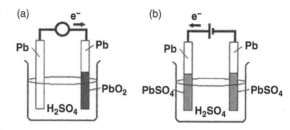

Figure 12.3 (a) Discharge and (b) recharge processes in a lead–acid battery.

If the external potential applied exactly equals the cell potential, then the system is at perfect equilibrium, since the flow of electrons is prevented by exactly balanced potentials. This is the situation that was called before a "reversible" condition. At this point, the reaction is stopped dead in its tracks, and no current flows. Thus, no resistive heating in the wire occurs due to current flow, and the work obtained is given by the free enthalpy, see Section 12.4.

We now turn to the determination of the cell potential of the reaction (in the spontaneous direction). The SRP for the two half-reactions may be found in a comprehensive table, or on-line.

$$PbO_2 + 2e^- + 4\,H^+ \rightarrow Pb^{2+}(aq) + 2\,H_2O \quad \varepsilon^\circ = 1.46\,[v] \tag{12.61}$$

$$Pb \rightarrow Pb^{2+}(aq) + 2e^- \quad \varepsilon^\circ = 0.13\,[v] \tag{12.58}$$

which gives for the total reaction

$$PbO_2 + Pb + 4\,H^+ \rightarrow 2\,Pb^{2+}(aq) + 2\,H_2O \quad \Delta\varepsilon^\circ = 1.59\,[v] \tag{12.62}$$

a cell potential of 1.59 [v], a far cry from the 12 volt of a car battery. How is that possible?

First of all, let us consider that 1.59 [v] represents $\Delta\varepsilon^\circ$, the cell potential at standard conditions. That requires that the Pb^{2+} ion concentrations in both the half-reactions are 1.0 [M]. However, if the counterion is sulfate, this is impossible, since lead sulfate is a sparingly soluble salt ($K_{sp} = 1.3 \cdot 10^{-8}$, see Section 9.9); thus, in a saturated solution, the lead ion concentration is much less than 1.0 [M].

Thus, we use half-reactions that explicitly take into account the counter ions:

$$PbO_2 + SO_4^{2-} + 2e^- + 4\,H^+ \rightarrow PbSO_4(s) + 2\,H_2O \quad \varepsilon^\circ = 1.69\,[v] \tag{12.63}$$

$$Pb + SO_4^{2-} \rightarrow PbSO_4(s) + 2e^- \quad \varepsilon^\circ = 0.35\,[v] \tag{12.64}$$

Then, the total reaction is

$$PbO_2 + Pb + 4\,H^+ + 2\,SO_4^{2-} \rightarrow 2\,PbSO_4 + 2\,H_2O \quad \Delta\varepsilon^\circ = 2.04\,[v] \tag{12.65}$$

Thus, we see that the potential obtained using lead sulfate, rather than $Pb^{2+}(aq)$, is nearly half a volt higher than that calculated for 1.0 [M] lead ion concentration (Eq. 12.57).

However, this voltage is still far less than the 12 volt specified for a car battery. How do we get 12 volt from an electrochemical cell whose potential is only 2[v]? This is achieved by connecting six lead acid cells in series. Thus, the voltages of the six cells add up to a little over 12 [v] to produce the voltage required to run all the electric components, such as the starter motor, in a car.

As we did for the AA dry cell, we now turn to an energy storage aspect of a car battery. When you go to buy a car battery, it may say something like "100 amp hours" capacity on the package, implying that a current of 100 [A] can be drawn for one hour from the battery before it is discharged. Using Eqs. 12.56 and 12.34 as before, we calculate the charge and energy E_{el} stored in the battery as

$$Q = 100\,[A] \cdot 3600\,[s] = 360{,}000 \quad [C = As] \tag{12.66}$$

With the voltage calculated above, this gives the electric energy stored as

$$E_{el} = \varepsilon\ Q = 12 \cdot 360{,}000 = 4{,}300\,[kJ] = 4.3\,[MJ]$$

$$= 4.3/3.6 \left[\frac{MJ}{\frac{MJ}{kWh}}\right] \approx 1.2\,[kWh] \tag{12.67}$$

using the energy conversion discussed in Section 11.2.1. How much lead sulfate must be converted to lead and lead dioxide to store this energy? The charge (Eq. 12.66) can be converted into moles of electrons, n:

$$n = Q/F = 360{,}000/96{,}485 \left[\frac{C}{\frac{C}{\text{mol of e}}}\right] = 3.73\,[\text{mol of e}^-] \tag{12.68}$$

Since two moles of electrons are transferred in both the cathode and anode reaction, the number of moles of lead sulfate converted to lead and lead dioxide is about 1.86 [mol], corresponding to ca. 0.38 [kg]. This accounts for the fact that lead–acid batteries are big and heavy (a 100 [ah] battery weighs about 20 [kg]), since the amount of lead converted is substantial. The fact that they are full of sulfuric acid and lead makes them environmentally quite unfriendly, and you should refrain from hugging them, since they leave ugly holes burned into skin and cloths if there is the slightest leak of sulfuric acid.

Since electric and hybrid cars have become quite commonplace, let us embark on a short detour and use the information calculated above and estimate how far a car powered by one lead–acid battery could go. Let us assume that a small-sized, gas-powered car engine (like in a VW buggy) produces about 50 horsepower [hp]; 1 [hp] very roughly corresponds to an electric power of 1 [kW]. So, to produce an electric vehicle that can compete with a gas-powered car, one would envision an electric motor with at least 50 [kW] power output. If a fully charged lead–acid battery provides 1.2 [kWh] (see Eq. 12.67) of energy, the electric car could go for about

$$1.2\,[\text{kWh}]/50\,[\text{kW}] = 0.024\,[\text{h}], \text{ or about } 1\tfrac{1}{2} \text{ minutes} \tag{12.69}$$

Using a stick-shift car like the VW buggy as an example: with the gas engine stalled, one could move a buggy for a few hundred yards using the electric starter motor to move the car out of harm's way... That, however, does not imply that a VW buggy was a practical, electric vehicle, and new batteries had to be developed that had higher energy storage, and much lower weight.

12.3.3 Lithium-ion Battery

The requirements listed in the last sentence led to the development of the lithium-ion battery. The capacity of the lithium-ion battery in a Tesla 3 is about 80 [kWh], or the equivalent of over 60 lead–acid batteries (which would have added 1200 kg to the car weight). The lithium-ion battery, in contrast, weighs in at about 480 [kg], making electric and hybrid vehicles practical.

The anode in a lithium-ion battery is a lithium–carbon intercalation compound that can be written as LiC_6. Here, Li^+ ions are embedded between sheets of graphite, and the electron given up by the Li atom is accepted into the delocalized π-electron cloud of the graphite. The lithium content is so high that the molecular formula LiC_6 is justified; that is, there is, on average, one lithium ion for each graphite ring. The oxidation numbers, therefore, are $+1$ for lithium, and $-1/6$ for carbon. The SRP for the Li^+ ion in aqueous solution is -3.04 [v] (see Table 12.1); however, it is somewhat less for the oxidation of LiC_6 (written as the oxidation reaction)

$$LiC_6 \rightarrow Li^+ + C_6 + e^- \quad \varepsilon^\circ = 2.84\,[\text{v}] \tag{12.70}$$

The cathode reaction during the discharge of a lithium-ion battery is

$$CoO_2 + Li^+ + e^- \rightarrow LiCoO_2 \quad \varepsilon^\circ = 0.56\,[\text{v}] \tag{12.71}$$

Here, Co (IV) oxide is reduced to $LiCoO_2$ (usually referred to as lithium-cobalt oxide) in which cobalt has an oxidation number of 3. Thus, the overall cell reaction can be written as

$$LiC_6 + CoO_2 \rightarrow C_6 + LiCoO_2 \quad \Delta\varepsilon^\circ = 3.4\,[\text{v}] \tag{12.72}$$

This cell produces a relatively high potential, which positively affects the energy storage (see Eq. 12.35). Furthermore, the energy storage density is very high, allowing relatively light-weight batteries to store large amount of energy. In addition, many electric vehicles employ regenerative braking technology: when braking, the kinetic energy of a moving vehicle is not wasted as heat at the brake pad and rotor, but is converted back to electric energy and used to recharge the battery. This can be accomplished by running the electric motor as a generator that converts some of the kinetic energy of the car back to electric energy.

12.4 Relationship Between Cell Potential and Free Enthalpy

We have seen before, in Section 11.9, that ΔG is the maximum amount of work one can obtain from a chemical reaction. When a reaction is carried out at constant temperature and pressure, and no other work than electrical work is extracted from a reaction, the maximum of electric work that can be obtained from an electrochemical reaction is just ΔG:

$$\Delta G = -w_{max} = -n \cdot F \cdot \Delta \varepsilon \tag{12.49}$$

Furthermore, at standard conditions (that is, at all concentrations at 1.0 [M], and 25 [°C])

$$\Delta G° = -n \cdot F \cdot \Delta \varepsilon° \tag{12.50}$$

As stated before, $\Delta G°$ denotes the maximum amount of energy that can be converted to useful work at standard state conditions; ΔG denotes the maximum amount of work at any other conditions. The rest of the energy change that accompanies a process is lost due to degradation of energy from useful to useless form (heat).

Next, we will discuss the relationship between the spontaneous direction of an electrochemical reaction and the sign of $\Delta \varepsilon°$. You may find the statement that for a positive $\Delta \varepsilon°$, the reaction occurs spontaneously. Indeed, the standard cell potential for the reaction discussed in the beginning of this chapter

$$2Al(s) + 3Cu^{2+}(aq) \rightarrow 2Al^{3+}(aq) + 3Cu(s) \tag{12.2}$$

is $\Delta \varepsilon° = 2.00$ [v] (see Eq. 12.43), and we have seen that this reaction is spontaneous, since solid copper is formed and aluminum dissolves. Furthermore, we have seen in the discussion of the lead–acid battery (Eq. 12.65) that the spontaneous discharge reaction

$$PbO_2 + Pb + 4H^+ + 2SO_4^{2-} \rightarrow 2PbSO_4 + 2H_2O \tag{12.65}$$

has a positive standard cell potential of $\Delta \varepsilon° = 2.04$ [v]. Thus, one may be tempted to argue that $\Delta \varepsilon°$ can be used as an indicator of the spontaneity of the reaction and one may think that a reaction occurs spontaneously if $\Delta \varepsilon°$ is positive. However, we have argued before, in Section 11.9 and 11.10, that the standard free enthalpy, $\Delta G°$, is not an indicator of spontaneity, only ΔG is. However, in Section 11.10 we showed that when we use standard states as conditions, the reaction will occur to equilibrium that is determined by $\Delta G°$. Since $\Delta G = \Delta G°$ if the starting conditions are the standard state, $\Delta G°$ determines the spontaneous reaction in this case. The same is true for the cell voltage: if we have both the reactions and products at standard state conditions, $\Delta G°$ and $\Delta \varepsilon°$ do indicate in which direction the standard state mixture will react to reach equilibrium. Thus, $\Delta G°$ and $\Delta \varepsilon$ do determine the direction of spontaneity *if and only if all reactants and products are at standard state*. In this case, we may write:

$$\begin{aligned} &\Delta \varepsilon° > 0 \text{ or } \Delta G° < 0 \quad \text{reaction occurs as written} \\ &\Delta \varepsilon° = 0 \text{ or } \Delta G° = 0 \quad \text{equilibrium} \\ &\Delta \varepsilon° < 0 \text{ or } \Delta G° > 0 \quad \text{reaction occurs in the opposite direction} \end{aligned} \tag{12.73}$$

If the starting conditions are not the standard state, $\Delta G°$ indicates the equilibrium position according to Eq. 11.59 which can be reached, depending on the starting conditions, in the forward or backward direction, as indicated by Figure 11.9.

12.5 Concentration and Temperature Dependence of EMF

We defined the relationship between the standard cell potential and the standard free enthalpy of the reaction by

$$\Delta G° = -n \cdot \Delta \varepsilon° \cdot F \tag{12.50}$$

Similarly, the free enthalpy and the cell potential at nonstandard conditions are related by

$$\Delta G = -n \cdot \Delta \varepsilon \cdot F \tag{12.49}$$

Since ΔG and $\Delta G°$ are related by

$$\begin{aligned} \Delta G - \Delta G° &= RT \ln Q \text{ or} \\ \Delta G &= \Delta G° + RT \ln Q \end{aligned} \tag{11.96}$$

we may write an equation that relates the cell potential at standard conditions to that at nonstandard conditions:

$$-n \cdot \Delta\varepsilon \cdot F = -n \cdot \Delta\varepsilon° \cdot F + RT \ln Q \tag{12.74}$$

$$\Delta\varepsilon = \Delta\varepsilon° - (RT/nF) \ln Q \tag{12.75}$$

This equation is known as the Nernst equation. Q is, as always, the quotient of concentration at initial (nonequilibrium) conditions. At 25 [°C], and converting the natural to the decadic logarithm, we write the expression (R T/F) as

$$(RT/F) = 2.303 \cdot 8.3 \cdot 298/96{,}485 \left[\frac{\frac{J}{mol\,K} K}{\frac{C}{mol}} = \frac{J}{C} = v \right] = 0.059\,[v] \tag{12.76}$$

where we used $\left[\frac{J}{C} = v\right]$ from Eq. 12.35.

Thus,

$$\Delta\varepsilon = \Delta\varepsilon° - (0.059/n) \log Q \tag{12.77}$$

Eq. 12.77 allows us to calculate how the potential of a cell varies with concentrations of the reactants. Take, for example, a cell such as the one given by Eq. 12.1:

$$2Al(s) + 3Cu^{2+}(aq) \rightarrow 2Al^{3+}(aq) + 3Cu(s) \quad \Delta\varepsilon° = 2.00\,[v] \tag{12.1}$$

If this reaction is carried out with concentrations of the electrode solutions different from 1.0 [M], for example, with

$$[Al^{3+}(aq)] = 0.1\,[M] \text{ and} \tag{12.78}$$
$$[Cu^{2+}(aq)] = 2.0\,[M],$$

the cell potential will be

$$\Delta\varepsilon = 2.00 - (0.059/6) \log Q \tag{12.79}$$

with Q given by

$$Q = [Al^{3+}(aq)]^2/[Cu^{2+}(aq)]^3$$
$$= (0.1)^2/2^3 = 0.01/8 = 0.00125 \tag{12.80}$$

$$\Delta\varepsilon = 2.00 - (0.059/6)(-2.90) = 2.03\,[v] \tag{12.81}$$

Interestingly, one can construct electrochemical cells where the cathode and the anode half-reaction are the same, but the concentrations of the ions are different. Take, for example, an electrochemical cell where both half-reactions consist of a copper electrode, immersed in a solution of copper (II) chloride, but where the ion concentrations are different:

$$Cu(s) \rightarrow Cu^{2+}(aq, 0.1M) + 2e^- \tag{12.82}$$

$$Cu^{2+}(aq, 1.0\,M) + 2e^- \rightarrow Cu(s) \tag{12.83}$$

The overall reaction is

$$Cu(s) + Cu^{2+}(aq, 1.0\,M) \rightarrow Cu(s) + Cu^{2+}(aq, 0.1M) \tag{12.84}$$

Since the cathodic and the anodic reactions have the same potential at standard conditions, $\Delta\varepsilon° = 0$, and the Nernst equation reduces to

$$\Delta\varepsilon = 0 - (0.059/2) \log[0.1/1] \tag{12.85}$$
$$= (-0.0295)(-1) \approx 0.03\,[v] \tag{12.86}$$

Thus, the reaction occurs in the direction as written, namely to equalize the concentrations in both solutions. Thus, solid copper will plate out of the 1.0 [M] solution, and solid copper will dissolve in the 0.1 [M] solution to reduce the concentration gradient or concentration difference, in the half-cells.

Example 12.5 Estimate the drop in voltage of a car battery when operated in mid-winter at −25 [°C]. Use the cell potential calculated in Eq. 12.65 and assume that the concentrations of sulfuric acid is about 4.5 [M] and does not change with temperature.

Answer: The reduction in voltage (about 0.02 [v]) is quite small. See answer booklet for details

Example 12.6 What are the units of electrical

(a) potential
(b) current
(c) charge
(d) power
(e) energy

Answer: See answer booklet

Example 12.7 What is the relation between

(a) power, current, and potential (include proper units)
(b) power and energy

Answer: See answer booklet

Further Reading

General and historical aspects: https://en.wikipedia.org/wiki/Electrochemistry
Alkaline battery: https://en.wikipedia.org/wiki/Alkaline_battery
Lithium-ion battery: https://en.wikipedia.org/wiki/Lithium-ion_battery

13

Chemical Kinetics: Rates of Reactions and Reaction Mechanisms

13.1 Scope of Kinetics Discussion

Chemical kinetics is the branch of chemistry dealing with the rate of chemical reactions, that is, how fast a chemical reaction occurs. The rate of a reaction can be thought of as the rate of disappearance of reactants (change of concentration of reactants with time) or the rate of formation of products (change of concentration of products with time). From everyday experience, we know that reactions can occur at a vastly different time scale: think of an explosion at one hand, where reactants are converted to products in a fraction of a second, or a slow and steady process like corrosion, where products are formed over the time scales of weeks or months.

Earlier in this course, we have dealt with identifying reactants and products, and the mole and mass relationships between them (stoichiometry). In the equilibrium chapter, we have concerned ourselves with the question of the extent of a chemical reaction, that is, whether all reactants disappear to form products or whether or not some of the products will react backward to recreate reactants. The discussion of chemical equilibrium subsequently led us to question what driving force makes a reaction go – and we introduced the concepts of free energy and entropy. However, in all these discussions, we have left out the question of how fast or how slowly a reaction proceeds, and what we can do to influence this rate of a chemical reaction.

However, the determination of the rate of a chemical reaction – although it is related to the equilibrium concepts discussed in the appropriate chapters – is only one reason to concern ourselves with kinetics. The other reason is that kinetics teaches us how chemical reactions occur – that is, it teaches us the *pathway* of a reaction. This is particularly important if we wish to influence a reaction – prevent it, favor it, speed it up, and so forth.

One problem we encounter for a given chemical reaction represented by an equation

$$A + B \rightarrow C \tag{13.1}$$

is that we do not know whether or not this is the way the reaction actually occurs. It may be that the reaction actually involves an intermediate compound (or compounds):

$$A + B \rightarrow D \rightarrow C \tag{13.2}$$

that we may or may not be able to observe, but that influences how fast a reaction occurs.

That leads us to the ideas of the pathways of chemical reactions, reaction intermediates, and the concepts of multistep chemical reaction. These ideas will be the subject of discussion in the next sections.

13.2 Elementary Steps and Chemical Reactions

13.2.1 Kinetic Model of Chemical Reactions

Intuitively, it makes sense to think that chemical reactions – in the gaseous or condensed phases – occur when molecules and/or atoms collide with each other. We remember from our discussions of gases that in a gaseous sample, atoms or molecules continuously and very frequently bump into each other, and into the walls of the container. These latter collisions are the manifestation of the pressure the gas exerts, whereas the collisions with other gaseous particles account for the rapid mixing and temperature equilibration of the sample, as well as processes like diffusion and effusion. In fact, in a gaseous sample of 1 mole of gas at standard temperature and pressure (STP), the number of collisions per second is many orders of magnitude larger than Avogadro's number! As the pressure or the concentration of a gaseous sample decreases, the number of collisions decreases as well. In outer space, where the concentration of matter may be as low as one molecule per cm^3, it may take seconds or minutes before a collision between particles occurs.

Let us take one of the chemical reactions occurring in outer space as an example. The interstellar space contains a number of small molecules, radicals, and molecular ions. One of these species is a radical that was identified by its spectral signatures (see Chapter 15), which does not occur on earth very much (because it is too reactive), namely the hydroxyl radical, OH·. Here, the dot refers to a single, unpaired electron. Remember, the hydroxide ion, OH$^-$, is fairly common and stable. Take one electron away and you get a species that is very reactive, because it violates the octet rule and has an unpaired electron.

When two hydroxyl radicals collide, in outer space or in a laboratory setting, a reaction may occur:

$$OH\cdot + OH\cdot \rightarrow H_2O_2 \tag{13.3}$$

The rate of the formation of H_2O_2 can be expressed as $\Delta[H_2O_2]/\Delta t$, i.e. the change in concentration of H_2O_2 with time (the square brackets, as usual, denote the concentration of a species). One could then argue that this rate of formation of H_2O_2 depends on the number of collisions between the OH· radicals.

There are detailed derivations, based on relatively simple classical physical arguments, to calculate the numbers of collisions between gaseous molecules. These depend on the concentration (or pressure) of particles, the temperature, and other parameters. It may be sufficient here to state that the number of collisions, n_c, between atoms of a mixture of gaseous A and B is proportional to the concentrations of A and B:

$$n_c \propto [A][B] \tag{13.4}$$

Thus, it appears perfectly logical to assume that the rate of the chemical reaction between two hydroxyl radicals to form hydrogen peroxide depends on the rate at which they collide:

$$\text{Rate of formation of } H_2O_2 = \Delta[H_2O_2]/\Delta t \propto [OH\cdot][OH\cdot] \tag{13.5}$$

Whenever we have a proportionality such as this one, we may write it as a proper equation by introducing a proportionality constant and write

$$\text{Rate of formation of } H_2O_2 = k[OH\cdot][OH\cdot] \tag{13.6}$$

where k is known as the rate constant. More about it will be discussed later. So, it appears that this reaction occurs faster (becomes more probable and produces products faster) as the concentration of hydroxyl radicals increases. This makes sense: the more OH· are around, the more they bump into each other, and the more often they react. So, we see that a reaction rate depends on concentrations. In other words, more H_2O_2 is formed per time when more OH· radicals are around.

However, it is not possible to write an equation for the rate of formation of product by just looking at the stoichiometric equation, since a stoichiometric equation only tells us about the ratios of reactants and products, but not of the reaction path. Take the formation of ozone in the atmosphere for an example. It simply reads

$$3O_2 \rightarrow 2O_3 \tag{13.7}$$

implying that stoichiometrically 3 moles of oxygen are needed to form 2 moles of ozone. However, this reaction occurs along a *multistep* pathway, producing some intermediate products that further react to produce the products. Such nasty behavior of chemical reactions was hinted above (end of Section 13.1) and is one of the major reasons why we discuss chemical kinetics, since this field helps us determine the reaction pathway. In the particular case of the formation of ozone, the first step of the reaction occurs according to:

$$O_2 \rightarrow 2O \tag{13.8}$$

This reaction is caused by the absorption of an ultraviolet photon from the sun's radiation. Oxygen atoms are called a reaction intermediate (or intermediate product). The oxygen atoms subsequently react according to

$$O + O_2 \rightarrow O_3 \tag{13.9}$$

Multiplying the second equation by 2 (since we created two oxygen atoms in Eq. 13.8) and adding the equations yields

$$\begin{aligned} O_2 &\rightarrow 2O \\ 2O + 2O_2 &\rightarrow 2O_3 \\ \hline 3O_2 &\rightarrow 2O_3 \end{aligned} \tag{13.10}$$

So, here we have a multistep equation, but we would not know that from the inspection of the total balanced equation, which simply tells us the stoichiometric ratios of reactants and products. The atomic oxygen does not appear in the overall balanced equation and is called a "reaction intermediate." Each step in a multistep chemical reaction is called an "elementary step," and the sum of these must be equal to the overall balanced chemical equation.

13.2.2 Basics of Chemical Kinetics: Rate Law and Rate Constant

Let us discuss next the rate equation, that is, the equation that determines how fast a chemical reaction proceeds. This is accomplished first for an elementary step. An elementary step is a reaction that occurs as written, that is, without an intermediate reaction, such as Eq. 13.3:

$$OH\cdot + OH\cdot \rightarrow H_2O_2 \tag{13.3}$$

or in a more generalized form

$$A + B \rightarrow C \text{ or even } A \rightarrow C$$

For such an elementary step, the rate of formation of C, measured in units of concentration change per time [mol/(L s)], is given by

$$\text{Rate of formation of C} = k\,[A][B] \tag{13.11}$$

where k is the rate constant introduced above, and [A] and [B] are the instantaneous concentrations, in [mol/L], of the reactants. The rate of formation of C is written as the change in concentration, $\Delta[C]$, in a given time period, Δt. So, if 10^{-3} [mol/L] of C is formed per second, we would say that the rate of formation of C in this elementary step is 10^{-3} [mol/(L s)]. We can define a "rate equation"

$$\frac{\Delta[C]}{\Delta t} = k[A][B] \tag{13.12}$$

A number of important items need to be discussed about rate equations. First of all, as pointed out, the arguments presented here only hold if the reaction, indeed, is an elementary step. Next, the rate constant k relates the instantaneous concentrations of the reactants to the observed rate. We can think of k as a fractional number, which is determined by the probability that a collision between A and B actually leads to a reaction. Let us go back to the example of the OH· radical reaction introduced above, for which the rate equation would be

$$\frac{\Delta[H_2O_2]}{\Delta t} = k[OH\cdot][OH\cdot] \tag{13.13}$$

The product of the concentration on the right, [OH·] [OH·], is simply a number proportional to the number of collisions between OH· radicals. So, one expects that higher hydroxyl radical concentrations lead to a higher reaction rate, hence the proportionality between reaction rate and concentration. However, not every collision will result in a successful chemical reaction. The molecules may collide so weakly that a reaction cannot occur. Also, the geometry of the collision may be such that the reaction is impossible (more on this in Section 13.5). Thus, we can think of the rate constant k to denote the probability that a collision leads to the desired chemical product. We refer to the expression

$$\Delta[C]/\Delta t$$

as the rate of production of C and to k as the rate constant. As pointed out before, the rate of formation of C has the units of [mol/(L s)] and varies constantly, since the concentrations of the reactants constantly change. The rate constant, however, is a constant as the name implies: it denotes the fraction of interactions between reactants that leads to formation of a product and is independent of the concentrations. k is dependent on the temperature, however, as will be discussed later. The units of k, for this given reaction, are [L/(mol s)]. This can be derived from the rate equation assuming that both sides have to have units of [mol/(L s)]. However, as we shall see shortly, the units of the rate constant are not always [L/(mol s)] but depend on the particular reaction (more specifically, the order of a reaction, see Section 13.3).

The rate equation has another important property, which we can deduct from the stoichiometry of reactants and products. For the elementary step

$$A + B \rightarrow C$$

the collision of one A particle with one B particle yields exactly one C particle. Thus, the rate of appearance of C must be tied to the rate of disappearance of A and B. A little reflection shows that the rate of disappearance of A (or B) is equal to the rate of appearance of C. If we utilize the sign of a concentration change as an indication of appearance or disappearance, we may write

$$\frac{\Delta[C]}{\Delta t} = -\frac{\Delta[A]}{\Delta t} = -\frac{\Delta[B]}{\Delta t} \tag{13.14}$$

Example 13.1 For the gas phase reaction $A + 2B \rightarrow C$, relate the rate of formation to the rate of disappearance of A and B.

Answer: $\frac{\Delta[C]}{\Delta t} = -\frac{\Delta[A]}{\Delta t}$ and $2\frac{\Delta[C]}{\Delta t} = -\frac{\Delta[B]}{\Delta t}$

13.2.3 Time Dependence of the Reaction Rate

To illustrate the points made in the previous section about rate laws and rate constants, let us take a simple one step reaction (or an elementary step)

$$A \rightarrow C$$

(for example, the radioactive decay $^{14}C \rightarrow {}^{14}N$, see Chapter 14) and follow the concentrations of reactants and products as the reaction proceeds, for a given initial concentration of A. We do that simply by plugging the instantaneous concentrations into the rate equation

$$\frac{\Delta[C]}{\Delta t} = k[A] \tag{13.15}$$

and calculating the rate and remaining concentrations at 1 [s] intervals (we make an assumption, to be detailed later, that the reaction only goes forward, that is, that the product, C, once formed, will not undergo any reactions). We pick a number for the rate constant k:

$$k = 0.1\,[1/s] \tag{13.16}$$

(note that the rate constant for this reaction has different units than the rate constant discussed earlier). Let us start with an initial concentration of [A] = 0.1 [mol/L] and [C] = 0. The initial rate of formation of C is given by

$$\frac{\Delta[C]}{\Delta t} = k[A] = 0.1 \cdot 0.1\,[(1/s) \cdot (mol/L)] = 0.01\,[mol/(L\,s)] \tag{13.17}$$

After 1 s, 0.01 [mol] of C has been produced. Since the rate of disappearance of A is tied to the rate of appearance of C by

$$\frac{\Delta[C]}{\Delta t} = -\frac{\Delta[A]}{\Delta t} \tag{13.18}$$

we see that after 1 s, 0.01 [mol/L] have disappeared such that the concentration of A after 1 s is

$$[A] = 0.1 - 0.01 = 0.09\,[mol/L].$$

Now, since A is reduced from 0.1 to 0.09 [mol/L], the reaction rate will slow down. In the next second, the rate of formation of C (which is the same as the rate of disappearance of A) is given by

$$\frac{\Delta[C]}{\Delta t} = k[A] = 0.1 \cdot 0.09 = 0.009\,[mol/(L\,s)] \tag{13.19}$$

So, after 2 s, there would be 0.09 − 0.009 = 0.081 [mol/L] of A left and 0.019 [mol] of C formed. A plot of the concentrations of A with time is shown in Figure 13.1. The concentration of C starts at zero and approaches a constant value, just as the concentration of A approaches zero concentration.

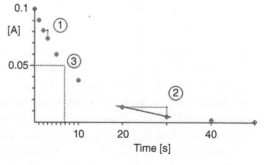

Figure 13.1 The dependence of the concentration of [A], and the reaction rate, with time for a first-order chemical reaction $A \rightarrow C$.

In Figure 13.1, it is apparent that the concentrations vary with time. Inspection of Figure 13.1 also reveals that the reaction rate changes with time: the rate of disappearance is about 0.06 [mol/L] per second between $t = 2$ and $t = 3$ (see dotted lines marked "1"). The same change in concentration takes about 10 s between $t = 20$ and $t = 30$ (point 2): the reaction has slowed down to about 1/10 of the original rate because the concentration of the reactant is only about 1/10 of the original value. Remember, though, that the rate constant does not change over the course of the reaction.

13.2.4 Integrated Rate Law

In the previous section, we calculated the concentrations of A and C at a given time by substituting the concentration at time zero (the start of the reaction), and determining the amount of material that had reacted in the first second. We then subtracted this amount from the initial concentrations and worked our way down to subsequent time intervals. There is a more elegant way to achieve the same information, which is in effect the way the curve in Figure 13.1 was derived. This new way to describe the reaction rate is known as the integrated rate equation. This equation can be derived using a little calculus.

Let us consider the same chemical reaction discussed above, namely

$$A \rightarrow C$$

We have written the rate either in terms of the formation of C or the disappearance of A as

$$\frac{\Delta [C]}{\Delta t} = -\frac{\Delta [A]}{\Delta t} = k[A] \tag{13.18}$$

Here, Δt is a finite time interval, denoted by the horizontal dashed lines in Figure 13.1, points 1 and 2, and the reaction rates we calculated before were the averaged values during a given time interval. If we wish to know the truly instantaneous values of the rate and concentrations, we can accomplish that by shrinking the time interval Δt to zero. This trick should sound familiar from the basic concepts of differential calculus. We denote the infinitesimally small time interval as dt, and the infinitesimally small change in concentration of A as dA. Then

$$-\frac{d[A]}{dt} = k[A] \tag{13.20}$$

which can be rewritten as

$$\frac{d[A]}{[A]} = -k\,dt \tag{13.21}$$

Integration of this expression between time zero ($t' = 0$) and a later time t ($t' = t$) and an original concentration $[A]_0$ and a concentration $[A]_t$ at time t yields:

$$\ln \frac{[A]_t}{[A]_0} = -kt \text{ or } \ln \frac{[A]_0}{[A]_t} = kt \tag{13.22}$$

or

$$\log \frac{[A]_0}{[A]_t} = k\,t/2.303 \tag{13.23}$$

or

$$\log [A]_t = -(kt/2.303) + \log [A]_0 \tag{13.24}$$

(All these mathematical steps are explained in Sections 1.6, 1.8, and 1.9.) The application of the integrated rate law allows the calculation of the concentration of the remaining reactant, or the product formed, at any time. Incidentally, since in the original rate equation the rate depends linearly on the concentration [A], we call the integrated rate law also the first-order integrated rate law (we shall discuss later some situations in which the reaction rate does not depend linearly on the concentration of the reactant). For example, if you wished to confirm the amount of A left for the reaction plotted in Figure 13.1 after 15 s have passed, you would solve

$$\log [A]_{t=15} = -(kt/2.303) + \log [A]_0 \tag{13.20}$$

with $[A]_0 = 0.1$ and $\log [A]_0 = \log (0.1) = -1.0$

$$-\frac{kt}{2.303} = -0.1 \cdot \frac{15}{2.303} = -0.651 \tag{13.21}$$

and $\log [A]_{t=15} = -0.651 - 1.0 = -1.651$ (13.22)
Therefore,

$$[A]_{t=15} = 10^{-1.651} = 0.022 \tag{13.23}$$

Example 13.2 Consider an elementary step A → C with a rate constant $k = 0.3 \ [s^{-1}]$.

(a) Calculate the ratio $[A]_o/[A]_t$ after 10 s.
(b) Calculate the ratio $[A]_o/[A]_t$ after 100 s.
(c) Calculate the time at which half the reactants are left.

Answer: (a) $\dfrac{[A]_0}{[A]_{10}} = 20.1$, (b) $\dfrac{[A]_0}{[A]_{100}} = 1.06 \cdot 10^{13}$, (c) $t = 2.3 \ [s]$

On the other hand, if the concentration at time zero and the concentration at a later time are known, it is possible to calculate back when time zero was, that is, when the reaction started. This important application of chemical kinetics is used in radiocarbon dating. Radiocarbon dating uses the principle that a carbon isotope, ^{14}C, decays radioactively into another element, namely ^{14}N, according to $^{14}C \rightarrow {}^{14}N$. We shall discuss this nuclear reaction in Chapter 14. The decay of ^{14}C proceeds according to the first-order rate equation given earlier. The principle of this method is based on the observation that any living tissue, be it from plants or animals, metabolizes ^{14}C contained naturally in trace amounts in carbon dioxide as $^{14}CO_2$. Thus, we all contain an equilibrium concentration of ^{14}C, as long as we are alive and metabolizing. After the death of plant or animal tissue, this metabolism stops, and the concentration of ^{14}C drops off following the first-order integrated rate equation given earlier. It is easy to measure the concentration of ^{14}C in tissues, since this isotope is radioactive and gives off radiation that can be detected. A comparison of the radioactivity of a prehistoric sample and a present piece of tissue gives one the values of $[^{14}C]_o$ and $[^{14}C]_t$, from which t, the time since the end of metabolism, can be calculated provided that the rate constant is known. Example 3.3 explicitly deals with this problem.

Next, the concept of half-life of a reaction will be introduced. We call the half-life of a reaction the time, $t_{(1/2)}$, required for the original concentration, $[A]_o$, to drop to half of its original value. In other words, the reaction has chewed up half of the starting materials:

$$[A]_{t_{(1/2)}} = \tfrac{1}{2}[A]_o \tag{13.24}$$

Substituting these concentrations into the integrated rate equation for first-order reactions

$$\ln\left[\frac{[A]_o}{[A]_{t_{(1/2)}}}\right] = k \, t_{(1/2)} \tag{13.25}$$

gives

$$\ln\left[\frac{[A]_o}{\tfrac{1}{2}[A]_o}\right] = \ln 2 = k \, t_{(1/2)} \tag{13.26}$$

from which we obtain a relationship between the rate constant and the half-life time:

$$t_{(1/2)} = 0.693/k \tag{13.27}$$

For first-order reactions, it is easy and convenient to determine the rate constants just by measuring the time required for the concentration of the starting materials to drop to half its original value. Referring to Figure 13.1 once more, we see that it takes about 6.9 s (0.693/0.1) for the reaction to reach the halfway point (point 3 in Figure 13.1). Since the reaction rate has slowed to half its original value at this point, the next half-life will consume ½…½ = ¼ of the reactants, leaving 25% of the original concentration of A.

Finally, a couple of words about chemical reactions that are not first order in a given reactant. For these reactions, we find rate equations of different forms. For example, the rate equation for the reaction A → D could be of the form

$$\frac{\Delta[D]}{\Delta t} = -\frac{\Delta[A]}{\Delta t} = k[A]^2 \tag{13.28}$$

This equation describes a chemical reaction in which the concentration of a reactant appears to the second power. We call this a "second-order reaction," or a reaction that is "second order in [A]." For such reactions, totally different integrated rate equations hold. We shall revisit this aspect in Section 13.4.

Example 13.3 ^{14}C occurs naturally at a certain level and is incorporated into plants by photosynthesis of $^{14}CO_2$. By consuming plant materials, which are partially labeled with ^{14}C, it is incorporated into a certain equilibrium concentration in human tissue. After death, the ^{14}C concentration declines with first-order kinetics.
During your next archeological field trip to Olduvai Gorge in Africa, you find an old thighbone of an early ancestor of ours (a relative of Lucy??). A ^{14}C dating experiment shows that the level of ^{14}C in the fossil is 0.015% of the original level, which we assign to be 100%. The half-life of ^{14}C, which decays according to first-order kinetics, is $t_{(1/2)} = 5730$ years. When did the thighbone's owner die?

Answer: $t = 72\,800$ [years]. Definitely not a relative of Lucy, who lived 3 million years ago.

13.3 Rates of Multistep Reactions and Equilibria

Many chemical reactions occur as a series of elementary steps that may involve reaction intermediates. In earlier discussions in this course, we have concerned ourselves mainly with the balanced stoichiometric equation and have not worried about the reaction intermediates and the pathway of the reaction. However, even reactions that appear perfectly simple and straightforward often involve an intermediate.

Take the reaction occurring in smog for an example:

$$NO_2 + CO \rightarrow NO + CO_2 \tag{13.29}$$

This reaction does not occur by simple collision of CO and NO_2 molecules, but *via* two elementary steps and the formation of a reaction intermediate as follows:

$$\left.\begin{array}{l} NO_2 + NO_2 \rightarrow NO + NO_3 \\ NO_3 + CO \rightarrow NO_2 + CO_2 \end{array}\right\} \tag{13.30}$$

The overall reaction is a sum of two elementary steps. Why does a reaction occur following an indirect path like the one indicated? The answer is that the direct reaction

$$NO_2 + CO \rightarrow NO + CO_2$$

may occur as well, but with a very low rate constant. As it turns out, the stepwise reaction is faster and, therefore, dominates the formation of products. Other reactions may compete as well, but it is the reaction with the highest rate that determines the major product formed.

The elementary steps are the actual chemical reactions that occur, and they are, of course, dependent on the collision of the two reactants. The rate of these elementary steps is, therefore, given by

$$\text{rate}_1 = \frac{\Delta[NO_3]}{\Delta t} = k_1[NO_2]\cdot[NO_2] \tag{13.31}$$

for the first step and

$$\text{rate}_2 = \frac{\Delta[CO_2]}{\Delta t} = k_2[NO_3]\cdot[CO] \tag{13.32}$$

for the second step.

The rate constants both have units of [L/(mol·s)]. It is, of course, very unlikely that these two consecutive elementary steps would have the same numerical value of rate constants. In general, rate constants can vary over many orders of magnitude, from reactions that take minutes or hours (the oxidation of elementary iron at room temperature to form rust) to fractions of a millisecond, in the detonation of an explosive. Thus, it is unlikely that k_1 and k_2 happen to be equal.

Then, there are two cases to be considered for our further discussion: one where the first step is significantly slower than the second step and another where the reverse situation is found.

In the example introduced so far, it turns out that the first step is slow:

$$\text{Rate}_1 = k_1[\text{NO}_2]\cdot[\text{NO}_2] \quad \text{slow}$$
$$\text{Rate}_2 = k_2[\text{NO}_3]\cdot[\text{CO}] \quad \text{fast}$$

Since the second step is faster than the first one, it will have the tendency to very rapidly use up the NO_3 produced by the first step. Thus, the overall rate depends on how fast NO_3 is produced. We therefore call the first step the rate-limiting or rate-determining step. A rate-limiting step can be visualized (nonchemically) by a bunch of people working on an assembly line. Let us say the third person on the line takes 2 min to perform his (or her) step, whereas it takes every other person only 1 min. Therefore, the total productivity depends on the rate-determining step, and no more than 30 units can be produced in each hour. There will be partially finished products piling up before the third person, and the persons behind him or her will idle half of the time since they cannot get sufficient parts to fully do their job. Thus, the overall reaction rate for the reaction

$$\text{NO}_2 + \text{CO} \rightarrow \text{NO} + \text{CO}_2 \tag{13.29}$$

obeys the rate law of the slow step

$$\text{overall rate law: } k_1[\text{NO}_2]^2 \tag{13.33}$$

Who would have expected this, and how do we know that this is true? To answer the first part: nobody would expect that, and the take-home message here is that it is not possible to determine the overall rate law of a chemical reaction by inspecting the balanced chemical equation. For elementary steps, on the other hand, the rate law always will be proportional to the concentrations of the reactants. The second question, "how do we know the form of the rate equation?" can be answered only by experiment, and we shall address this point later in this Section 13.4.

Now, consider a two-step reaction with the first step faster than the second, for example, the reaction between gaseous hydrogen and iodine at a temperature where iodine is in the gaseous state:

$$\text{H}_2 + \text{I}_2 \rightarrow 2\text{HI} \tag{13.34}$$

This reaction proceeds *via* the following elementary steps:

$$\text{I}_2 \rightarrow 2\text{I} \quad \text{with a fast rate determined by } k_1 \tag{13.35}$$

and a slow second reaction

$$2\text{I} + \text{H}_2 \rightarrow 2\text{HI} \quad \text{with a rate constant } k_3. \tag{13.36}$$

Since the first step produces more iodine atoms than the second (slow) reaction can consume, there is a build-up of I atoms. However, the reaction in Eq. 13.35 is an equilibrium reaction, and the reverse reaction of the first step becomes significant. I_2 molecules are regenerated according to:

$$2\text{I} \rightarrow \text{I}_2 \quad \text{reverse reaction, } k_2 \tag{13.37}$$

Thus, there are two competing reactions that reduce the I concentration: the formation of HI (Eq. 13.36) and the formation of I_2 (Eq. 13.37). At the beginning of the reaction, the concentration of I increases with time as the concentration of I_2 decreases (Eq. 13.35). Once enough atomic iodine has been produced, the reverse reaction (Eq. 13.37) becomes significant and we reach an equilibrium, in which the concentrations of I and I_2 do not change with time. This is shown in Figure 13.2. Previously, in Figure 13.1, we had assumed that there is no reverse reaction; therefore, the concentration of the starting material reached zero asymptotically. Now, in the case of a reverse reaction, both concentrations of reactants and products reach equilibrium values.

At equilibrium, the rate of the forward reaction and the rate of the reverse reaction just balance each other. So, at this point at equilibrium, the formation

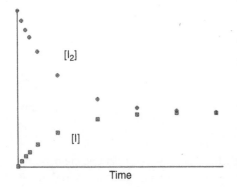

Figure 13.2 Concentrations of reactant I_2 and product I or the equilibrium reaction $\text{I}_2 \leftrightarrow 2\text{I}$.

of products and the disappearance of products to reform the reactants are exactly equal. Since both the forward and reverse reactions are elementary steps, we can write these rate equations for the two reactions as follows

Forward reaction rate $= k_1[I_2]$ (13.38)

Reverse reaction rate $= k_2[I]^2$ (13.39)

Since these rates are equal at equilibrium, we may write

$$k_1[I_2] = k_2[I]^2 \tag{13.40}$$

or

$$\frac{k_1}{k_2} = \frac{[I]^2}{[I_2]} = K_{eq} \tag{13.41}$$

which we recognize to be the equilibrium constant for the reaction $I_2 \rightarrow 2\,I$ (Eq. 13.35). Thus, the ratio k_1/k_2, the rate constants for the forward and reverse reaction, is the equilibrium constant! This is a general result: for a dynamic equilibrium as defined in Chapter 9, the forward and reverse reaction rates (not the rate constants) are equal; therefore, there are no changes occurring in the concentrations of reactants and products. For the $I_2 \rightarrow 2\,I$ equilibrium, the reverse reaction is controlled by rate constant k_2, which is smaller than k_1; the reaction will first occur in the forward direction. Once the concentration of I atoms becomes larger, the reverse reaction becomes significant.

Back to the $H_2 + I_2 \rightarrow 2\,HI$ reaction (Eq. (13.34)), we assume that reaction 13.35 achieves a "steady-state" condition in which iodine atoms, the reaction intermediate, reach a constant concentration with time. The I atoms are removed from the reaction mixture not only by the second step (the formation of the product), but also *via* the reverse reaction to recreate I_2.

Now, the slow and consequently rate-determining step obeys the rate equation:

rate $= k_3\,[I]^2[H_2]$ (13.42)

since it is an elementary step that depends on the collision of three particles, namely two I atoms and an H_2 molecule. The term $[I]^2$ can be obtained by solving Eq. 13.40 for $[I]^2$:

$[I]^2 = (k_1/k_2)[I_2]$ (13.43)

Finally, we substitute the value of $[I]^2$ into Eq. 13.42 and get for the overall rate

rate $= (k_1 k_3/k_2)[H_2][I_2]$ (13.44)

So, we have discussed both scenarios of a two-step reaction: if the first step is slow, as in the example in Eqs. 13.29 through 13.33

$NO_2 + CO \rightarrow NO + CO_2$

the overall reaction rate depends on the rate of the first step, the slow and rate-determining step that was found in this case to be of the form

Rate $= k[NO_2]^2$ (13.33)

If the second step is the slow step, as in the case of the H_2/I_2 reaction (Eqs. 13.37 through 13.37), we have to invoke a steady-state situation in which the first, fast step produces the reaction intermediate at such a high rate that the reverse reaction of this step has to be taken into account. Thus, we reach an equilibrium concentration of the reaction intermediate. Example 13.4 presents another two-step reaction with a fast first step.

Obviously, there is no way to predict the form of the rate law from the stoichiometry of the reaction. Furthermore, a rate law does not automatically determine the mechanism of a reaction. Take the hydrogen/iodine reaction: the rate law suggests that the reaction rate is linear in both the hydrogen and iodine concentration. So, one may believe that the reaction occurs by direct collisions between hydrogen and iodine molecules, and that, in fact, was believed for some time. However, a more detailed study showed that there is a reaction intermediate, the I atom, and the resulting rate expression contains the equilibrium constant for the I/I_2 reaction. Thus, we can summarize and say that a rate law never proves a reaction mechanism, but that certain mechanisms are consistent with an observed rate law.

Example 13.4 The decomposition of ozone to oxygen proceeds according to the following two-step mechanism:

$$O_3(g) \rightarrow O_2(g) + O(g) \text{ (fast, equilibrium constant K)}$$
$$\underline{O_3(g) + O(g) \rightarrow 2O_2(g)}$$
$$2O_3(g) \rightarrow 3O_2(g) \quad \text{(slow, rate constant } k_2\text{)}$$

Derive an expression for the rate of the overall reaction.

Answer: Rate = $k_2 K [O_3]^2/[O_2]$

13.4 Reaction Rates for Reactions That Are Nonlinear in Concentrations

That brings us back to the question posed earlier: how do we know what the rate law of a chemical reaction is? The answer to this question can only be found in experiments. In order to answer the question, one has to perform a number of different experiments, at the same conditions, but varying concentrations in a controlled manner. Take the reaction discussed earlier for an example:

$$NO_2 + CO \rightarrow NO + CO_2 \tag{13.29}$$

One finds experimentally that the overall reaction rate is independent of the CO concentration. This implies that the reaction rate, measured for example by the formation of CO_2, does not change at all when the CO concentration changes. This must imply that [CO] does not appear in the rate equation. What will happen when one doubles the NO_2 concentration between two consecutive experiments? It is found experimentally that the reaction rate (the rate of production of CO_2) goes up by a factor of 4. Thus, one concludes that [NO_2] appears to the second power in the rate law.

In general, if we want to determine the rate law for a generic reaction

$$A + B \rightarrow C$$

we write a trial rate equation of the form

$$\text{Rate} = \frac{\Delta[C]}{\Delta t} = k[A]^x[B]^y \tag{13.45}$$

where the x and y are **not** the stoichiometry coefficients, but unknowns to be determined experimentally. One accomplishes this by measuring the reaction rate (for example, the rate of formation of C) for initial concentrations [A] and [B]. Then, one varies these concentrations and re-measures the rate of the reaction. Assume, we leave [B] constant and double the concentration of A:

if reaction rate does not change	$x = 0$
if reaction rate doubles	$x = 1$
if reaction rate quadruples	$x = 2$

A similar set of experiments is carried out leaving A constant and varying B. The resulting exponents need not even be integers, and rate laws with fractional exponents are known to occur.

Two more aspects of the rate laws need to be mentioned here. First, the measurement of the actual reaction rates is by no means trivial. In the case of the disappearance of NO_2, it may be relatively simple to measure a rate because this species is strongly colored (which is a typical property of radical molecules). The disappearance of NO_2 can be monitored by the bleaching of the dark brown color of NO_2. Simple mathematical formulas exist that relate the intensity of the color to the concentrations. Often, however, it is much more difficult to determine the reaction rates. Typically, a small aliquot of the reaction mixture is removed at fixed times, say every 10 [s], and analyzed for the concentration of a reaction or product by standard analytical means. The rates are, of course, given by the time variation of the concentrations.

Secondly, we need to introduce more nomenclature and a few more equations. If a reaction depends on the second power of a reactant, we say that the reaction is of second order in this reactant. Using the examples above, the rate equation for

$$NO_2 + CO \rightarrow NO + CO_2 \quad rate = k[NO_2]^2 \tag{13.33}$$

is second order in NO_2 and zeroth order in CO. The overall order of the reaction is the sum of the individual orders, and we see that the reaction is of second order overall.

The formation of HI is first order in both hydrogen and iodine, and second order overall.

$$H_2 + I_2 \rightarrow 2\,HI \quad rate = (k_1 k_3/k_2)[H_2][I_2] \tag{13.44}$$

The rate equations are different for zeroth-, first-, and second-order reactions:

$$-\frac{d[A]}{dt} = k[A]^0 \quad \text{or} \quad -d[A] = k\,dt \qquad [A]_t = [A]_0 - kt \quad \text{Zeroth order} \tag{13.46}$$

$$-\frac{d[A]}{dt} = k[A]^1 \quad \text{or} \quad -\frac{d[A]}{[A]} = k\,dt \qquad \ln\frac{[A]_0}{[A]_t} = k\,t \quad \text{First order} \tag{13.47}$$

$$-\frac{d[A]}{dt} = k[A]^2 \quad \text{or} \quad -\frac{d[A]}{[A]^2} = k\,dt \qquad \frac{1}{[A]_t} - \frac{1}{[A]_0} = kt \quad \text{Second order} \tag{13.48}$$

The integrated rate laws for zeroth and second order can be obtained by integrating the three expressions on the left of 13.46 to 13.48 to give the different integrated rate laws. It will be left as an exercise to derive the equations for the reaction half-life for the zeroth and second-order reactions.

Example 13.5 Derive the equations for the reaction half-life for zeroth- and second-order reactions.

Answer: For a zeroth-order reaction, $t_{1/2} = \dfrac{[A]_0}{2k}$

For a second-order reaction, $t_{1/2} = \dfrac{1}{k[A]_0}$

Example 13.6 Consider the gas phase reaction

$aA + bB \rightarrow cC$.

The rate of the reaction doubles, if [B] is doubled, and does not change if [A] is doubled. Derive an expression for the rate law of this reaction.

Answer: Rate = $k\,[B]$

13.5 Reaction Path and Catalysis

The number of collisions between gaseous molecules depends on their concentrations (pressures) and the temperature, since gas particles move faster at higher temperatures. However, the temperature effects the collisions in a different way as well, namely through the kinetic energy. We remember that the absolute temperature of a gas is proportional to the average kinetic energy of the gaseous particles. So, as the temperature increases, the collisions become much more energetic. This is an important point, because when atoms or molecules collide, their prime reaction is to get away from the other species as fast as possible without getting involved. In other words, at low energies the particles behave like billiard balls and may collide and bounce off each other without reaction. This is so since the outsides of atoms and molecules are made up of the electrons in bonding and nonbonding orbitals that extend into space from the center of the atoms. Thus, the collision must be sufficiently energetic to overcome these repulsive forces. This repulsive energy that needs to be overcome is known as the activation energy. If we plot the energy (or enthalpy) of reactants and products along the pathway of the reaction, we may represent this activation energy as the hump between reactants and products, as shown in Figure 13.3.

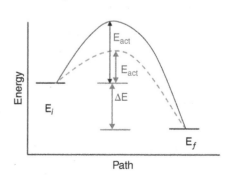

Figure 13.3 Activation energy for a reaction proceeding from an initial energy state E_i to a final energy state E_f via two different pathways.

The activation energy E_a affects the rate constant k of a given chemical reaction. We mentioned earlier that rate constants are independent of concentrations but are affected by temperature. This dependence is of the form

$$k = A_r \, e^{-\frac{E_a}{RT}} \quad \text{or} \quad k = A_r \, 10^{-\frac{E_a}{2.303\,RT}} \tag{13.49}$$

Eqs. 13.49 is known as the Arrhenius equation and describes how the temperature affects reaction rates. In Eq. 13.49, A_r is referred to as the pre-exponential factor (how imaginative) that takes values between 0 and 1 and determines what factors other than the temperature determine whether or not a collision between two reactants is leading to the formation of the products. It turns out that only a small fraction of collisions has the proper arrangement of the colliding particles to lead to the formation of the products. Think about it this way: in the reaction described by Eq. 13.3, a collision between the OH· radicals may be successful if the two oxygen atoms are facing each other, but unsuccessful if the two hydrogen atoms face each other.

The activation energy, E_a, can be determined by the slope of a plot of $\log k$ vs. $1/T$. We have encountered a similar situation before when we found that the vapor pressure of a liquid is given by

$$p_{\text{vap}} \propto 10^{-\frac{\Delta H_{\text{vap}}}{2.303\,RT}} \tag{9.10}$$

The enthalpy of vaporization of a liquid could be obtained by plotting the logarithm of the vapor pressure vs. $1/T$, see Figure 9.5. Similarly, we can obtain a value for the activation energy of a reaction by plotting the logarithm of the rate constant of a reaction vs. $1/T$. The (negative) slope of this plot is proportional to the activation energy, as shown by Eq. 13.49.

Example 13.7 For a chemical reaction with an activation energy of 50 [kJ/mol], calculate the increase in reaction rate caused by an increase in temperature from 25 [°C] to 45 [°C].

Answer: The reaction rate increases by a factor of 3.6.

Sometimes it is desirable to speed up the reaction rate by increasing the rate constant of chemical reactions, either to make more money (if you are in the chemical industry) or to escape a charging lion, if you are a zebra in the prairie and need massive amounts of ATP in your muscle to run. Raising the temperature will speed up chemical reactions but may either prove expensive or, as discussed for the formation of ammonia from the elements (see Section 9.5), may shift the equilibrium in an undesirable direction.

If you are the zebra in the prairie or any other warm-blooded creature running away from a predator, raising the temperature may not be a useful approach (unless you want to present yourself to the lion already cooked). There is another way to achieve raising a reaction rate without varying the temperature, and this is by providing a different pathway for the chemical reaction *via* catalysis. A catalyst is an additional reactant introduced into the chemical reaction that is regenerated at the end of the reaction. It may be thought of as forming a reaction intermediate, and therefore does not show up in a stoichiometric equation. How does a catalyst speed up a chemical reaction? It does so by providing another path for the reaction that lowers the activation energy of the reaction by making one or more of the reactants more likely to react. A prototypical reaction to study catalysis is the formation of water from a mixture of oxygen and hydrogen gas according to

$$2\,H_2(g) + O_2(g) \rightarrow 2\,H_2O(g) \tag{13.50}$$

Although the free enthalpy of this reaction predict that the equilibrium lies far on the side of the product, a mixture of hydrogen and oxygen does not produce any observable amount of water at room temperature. If, however, a spark or flame is used to ignite this mixture of gases, a spectacular explosion will occur and produce gaseous water with a large release of enthalpy. Thus, one concludes that the activation energy of the reaction in Eq. 13.50 is so high that the reaction rate at room temperature is practically zero. However, by (locally) providing with a flame or spark the energy to overcome the activation energy, the reaction will occur very fast, with the reaction enthalpy released at the onset of the reaction providing sufficient energy to maintain the reaction until all reactants are used up. However, the reaction will proceed rapidly and quietly if a pellet of spongy platinum is added to the reaction mixture. In this case, one writes Eq. 13.50 as

$$2\,H_2(g) + O_2(g) \xrightarrow{\text{Pt}} 2\,H_2O(g) \tag{13.51}$$

How does the platinum catalyze the reaction? It is well known that hydrogen gas is readily adsorbed on the surface of platinum metal. This process can be described as a weak chemical bond forming between the hydrogen molecules and the Pt atoms, as shown in Figure 13.4. The Pt—H bond weakens the H—H bond to such an extent that collisions of oxygen molecules with the hydrogen on the Pt surface causes the reaction between oxygen and hydrogen to occur rapidly, but without the explosion.

Figure 13.4 Schematic of hydrogen adsorption to a Pt surface.

Biological catalysts, also known as enzymes, accomplish the lowering of the activation energy by binding to the reactants (also referred to as the substrates) and surrounding the bonds that need to be formed or broken with groups to facilitate the process. If a polar bond breaks during a reaction, an enzyme may surround this bond with polar or charged centers to weaken the bond, thereby providing a new and improved pathway compared to the one in the absence of the enzyme. The lowered activation energy, in turn, speeds up the chemical reaction according to $e^{-\frac{E_a}{RT}}$. The pathway with lowered activation energy for the catalyzed process is shown by the dotted gray line in Figure 13.3, along with the lowered activation energy (also in gray).

Example 13.8 The half-life of the first-order radioactive decay of plutonium is 12 400 years. After how many years will 1 ton (1000 kg) of Pu have decayed to 1 g of Pu?

(a) $1.2 \cdot 10^4$ years
(b) $1.2 \cdot 10^5$ years
(c) $2.5 \cdot 10^5$ years
(d) None of the above

Answer: (c) Details in answer booklet

Example 13.9 Using a program such as EXCEL, plot the $[A]_t$ dependence with time for a first- and second-order gas phase reaction see Figure 13.5. Use $[A]_0 = 0.1$ [atm] and $k = 0.3$, and follow the reactions for 60 [s].

Answer: Details in answer booklet

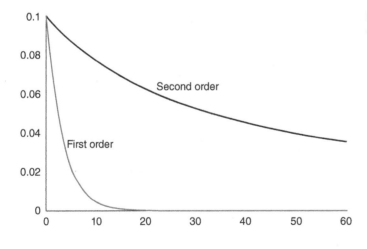

Figure 13.5 Dependence of the concentration [A] for first- and second-order reaction.

Example 13.10 Acetone reacts with bromine according to:

$$CH_3COCH_3 + Br_2 \rightarrow CH_3COCH_2Br + H^+ + Br^-$$

The rate of disappearance of bromine was measured in a series of experiments for several different concentrations of acetone, bromine, and hydrogen ions at a single temperature:

Experiment Nos.	[CH$_3$COCH$_3$]	[Br$_2$]	[H$^+$]	Rate of disappearance of Br$_2$
1	0.30	0.050	0.050	$6.0 \cdot 10^{-5}$
2	0.30	0.10	0.050	$6.0 \cdot 10^{-5}$
3	0.30	0.050	0.10	$1.2 \cdot 10^{-4}$
4	0.40	0.050	0.050	$8.0 \cdot 10^{-5}$

What is the rate law of the reaction?

Answer: The rate law is $-\Delta[\text{Br}_2]/\Delta t = k\,[\text{CH}_3\text{COCH}_3][\text{H}^+]$

Further Reading

https://chem.libretexts.org/Bookshelves/General_Chemistry/Map%3A_Chemistry_-_The_Central_Science_(Brown_et_al.)/14%3A_Chemical_Kinetics and subsections therein

14

Nuclear Reactions

14.1 Nuclear Reactions and Transmutations

So far in the discussion of chemistry in this book (with the exception of a short excursion in Chapter 13), we have assumed that chemistry deals with the rearrangement of electrons and that the nucleus of atoms does not undergo any changes. As far as chemistry is concerned, this assumption is correct. However, the nuclei of atoms themselves can undergo reactions that occur independently of the chemical reactions caused by the rearrangement of electrons. When nuclear reactions occur, the number of protons and/or neutrons in the nucleus changes; therefore, elements are converted into other elements or other nuclides. A nuclide is a species characterized by the atomic number Z (which also denotes the number of protons in the nucleus) and the number of neutrons N (we use an italic, capital letter N for the number of neutrons). For example, isotopes (introduced in Chapter 2) are nuclides with equal Z but different values of N. An example would be $^{13}_{6}C$ and $^{12}_{6}C$, for which the number of protons is $Z = 6$, but the number of neutrons is $N = 7$ and $N = 6$, respectively. Isotones are nuclides where the number of neutrons is equal, but the number of protons differs, for example $^{13}_{6}C$ and $^{14}_{7}N$. Nuclides can be stable, such as $^{13}_{6}C$ and $^{12}_{6}C$, or unstable, i.e. they undergo nuclear reactions. An unstable nuclide discussed earlier is $^{14}_{6}C$, which undergoes β-decay (see Section 14.2.2) to form $^{14}_{7}N$ with a half-life of 5730 years; obviously, it is a slow process. Other nuclide reactions might have half-lives of fractions of a second. Nuclear reactions require or produce energies far above those of chemical reactions.

In general, the process of one element turning into another element is referred to as "nuclear transmutation." Such a process may occur naturally, such as the very slow decay of $^{14}_{6}C$ to $^{14}_{7}N$ or the decay of uranium $^{238}_{92}U$ to lead (see Sections 14.3.1 and 14.3.5). Other transmutations are caused by human intervention: when atomic nuclei are bombarded with neutrons or α-particles (see below), nuclei can be changed and transformed from one element into another. The dream of alchemists of the Middle Ages of converting certain metals into gold has, thus, been realized; however, it is unlikely that the conversion of donkey dung into gold (another example of their dreams) will be feasible.

14.2 The Structure of Atomic Nuclei

The very basic view of the atomic nucleus introduced earlier (Chapter 2) stated that nuclei are composed of protons and neutrons, with the neutrons required to overcome the enormous repulsion two (or more) positively charged protons exert on each other. Furthermore, we assumed that in each successive element in the periodic chart, the number of protons increases by one; this increase in the number of protons is accompanied by an increase in N, the number of neutrons, but this latter increase is not linear, and heavy nuclei have far more neutrons than protons. Take the aforementioned $^{238}_{92}U$ isotope for an example: it has 92 protons and 146 neutrons. However, the added protons and neutrons, as we proceed through the periodic chart, are not just arbitrarily squashed in the nucleus but follow certain rules about nuclear energy levels.

As pointed out in the last paragraph of Chapter 5, these nuclear energy levels are summarized in a "nuclear shell model" that predicts the stability of nuclei as a function of the number of protons and neutrons. Since both protons and neutrons have spins, similar to the electronic spin, the nuclear "Aufbau" principle mimics the electronic Aufbau principle. The nuclear shell model predicts a maximum number of protons that may be found in the nucleus, as well as a maximum number of neutrons. In addition, the model describes which nuclei are energetically more stable and predicts the nuclear binding energy that determines which elements can be produced by nuclear fusion and fission (see Section 14.4).

Understanding Essential Chemistry, First Edition. Max Diem.
© 2025 John Wiley & Sons, Inc. Published 2025 by John Wiley & Sons, Inc.

In the nuclear shell model, the lowest shell (corresponding to the s orbitals in the electronic Aufbau principle) can hold a maximum of two protons and two neutrons, a situation that is found in the $^4_2$He atom. The next shell can contain six protons and six neutrons, and the atom that has the two lowest energy shells filled is the $^{16}_8$O atom. The third shell can contain 12 protons and 12 neutrons, and the element with all three shells filled to maximum capacity is $^{40}_{20}$Ca. These elements, with an equal number of protons and neutrons, and totally filled nuclear shells, represent particularly stable nuclei.

When nuclides have values of Z and N that do not obey these rules, they are less stable and may reach arrangements of lower energy by emitting particles from the nucleus. These processes give rise to what we refer to as radioactive decay and decay Series (see Section 14.3).

Before these nuclear reactions can be discussed, we need to introduce the nomenclature of nuclear particles as they are used in this chapter. An α-particle is defined as the naked nucleus of a helium atom:

$$\text{He}^{2+} = {}^4_2\alpha \tag{14.1}$$

where the superscript denotes the mass number ($Z+N$) and the subscript the number of protons. A neutron, therefore, is represented by

$$^1_0 n \tag{14.2}$$

a proton by $^1_1 p$ (14.3)

and an electron by $^{\ \ 0}_{-1} e$ (14.4)

(You may wonder why electrons are listed as "nuclear particles," since we know that the electrons are not found in the nucleus. This aspect will be clarified in Section 14.3.2). The electron specified in Eq. 14.4, $^{\ \ 0}_{-1}e$, is also referred to as a β-particle.

14.3 Radioactive Decay and Decay Chains

14.3.1 α-Decay

It was mentioned in Section 14.1 that uranium undergoes a radioactive decay into thorium. In order to write a balanced "chemical" equation for such a decay,

$$^{238}_{92}\text{U} \rightarrow {}^{234}_{90}\text{Th} + ?? \tag{14.5}$$

we see that the mass number changes by 4, and the atomic number by 2 units. Thus, the mass difference is carried away by an α-particle:

$$^{238}_{92}\text{U} \rightarrow {}^{234}_{90}\text{Th} + {}^4_2\alpha \tag{14.6}$$

We consider Eq. 14.6 to be a balanced equation since the number of protons on both sides is the same ($92 = 90 + 2$) and the mass numbers add up ($238 = 234 + 4$). Notice that we have ignored the fact that two valence electrons are not accounted for, but as we know, their mass is negligible compared to the nuclear masses involved, and the notation of nuclear reactions does not account for valence electrons. α-Decay is observed for heavy elements only and is a very common occurrence in the natural decay chain of uranium (see Section 14.3.5). The α-particle typically leaves the nucleus at a high velocity ($\sim 1.5 \cdot 10^7$ [m/s] or about 5% of the velocity of light). Despite the enormous energy of α-particles, they can be shielded readily by even a sheet of paper but can cause havoc when interacting with molecules.

Example 14.1 Calculate the energy of α particles in [kJ/mol], using their masses and the velocity given above. Compare this energy to the energy of a chemical reaction.

Answer: $4.5 \cdot 10^5$ [MJ/mol], as compared to about 1 [MJ/mol] for chemical reactions

14.3.2 β-Decay

Although not explicitly identified as such in Chapter 13 and Section 14.1, the radioactive decay of ^{14}C to ^{14}N is a β-decay:

$$^{14}_6\text{C} \rightarrow {}^{14}_7\text{N} + ?? \tag{14.7}$$

Here, it is much more difficult to decide what is missing in Eq. 14.7. Obviously, the number of protons increases from 6 to 7 in the reaction, but the total number of nucleons stays the same. This can happen only if one of the neutrons is converted to a proton. How can this be?

The answer to this problem lies in the fact that neutrons in a nucleus can decay into a proton and electron:

$$^1_0n \rightarrow {}^1_1p + {}^{\ 0}_{-1}e \tag{14.8}$$

This equation needs some discussion[1]. First, the equation is balanced in the sense of Eqs. 14.2 through 14.4 in that the masses and charges of particles add up: the neutron has neutral electric charge because the positive and negative charges of the proton and electron exactly balance. The number of protons (the subscripts) also balances, since the neutron has no protons, the proton (obviously) has one proton, and the electron has (−1) protons. Within the nucleus, neutrons may be considered stable; however, free neutrons decay according to Eq. 14.8 with a half-life of about 10 min. Combining the information of Eqs. 14.7 and 14.8, we obtain for the β-decay of ^{14}C:

$$^{14}_{6}C \rightarrow {}^{14}_{7}N + {}^{\ 0}_{-1}e \tag{14.9}$$

In Chapter 2, we listed the masses of the proton, neutron, and electron as $1.67262 \cdot 10^{-27}$ [kg], $1.67493 \cdot 10^{-27}$ [kg], and $0.00091 \cdot 10^{-27}$ [kg], respectively. Eq. 14.8 stipulates that the mass of the neutron, therefore, should be $(1.67262 + 0.00091) \cdot 10^{-27} = 1.67353 \cdot 10^{-27}$ [kg]; however, the accepted mass of the neutron is $0.0014 \cdot 10^{-27}$ [kg] higher than the combined mass of proton and electron. Some of this mass difference is converted to energy according to Einstein's mass–energy relation

$$E = m\ c^2 \tag{14.10}$$

and accounts for the enormous energies set free in nuclear reactions. This energy is about $1.5 \cdot 10^4$ [MJ/mol] for the β-decay of ^{14}C according to Eq. 14.9.

As opposed to α-decay that is restricted to heavy nuclei, β-decay occurs both in heavy and light nuclei. We have already discussed an example of β-decay in light nuclei above. An example of β-decay in a heavy nucleus is that of thorium ^{234}Th produced by the reaction described in Eq. 14.6:

$$^{234}_{90}Th \rightarrow {}^{234}_{91}Pa + {}^{\ 0}_{-1}e \tag{14.11}$$

in which the product, protactinium, Pa, has one more proton than the thorium atom. The overall mass number stays the same in Eq. 14.11; thus, the emitted particle must have mass number of zero and a neutron in the nucleus decays into a proton and an electron.

As mentioned above, the emitted electron in β-decay has high kinetic energy and, being an electron, can ionize atoms or molecules it encounters. In particular, ionization of a cell's DNA can cause permanent damage to DNA. Therefore, the health effects of β-radiation are substantial, and any handling of β-emitters requires shielding. However, a few centimeter thicknesses of plastic or paraffin is sufficient to capture the emitted electrons.

14.3.3 γ-Emission (γ-Decay)

Many of the α- or β-decay reactions discussed above produce nuclei in a highly excited state. It was pointed out before (Section 14.2) that nuclei themselves have different energy levels, just as electrons in an atom can be found in excited energy states. Highly excited nuclei can emit their excess energy as photons. However, these photons are very energetic and typically have wavelengths of about 10^{-11} [m] or frequencies of about $3 \cdot 10^{19}$ [Hz] and are referred to as γ-rays. For comparison, a mid-range X-ray photon has a wavelength of about 10^{-9} [m]. In nuclear fission processes, where large nuclei are split into smaller nuclei (for example, U into Ba and Kr, see Section 14.4), the energy released very commonly leaves the nuclei in highly excited states and γ-emission is very common. These γ-rays are the most dangerous radiation from nuclear reactions, since these photons have sufficient energy to break just about all chemical bonds and thus are ionizing radiation. Thick shielding using high atomic number atoms (such as lead) can protect from these photons.

---

1 Eq. 14.8 actually omits a neutrino-type particle that accounts for spin conservation, but is omitted here in this introductory discussion of nuclear reactions.

14.3.4 Positron Emission

In Section 14.3.2, we introduced the concept that neutrons can decay into a proton and an electron, according to Eq. 14.8:

$$_0^1 n \rightarrow\ _1^1 p +\ _{-1}^0 e \tag{14.8}$$

The reverse process, a proton decaying into a neutron, can also occur:

$$_1^1 p \rightarrow\ _0^1 n +\ ?? \tag{14.12}$$

For this to happen, a positively charged electron would have to be created. What is a positively charged electron, and does it exist? Indeed, it does. The positively charged electron is known as a positron (with a symbol $_{+1}^0 e$) and is a form of antimatter. In antimatter, the charge of the antielectron (positron) is +1, and the antiproton has a charge of −1. The mass of a positron is exactly the same as the mass of an electron, and the mass of a proton is exactly the same as that of an antiproton. A positron and an antiproton can form antihydrogen atoms. However, matter and antimatter cannot coexist: they annihilate to produce energy in the form of photons:

$$_{-1}^0 e +\ _{+1}^0 e \rightarrow 2\gamma \tag{14.13}$$

with the total mass of the particles converted to energy. Similarly, a proton and an antiproton will annihilate.

Example 14.2 Calculate the energy and wavelengths of the photons created in the reaction described by Eq. 14.13

Answer: $E = 8.19 \cdot 10^{-14}$ [J]; $\lambda = 2.4$ [pm]

Thus, the process introduced in Eq. 14.12, the decay of a proton into a neutron, can be completed

$$_1^1 p \rightarrow\ _0^1 n +\ _{+1}^0 e \tag{14.14}$$

Nuclear reactions that proceed with positron emission are, for example,

$$_{11}^{22}\mathrm{Na} \rightarrow\ _{10}^{22}\mathrm{Ne} +\ _{+1}^0 e \tag{14.15}$$

Notice that in positron emission, an element transmutes into an element of lower atomic number, since a proton decays to a neutron, while the mass number remains unchanged. An important positron emitter is $_9^{18}\mathrm{F}$.

$$_9^{18}\mathrm{F} \rightarrow\ _8^{18}\mathrm{O} +\ _{+1}^0 e. \tag{14.16}$$

It has a half-life of 109 min and can be substituted for a hydroxyl group in glucose, $C_6H_{12}O_6$, to form $C_6H_{11}O_5{}^{18}F$, which is uptaken and metabolized by mammalian cells very similarly to the way glucose is metabolized. Whatever the chemistry is for the fluorinated glucose, it still undergoes the positron decay described by Eq. 14.16 and can help localize specific sites of accumulation. Cancer cells, in general, uptake and metabolize glucose more rapidly than normal cells; therefore, cancerous areas will have higher positron production than normal cells. Since the positrons produced nearly immediately react with electrons in the tissue, two gamma photons are emitted that readily can be detected. Since the two photons created by the electron/positron annihilation are emitted collinearly but at 180° with respect to each other, an array of detectors surrounding the site of gamma ray emission can localize the site of the origin of the gamma ray to within a few millimeters by the time delay at which the gamma photons reach the detectors.

14.3.5 Nuclear Decay Chains

Two important naturally occurring nuclear decay chains will be introduced in this section. A nuclear decay chain is a set of α- and β-decay reactions that start with naturally occurring nuclides and proceed to a stable element. One of these chains starts with the uranium isotope $_{92}^{238}\mathrm{U}$ that, as discussed earlier (Eq. 14.6), undergoes α-decay to form $_{90}^{234}\mathrm{Th}$. $_{92}^{238}\mathrm{U}$ is the heaviest element found naturally on the earth (with the possible exception of trace amounts of plutonium, see Section 14.4) and is one of the primordial constituents of the earth. Its α-decay is very slow, indeed, with a half-life of $4.5 \cdot 10^9$ years (see Figure 14.1). Given that the estimated age of the earth is about $4.5 \cdot 10^9$ years, we can state that at this point in time, about half the original amount of uranium has decayed into thorium. This α-decay has produced a large percentage of all helium found on the earth and is also responsible for nearly half of the radioactive heat production that keeps the core of the earth molten.

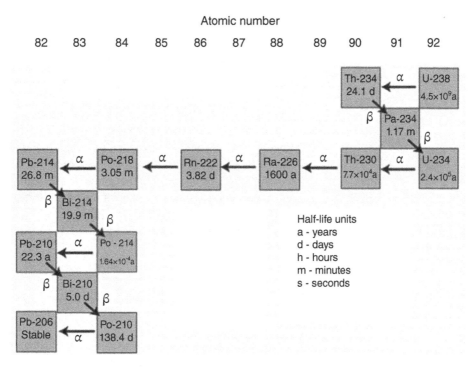

Figure 14.1 Nuclear decay chain of ^{238}U to ^{206}Pb. *Source:* Adapted from US Geological Survey.

The $^{234}_{90}$Th produced by the α-decay of $^{238}_{92}$U is unstable and decays *via* β-emission to palladium isotope, $^{234}_{91}$Pd with a half-life of 24.1 days (see Figure 14.1). After a second β-decay, we return to the element uranium, albeit to a lighter isotope, $^{234}_{92}$U. After this step, five consecutive α-decays produce the unstable lead isotope $^{214}_{82}$Pb. Two β-decays, one α-decay, two more β-decays, and a final α-decay finally produce the stable lead isotope, $^{206}_{82}$Pb. This decay chain is appropriately called the U-238 decay chain.

The other natural nuclear decay chain (the Thorium-232 decay chain) starts with $^{232}_{90}$Th and ends with the stable lead isotope $^{208}_{82}$Pb. Details of this chain will be the subject of Example 14.3.

Example 14.3 The Thorium-232 decay chain starts with $^{232}_{90}$Th. The following decays occur in this chain, starting with $^{232}_{90}$Th: α, β, β, α, α, α, α, β, α, β. Identify the intermediate and the final nuclides.

Answer: See answer booklet

Example 14.4 Which step in the decay scheme in Figure 14.1 is the rate-determining step?

Answer: Step 1 has the longest half-life; therefore, it is the slowest and rate-determining step.

Example 14.5 The nuclide $^{38}_{19}$K undergoes decay by positron emission. What is the nuclide produced?

Answer: $^{38}_{18}$Ar

14.3.6 Nuclear Dating

All nuclear decay reactions discussed so far, whether by α-, β-, or positron decay, follow first-order kinetics in the reactant. Thus, the rate laws derived for first-order kinetics hold:

$$-\frac{d[A]}{dt} = k\,[A] \tag{13.20}$$

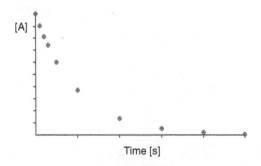

Figure 14.2 Exponential decay of the concentration of [A] as a function of time.

$$\ln\frac{[A]_0}{[A]_t} = k\,t \qquad (13.22)$$

$$\log\frac{[A]_0}{[A]_t} = k\,t/2.303 \qquad (13.23)$$

$$t_{(1/2)} = 0.693/k \qquad (13.27)$$

Therefore, the nuclear decay of any of the nuclides discussed will occur with an exponential decay as shown in Figure 14.2. Notice that we do not have to account for a back-reaction to reach an equilibrium condition as shown in Figure 13.2 since nuclear reactions are irreversible. We can, therefore, use the concentration of a nuclear reactant at time t, $[A]_t$, to calculate the concentration of A at any time before and after the present time t if the half-life, or the reaction rate, is known.

One example of nuclear dating, also known as radiometric dating, was discussed earlier (see Example 13.3) where the nuclear reaction was the β-decay of ^{14}C to ^{14}N according to

$$^{14}_{6}C \rightarrow {}^{14}_{7}N + {}^{0}_{-1}e$$

which has a half-life of 5730 years. Incidentally, ^{14}C is produced in the upper atmosphere by neutron capture (see Section 14.4) of $^{14}_{7}N$ according to

$$^{14}_{7}N + {}^{1}_{0}n \rightarrow {}^{14}_{6}C + {}^{1}_{1}p \qquad (14.17)$$

The ^{14}C is incorporated into plant tissue by photosynthesis and reaches a constant level as long as the plant is alive. After the plant stops photosynthesis, for example after it is consumed by you or a cow, the ^{14}C level in the plant or animal tissue decreases exponentially, as shown in Figure 14.2. ^{14}C dating can be used to determine the age of materials up to about 70 000 years.

Example 14.6 Why, in your opinion, is ^{14}C dating limited to time periods of about 70 000 years?

Answer: See answer booklet

When ^{14}C dating is no longer an option to determine the age of any materials, the radioactive decay processes with a longer half-life can be used. In geochemistry and geology, for example, the nuclear decay of nuclides such as ^{238}U to ^{206}Pb (see Figure 14.1) can be used to date very old rock formations since the half-life of this process is about 4.5 billion years or about the age of the earth.

Example 14.7 The total amount of ^{238}U presently on earth is estimated to be about $4 \cdot 10^{15}$ [kg]. What was the amount of ^{238}U on earth when it formed?

Answer: Twice this amount. See answer booklet.

14.4 Nuclear Fission and Nuclear Fusion

So far, we have discussed nuclear reactions that occur naturally, such as the different decay modes of nuclei. We also briefly touched upon a process (Eq. 14.17), neutron capture, which occurs in the presence of neutrons with suitable kinetic energy to enter a nucleus. Since the neutron is electrically neutral, it is not rejected by the highly positively charged nucleus of an atom and can penetrate it relatively easily. α-Particles, on the other hand, need much larger kinetic energies to penetrate a nucleus, but such processes can be readily carried out using modern technologies. Bombardment of nuclei with either neutrons or α-particles created the atoms in the periodic system past the element uranium.

When nuclei are bombarded with neutrons or α-particles, several different reactions can occur. Heavy nuclei, such as the uranium isotope ^{235}U, may break apart into smaller nuclei according to

$$^{235}_{92}U + {}^{1}_{0}n \rightarrow {}^{141}_{56}Ba + {}^{92}_{36}Kr + 3{}^{1}_{0}n \qquad (14.18)$$

This process is known as "nuclear fission," which will be discussed in more detail later (Section 14.4.2).

Light elements, on the other hand, may incorporate a neutron or α-particles in their nuclei and become other elements, like the example given in Eq. 14.17, or become heavier elements, like shown in Eq. 14.19:

$$^{12}_{6}C + ^{4}_{2}\alpha \rightarrow ^{16}_{8}O + \gamma \tag{14.19}$$

This process is known as nuclear fusion since two nuclei are fused together to produce a new, heavier element. Nuclear fusion is responsible for the formation of all heavy elements and will be discussed in Section 14.4.3. Thus, we see that the nuclei themselves have stabilities that are totally independent of the chemical conditions in which the corresponding atom is involved. This leads us to the discussion of nuclear binding energy or the energy involved in forming a nucleus from protons and neutrons.

14.4.1 Nuclear Binding Energy

The nuclear binding energy can be visualized by considering the exact masses of the nucleons forming an atomic nucleus. Take the oxygen isotope, $^{16}_{8}O$, for an example. It consists of eight protons and eight neutrons. The combined mass of these 16 nucleons, using the proton mass of $1.6726219 \cdot 10^{-27}$ [kg] and the neutron mass of $1.6749274 \cdot 10^{-27}$ [kg] amounts to $2.6780394 \cdot 10^{-26}$ [kg], whereas the mass of an oxygen atom is $2.656696 \cdot 10^{-26}$ [kg]. The mass difference, also known as the "mass deficit," $\Delta m = 2.134 \cdot 10^{-28}$ [kg], is proportional to the energy gained by forming oxygen from the nucleons. The proportionality constant is c^2, the square of the velocity of light, according to Einstein's mass–energy relation. Thus, this energy is

$$E = 1.908 \cdot 10^{11} \text{ [J/atom]} \tag{14.20}$$

This energy usually is expressed in units used in nuclear physics, MeV (mega electron volt), where

$$1 \text{ [eV]} = 1.602 \cdot 10^{19} \text{ [J]} \tag{14.21}$$

Then, the energy in Eq. 14.20 is commonly reported as

$$E = 119 \text{ [MeV/atom]} \tag{14.22}$$

Although the energy appears very small, it amounts to enormous values when converted to energy per mole of particles:

$$E = 1.908 \cdot 10^{11} \left[\frac{J}{atom}\right] \cdot 6.02 \cdot 10^{23} \left[\frac{atoms}{mol}\right] = 1.15 \cdot 10^{10} \text{ [kJ/mol]} \tag{14.23}$$

This nuclear binding energy explains why the oxygen nucleus is more stable than the 16 nucleons. The nuclear binding energy often is expressed in units of energy/nucleon; since there are 16 nucleons in an oxygen atom, the nuclear binding energy per nucleon is

$$E = 119/16 = 7.4 \text{ [MeV]} \tag{14.24}$$

in accordance with the value obtained from Figure 14.3. Notice that in this figure, the elements identified in Section 14.2 ($^{4}_{2}He$ and $^{16}_{8}O$) as having filled nuclear shells have high nuclear binding energies, indicating particularly stable nuclei.

Inspection of Figure 14.3 reveals that nucleus lighter than ^{56}Fe have lower nuclear binding energies; thus, fusion of these nuclei into heavier elements will release energy, a process referred to as nuclear fusion. Elements heavier than iron will release energy when the nuclei are broken into smaller nuclei, in a process referred to as nuclear fission. These two processes will be discussed next.

14.4.2 Nuclear Fusion

The nuclear reaction that powers life on earth, namely the nuclear fusion of hydrogen nuclei into helium nuclei that occurs in the sun, produces unimaginable amounts of energy. Looking again at Figure 14.3, we see that the difference in nuclear binding energy between H and He is about 7 or 28 [MeV] for the four nucleons in He. The sun converts about $6 \cdot 10^9$ [kg] of H nuclei into He nuclei every second, producing about $3.8 \cdot 10^{26}$ [J] of energy every second, which corresponds to a mass loss of $4 \cdot 10^9$ [kg], or 4 million tons/s. (Fear not, the sun's mass at present is about $2 \cdot 10^{30}$ [kg], so it can accommodate this mass loss for a while.)

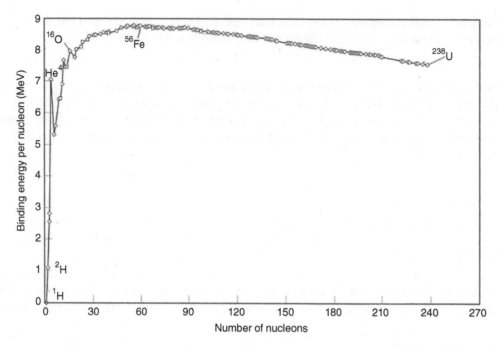

Figure 14.3 Nuclear binding energy per nucleon. *Source:* Adapted from https://en.m.wikipedia.org/wiki/File:Binding_energy_curve_-_common_isotopes.svg.

Heavier elements are produced in stars that are much more massive than our sun by successive addition of α-particles to $^{8}_{4}Be$ (which itself is produced by the fusion of two α-particles) to create $^{12}_{6}C$, $^{16}_{8}O$, $^{20}_{10}Ne$, etc., until $^{52}_{26}Fe$ is reached. This process is referred to as the "alpha process." Each successive addition of an α-particle releases about 7 [MeV] of energy; thus, these very massive stars can maintain nuclear fusion even when all the hydrogen is used up and converted to helium. Our sun does not have sufficient mass to enter the alpha process and nuclear fusion will come to an end once all the hydrogen is converted to helium.

Controlled nuclear fission to produce clean and nearly unlimited energy on earth has been the goal of intensive research in many sites in the world for about seven decades now. The process most likely to be harnessed is the fusion of two hydrogen isotopes, deuterium $^{2}_{1}H$, (often written as $^{2}_{1}D$), and tritium $^{3}_{1}H$, (often written as $^{3}_{1}T$), according to

$$^{2}_{1}H + ^{3}_{1}H \rightarrow ^{4}_{2}He + ^{1}_{0}n \tag{14.25}$$

releasing an energy of 17.6 [MeV]. To initialize a nuclear fusion reaction, the reactants have to be heated to temperatures of about 100 million degrees and kept at high pressures. This is accomplished by confining the reactants, in the form of a plasma, in a doughnut-shaped magnetic chamber known as a Tokamak. Very recently, fusion processes have been maintained for many seconds, producing more power than required for the heating of the plasma and the creation of the magnetic fields. The reactants in Eq. 14.25 are naturally occurring deuterium and tritium, which can be produced by irradiation of Li atoms by neutrons, according to

$$^{6}_{3}Li + ^{1}_{0}n \rightarrow ^{4}_{2}He + ^{3}_{1}H \tag{14.26}$$

Uncontrolled nuclear fission is also achieved in thermonuclear weapons, where a fusion reaction similar to Eq. 14.25 is initiated by the detonation of a nuclear fission device (see next section).

14.4.3 Nuclear Fission

Nuclear fission is a form of nuclear transmutation in which the nucleus of an element is transformed into other elements. An important nuclear fission reaction was introduced earlier, the fission of the ^{235}U isotope according to

$$^{235}_{92}U + ^{1}_{0}n \rightarrow ^{141}_{56}Ba + ^{92}_{36}Kr + 3^{1}_{0}n \tag{14.18}$$

but the products can vary and typically have mass numbers of 135 ± 15 and 95 ± 15. The nuclides produced in fission reactions generally are highly radioactive themselves and decay *via* the decay reactions discussed in Section 14.3 into more stable nuclides.

The fission reaction described by Eq. 14.18 produces three neutrons and releases about 200 [MeV] of energy per ^{235}U atom. A fraction of the neutrons released collide with other ^{235}U atoms to create what is termed a nuclear chain reaction. This chain reaction can lead to an uncontrolled reaction (i.e. a nuclear explosion) if the amount of ^{235}U exceeds what is known as the "critical mass." This critical mass is a little over 50 [kg] for ^{235}U. In order to control the nuclear chain reaction in a nuclear reactor for the commercial production of electricity, the number of neutrons being reabsorbed by ^{235}U atoms must be controlled such that the chain reaction proceeds in a manageable fashion. This is achieved by confining the "fuel," the ^{235}U, usually in the form of pellets of uranium oxides, in fuel rods that are submersed in a medium to carry away the heat produced, along with "moderators," a material that readily absorbs neutrons. The moderator rods, often made of graphite, also are submerged in the cooling liquid and adjusted such that the number of neutrons captured by the moderator allows for a controlled nuclear reaction. The heat generated is carried away by the cooling medium and transferred to water that evaporates into steam to drive steam-powered generators. This final step, of course, is the same for a coal, oil, or natural gas-fired power plant.

Although ^{235}U-based nuclear reactors can produce large amounts of energy, their practicality is limited due to the fact that ^{235}U occurs in naturally found uranium ores in a very small percentage (c. 0.7%). Thus, ^{235}U has to be enriched or purified for it to be useful as a nuclear fuel. This enrichment is carried out, among other methods, by converting the mixture of U isotopes to UF_6, which is a gaseous molecule. The two isotopic forms, $^{235}UF_6$ and $^{238}UF_6$, can be separated by the fact that their effusion rates differ ever so slightly, as discussed in Section 8.7. However, this process is very time- and energy-intensive and limits the use of ^{235}U-based nuclear reactors. Therefore, another nuclear fission reaction is frequently used in nuclear power plants, starting with ^{239}Pu. This nuclide can readily be produced in nuclear reactors (the so-called breeder reactors) by irradiating ^{238}U with neutrons to produce ^{239}U, which undergoes two successive β-emissions to turn into $^{239}_{94}$Pu.

Example 14.8 Write complete nuclear reaction equations for the process indicated above, leading from ^{239}U to ^{239}Pu.

Answer: See answer booklet

Upon capturing a neutron, $^{239}_{94}$Pu can undergo fission into several products along different pathways such as

$$^{239}_{94}\text{Pu} + ^{1}_{0}\text{n} \rightarrow ^{134}_{54}\text{Xe} + ^{103}_{40}\text{Zr} + 3\,^{1}_{0}\text{n} \tag{14.27}$$

that again produces a sufficient number of neutrons to maintain a chain reaction.

^{235}U- and ^{239}Pu-based nuclear reactors have produced enormous amounts of energy since their first commercial uses in the late 1950s and early 1960s. However, there are some serious problems with the technology. First of all, they produce copious amounts of highly radioactive waste products that need to be stored for millennia deeply underground in vitrified form. Second, nuclear reactors may, if improperly operated or maintained, pose a significant danger, as the three most publicized nuclear accidents demonstrate: the Three Mile Island (1979, Pennsylvania, USA), the Chernobyl (1986, Ukraine), and the Fukushima (2011, Japan) disasters. In each of them, a meltdown of the nuclear fuel rods occurred, and massive amounts of radioactive fission products breeched the confinement buildings. In addition, a large number of lesser nuclear accidents happened, many of them in military nuclear reactors that were often not revealed to the public until decades later. Although there is certainly the necessity to provide electrical power for both industrial and private use, the use of nuclear fission technology to provide this power is a double-edged sword and has to be carefully balanced against other technologies that have a zero-carbon footprint.

Further Reading

https://en.wikipedia.org/wiki/Radioactive_decay

15

Fundamentals of Quantum Chemistry, Spectroscopy, and Structural Chemistry

This is a chapter that intends to introduce you, the science students, to an area of science where mathematics, physics, and chemistry intersect and provide you with a broader (and perhaps more philosophical) view of chemistry in the context of mathematics. The chemistry course you are enrolled in most likely will never get to the subjects discussed in this chapter, so its intension is to take you one notch higher in the understanding of the essential chemistry principles. It also intends to answer the questions that might have popped up while studying the previous chapters: how do we know all this stuff? How do we know that the H—N—H bond angle in ammonia is smaller than the H—C—H bond angle in methane? How do we know that our model of atoms predicts the distribution of electrons correctly? How do we know that MO theory-based prediction of bonding and antibonding orbital energies is correct? How do we know that SF_4 is not a tetrahedral molecule?

The answers to such questions are, as in most cases in empirical sciences, "…from experiment." Therefore, this chapter introduces the experiments that underlie modern methods of structural analysis and discusses how the interpretation of experimental results has led to models that guided us to understand the structure of atoms, molecules, and materials, in general. This understanding, in turn, has led to advances in producing compounds that can fulfill specific goals, such as a medication that targets a distinct disease.

We start this chapter with a more sophisticated view of the quantum mechanical foundations of the hydrogen atom structure, and how understanding of the mathematical approach of quantum mechanics led to the formulation of the periodic chart of elements that contains much of the principles of chemistry. We then shall expand from the H atom to other atoms and describe how spectroscopy, in this case atomic absorption and emission spectroscopy, shed light on the structure of atoms heavier than hydrogen and enabled us to experimentally determine atomic energy levels. In this context, we shall explore modern methods in analytical spectroscopy: rather than using time-consuming methods of classical analytical chemistry – quantification of elemental species in a sample by separation and selected precipitation methods that were still taught as the "state-of-the-art" to the author in the late 1960s – modern spectroscopic methods such as induction-coupled plasma atomic emission (ICP-AE) spectroscopy will be introduced.

This will be followed by a discussion of molecular energy levels that are also predicted by quantum mechanics and are the foundation for many branches of molecular spectroscopy, such as electronic, vibrational, fluorescence, rotational, and spin spectroscopies, as well as another structural method, X-ray diffraction, which allows three-dimensional models of small and large molecules (such as proteins, enzymes, or DNA sequences) to be established. X-ray diffraction methods have even revealed the shape of electron distribution around atoms and have, thereby, confirmed the shape of atomic orbitals (see comments in Section 5.4) predicted by quantum mechanics. Among the most powerful methods of modern methods in analytical chemistry is mass spectrometry, which will be discussed after the spectroscopic methods.

15.1 Wavefunctions and the 1D and 2D Particle in a Box

Quantum mechanics is unique in the sciences in that it is based on postulates, rather than axiomatically on steps that can be derived from prior knowledge. One of these postulates is that everything about a system, including its present and future behavior, is contained in a wavefunction, ψ, where the square of the wavefunction expresses the probability of finding the system at a given time and space. These wavefunctions are different for the various problems in which we are interested; thus, the wavefunctions for electronic distributions in atoms or molecules or the rotation or vibration of molecular systems are all different.

The quantum mechanical method, in general, works as follows. We define a mathematical operator – a mathematical set of instructions – that describes the problem. For example, for the hydrogen atom, the operator states that the total energy of the hydrogen atom is the sum of the kinetic and potential energies of the electron in the electrostatic field of the nucleus. Then, one writes the kinetic energy in terms of the electron's momentum p and mass as $p^2/2m$ and replaces the momentum p by its quantum mechanical equivalent, namely the derivative of the wavefunction with respect to the spatial coordinates. Since the momentum operator appears squared in the expression for the kinetic energy, the quantum mechanical expression for the kinetic energy contains the second derivative with respect to the spatial coordinate.

Therefore, the operator – known as the Hamilton operator or simply as the "Hamiltonian" – is a differential operator (more on this in Section 15.2). When this operator is applied to the (still unknown) wavefunctions, it gives us a set of allowed energy states of the system, known as the energy "eigen" values (or specific values). Furthermore, in this process, the unknown wavefunctions associated with each energy eigenvalue are obtained. These functions are known as the eigenfunctions. This formulation of a problem as an operator/eigenvalue/eigenfunction is nothing specific to quantum mechanics but is common in a branch of mathematics known as linear algebra. Unfortunately, most chemistry undergraduates are not familiar with this mathematical approach, and therefore the mathematics of quantum mechanics appears somewhat mysterious to most of us.

Without going into the mathematical details, the linear algebra formalism will be introduced using a very simple quantum mechanical system, an electron trapped inside an energy well. Such an energy well might be a small (< 50 nm diameter) spot of a semiconductor, such as silicon, on a nonconducting substrate. The electrons in the conductivity band (see Section 7.1) of the silicon "dot", or "quantum dot", are a physical system that can be described very well by this electron-in-a-box model. To make things easy, let us consider a one-dimensional box first and make the length of the box L.

The potential energy walls outside are so high that the electron cannot escape. Classically, a particle in such an energy well appears as shown in Figure 15.1a. The particle can be anywhere inside the "box" and rests on the bottom of the box. However, a quantum mechanical system, such as an electron, behaves differently: since it has wave character. According to the Heisenberg uncertainty principle, it cannot be localized exactly and behaves like a standing wave, more like the string of a guitar. This is shown in Figure 15.1b by the lowest curve (marked $n = 1$), which represents the wavefunction of the electron inside the box. The square of this wavefunction, shown by the bottom curve in Figure 15.1c, represents the probability of finding the electron and is highest at the center of the box. Notice also that the wavefunction is not at the bottom of the box, but offset from the bottom (see later).

The "particle-in-a-box" wavefunction $\psi(x)$ is obtained mathematically by solving a simple differential equation (the particle-in-a-box Schrödinger equation), the solutions of which are of the form

$$\psi_n(x) = \sqrt{\frac{2}{L}} \sin\left(\frac{n\pi}{L}\right) x \tag{15.1}$$

In Eq. 15.1, L is the length of the box and n is a quantum number, running from 1 to infinity, since the differential equation which needed to be solved has infinitely many solutions. These functions $\psi_n(x)$ are called the eigenfunctions for the particle-in-a-box. For $n = 1$ (the ground state), Eq. 15.1 becomes

$$\psi_1(x) = \sqrt{\frac{2}{L}} \sin\left(\frac{\pi}{L}\right) x \tag{15.2}$$

This wavefunction is a simple sine wave with the amplitude $\sqrt{\frac{2}{L}}$, shown by the lowest curve (marked $n = 1$) in Figure 15.1b. The amplitude of the wavefunction is zero at $x = 0$, since $\sin(0) = 0$. It is also zero at $x = L$, since $\sin(\pi) = 0$. These two conditions result from the boundary condition when solving the differential equation of the system. The second eigenfunction would be

$$\psi_2(x) = \sqrt{\frac{2}{L}} \sin\left(\frac{2\pi}{L}\right) x \tag{15.3}$$

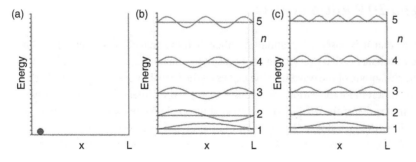

Figure 15.1 (a) A "classical" particle placed into a box. (b) A quantum mechanical particle in a one-dimensional box. Notice the eigenfunctions and corresponding eigenvalues. (c) Squared wavefunctions, giving the probability (amplitude) of the particle.

which is a sine wave of twice the frequency. It is still zero at $x = 0$ and $x = L$, but is also zero (has a "node") in the middle of the box, as shown by the second lowest curve in Figure 15.1b, marked $n = 2$. The next eigenfunction is

$$\psi_3(x) = \sqrt{\frac{2}{L}} \sin\left(\frac{3\pi}{L}\right) x \tag{15.4}$$

and so forth.

Each wavefunction or eigenfunction is associated with a particular energy eigenvalue, given by

$$E_n = \frac{n^2 h^2}{8mL^2} \tag{15.5}$$

In Eq. 15.5, h is Planck's constant and m the mass of the electron. Thus, we see that for

$$\begin{aligned} n = 1 \quad & \psi_1(x) = \sqrt{\frac{2}{L}} \sin\left(\frac{\pi}{L}\right) x \quad & E_1 = \frac{h^2}{8mL^2} \\ n = 2 \quad & \psi_2(x) = \sqrt{\frac{2}{L}} \sin\left(\frac{2\pi}{L}\right) x \quad & E_2 = \frac{4h^2}{8mL^2} \\ n = 3 \quad & \psi_3(x) = \sqrt{\frac{2}{L}} \sin\left(\frac{3\pi}{L}\right) x \quad & E_3 = \frac{9h^2}{8mL^2} \end{aligned} \tag{15.6}$$

These energy eigenvalues explain why the curves for the different values of n were drawn offset by n^2 in Figure 15.1b and c.

This simplest quantum mechanical system actually explains many features that will reoccur in future discussions of molecular systems, such as quantized energy states. These quantized energy states imply that quantum mechanical systems cannot assume arbitrary energy values that a ball on an inclined plane can assume, but only quantized energy levels like the ball on a staircase (refer to Figure 5.5 in Chapter 5). Furthermore, this model explains the color – the wavelength at which a molecule absorbs light – of certain compounds which may be described by an electron in a one-dimensional confinement. An example for such a molecule is retinal, shown in Figure 15.2. In this molecule, the chain of single and double bonds represents a system that can be described quite well by the particle-in-a-box model since the electrons involved in the π bonds are delocalized over the entire single bond–double bond distance (the "conjugated length"). The color of such a molecule is determined by an electron being promoted from one energy level to another. The energy difference between the initial and the final energy states of the electron is provided by light. This process requires a photon of a particular wavelength to be absorbed by the dye molecule, thereby removing a particular color of light from a broadband light source such as the sun or an incandescent light bulb. More about this is in Examples 15.1 and 15.2.

Example 15.1 Calculate the energy levels [in units of J] for an electron in a box for $n = 1, 2$, and 3. Let the dimensions of the "box" be 100 [pm] (the approximate length of a chemical bond). The mass of an electron is $m_e = 9.1 \cdot 10^{-31}$ [kg].

Answer: $E_1 = 6.0 \cdot 10^{-18}$ [J] $\quad E_2 = 2.4 \cdot 10^{-17}$ [J] $\quad E_3 = 5.4 \cdot 10^{-17}$ [J]

Example 15.2 Calculate the wavelength of a photon that can promote the electron in a box from the ground state ($n = 1$) and the first excited state ($n = 2$). Use the energy levels from Example 15.1.

Answer: $\lambda = hc/\Delta E = 11$ [nm], a deep UV photon

Example 15.3 Recalculate the energy difference between the ground and the first excited state for an electron in a box if the length of the box is 1 [nm].

Answer: $\lambda = hc/\Delta E = 1.1$ [µm], a near IR photon

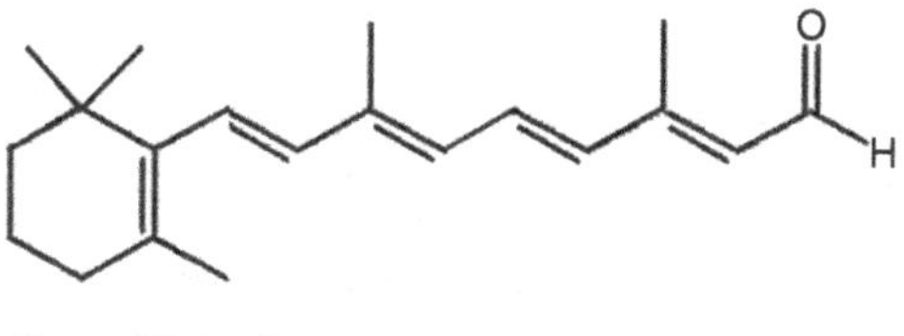

The difference between the particle- and the wave-like behavior of the particle in a box is summarized in Figure 15.1. The classical model predicts that the electron is found anywhere at the bottom of the energy well just as a marble would be in a box, whereas in the quantum mechanical model, the electron has wave character and is best described by a wavefunction. The square of this wave function indicates where the electron is

Figure 15.2 Structure of retinal, which is the light-absorbing moiety of the pigment rhodopsin in the human eye.

found most likely, see Figure 15.1c. Furthermore, the electron in the particle-in-a-box model can no longer assume arbitrary energy values, but only discrete values given by Eq. 15.5. Notice that the number of nodal points in the wavefunctions increases with increasing quantum number n.

Although the one-dimensional particle-in-a-box model has physical applications, it is useful to consider a two-dimensional case, such as a square quantum dot described above. Let us assume that the length L of the box is the same in both the x and y direction. Then the wavefunction of the electron $\psi_{n_x n_y}(x,y)$ is just the product of the two wavefunctions in the x and y direction:

$$\psi_{n_x n_y}(x,y) = \frac{2}{L} \sin \frac{n_x \pi x}{L} \sin \frac{n_y \pi y}{L} \tag{15.7}$$

The wavefunction can be represented as shown in Figure 15.3 for the cases $n_x = 1$ and $n_y = 1$ (Panel a), $n_x = 2$ and $n_y = 1$ (Panel b), and $n_x = 1$ and $n_y = 2$ (Panel c). The energy of the electron is

$$E_{n_x n_y} = \frac{(n_x^2 + n_y^2) h^2}{8mL^2} \tag{15.8}$$

These wavefunctions represent the standing waves on a square drum. Notice that the energy eigenvalues for these two cases shown in Figure 15.3b and c are the same:

$$E_{21} = E_{12} = \frac{5 h^2}{8mL^2} \tag{15.9}$$

When two or more energy eigenvalues for different combination of quantum numbers are the same, these energy states, E_{21} and E_{12}, are said to be *degenerate*. This is a common occurrence in quantum mechanics, as will be seen later in the discussion of the hydrogen atom where all three 2p orbitals, all five 3d orbitals, and all seven 4f orbitals are found to be degenerate. Please note that instead of nodal points (as shown in Figure 15.1), we now observe nodal lines, that is, lines that have zero amplitude of the wave function during an entire oscillation cycle. Furthermore, it is important to state that for degenerate states, any combination of the wavefunctions is also an allowed solution of the wave equation. To be specific, consider the degenerate states described by the wavefunctions $\psi_{2,1}(x,y)$ and $\psi_{1,2}(x,y)$. Then, any combination of the two wavefunctions of the form

$$a\psi_{2,1}(x,y) + b\psi_{1,2}(x,y) \quad 0 < a, b < 1, \ a + b = 1 \tag{15.10}$$

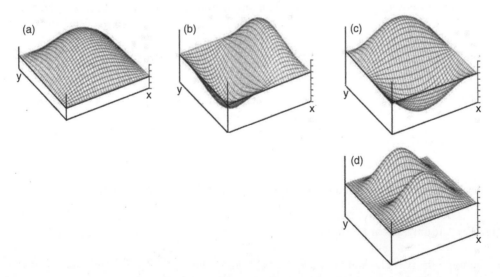

Figure 15.3 Panel (a): Wavefunction $\psi_{n_x n_y}(x,y)$ for a particle in a two-dimensional box with $n_x = n_y = 1$. Panel (b): Wavefunction $\psi_{n_x n_y}(x,y)$ for a particle in a two-dimensional box with $n_x = 2$ and $n_y = 1$. Panel (c): Wavefunction $\psi_{n_x n_y}(x,y)$ for a particle in a two-dimensional box with $n_x = 1$ and $n_y = 2$. Panel (d): Density plot $\psi_{n_x n_y}^2(x,y)$ for a particle in a two-dimensional box with $n_x = 1$ and $n_y = 2$. Modified from Breneman, G.L., J.Chem.Ed. (1990). For an animation of these wavefunctions, see http://www.acs.psu.edu/drussell/Demos/MembraneSquare/ Square.html. *Source:* Breneman G.L. (1990).

$k = 1, m = 0$ $k = 2, m = 0$ $k = 0, m = 1$

Figure 15.4 Plot of the square of the three lowest energy wavefunctions of an electron in a cylindrical well. *Source:* Electron density plots created by https://demonstrations.wolfram.com/ParticleInAnInfiniteCircularWell/. For an animation of these wavefunctions, see https://en.wikipedia.org/wiki/Vibrations_of_a_circular_membrane.

is also a solution to the mathematical problem. This "mixing" of wavefunction is called "hybridization," which we came across when combining atomic orbitals in Chapter 6.

Things get more realistic, but mathematically more complicated, when we confine the electron to a cylindrical well. Here, it is no longer useful to use the Cartesian coordinates x and y to define the position of the electron, but rather the radius r and its azimuth φ. The transformation from the Cartesian coordinates x, y to the radial, or polar, coordinates r, φ is discussed along with spherical polar coordinates, in Chapter 1, Section 7. The center of this new coordinate system is the center of the well with radius r. The angle φ varies from 0 to 2π. For this system, the wavefunctions and energies are described by two quantum numbers m, associated with the angle φ, and k, associated with the radius r.

The three lowest energy solutions of the electron in a cylindrical well system are shown in Figure 15.4. Panel (a) of Figure 15.4 shows the electron density plots (the squares of the wavefunctions) for the lowest energy state. The solution for this circular boundary case is the equivalent of the solution we saw in Figure 15.3a for the square case. However, in the circular case, the solution shown in Figure 15.4a has radial symmetry; that is, its wavefunction and density distribution are the same for any angle φ. The quantum numbers k and m are 1 and 0, respectively.

The next two solutions are really intriguing and pave the way toward our understanding of the hydrogen atom: The solution shown in Figure 15.4b still has radial symmetry and a maximum electron density at the center (at $r = 0$). However, it exhibits a ring of zero density (a nodal circle), followed by a second maximum in electron density. This density distribution is the two-dimensional equivalent of the "2s" orbital in the hydrogen atom. The quantum numbers k and m are 2 and 0, respectively. This distribution, with its radial symmetry, has no equivalent in the square box. The distribution shown in Figure 15.4c shows an electron density distribution that does not exhibit radial symmetry but has two maxima in the electron distribution similar to those shown in Figure 15.3d for the square box. The quantum numbers k and m are 0 and 1, respectively. This electron density corresponds to the 2p hydrogen orbitals.

Example 15.4 Describe the number and nature of the nodes of the wavefunctions for a
(a) particle in a one-dimensional box
(b) particle in a square box

Answer: (a) There are $(n - 1)$ nodal points; (b) there are $(n_x + n_y - 2)$ nodal lines.

15.2 Spherical Harmonics, Hydrogen Atom Wavefunctions, and Hydrogen Atomic Orbitals

The probability of finding an electron in a hydrogen is described as a three-dimensional problem, and the potential function is quite different from that of the particle-in-a-box systems described before. In the latter case, we assumed that the potential energy was zero inside the box and infinitely large outside the box. Therefore, the electron could not escape this confinement. In the hydrogen atom, the electron is subject to the electrostatic potential energy of the nucleus, which we described before by Coulomb's law as

$$U = k\, e_e\, e_N/r \tag{5.17}$$

Since the electronic charge is equal and opposite to the nuclear charge, we write Eq. 5.17 as

$$U \propto -e^2/r \tag{15.11}$$

A plot of this function is shown in Figure 15.5b. Notice that the potential energy in a hydrogen atom has spherical symmetry, that is, the potential energy, or the interaction between nucleus and electron, decreases with $1/r$ in whatever direction we consider. Thus, it makes sense that we describe the energy of the electron in a coordinate system that reflects this spherical symmetry, just as we described the coordinates of the electron in a circular well by polar coordinates (see Section 1.7). This leads us to the hydrogen-atom Schrödinger equation, where the term "Schrödinger equation" is used for any quantum mechanical expression where an energy operator (see Section 15.1) is applied to a set of wavefunctions to give us the energy eigenvalues of the system, and – in the process – defines the wavefunctions. The total energy operator - the Hamiltonian of the system – consists of kinetic and potential energy terms. The potential energy term is very simple and is defined by Eq. 15.11. The kinetic energy part of the Hamiltonian contains the second derivative of the wavefunction with respect to the spatial coordinates, as pointed out before in the discussion of the one-dimensional particle in a box. Thus, the kinetic energy for a three-dimensional problem contains the term

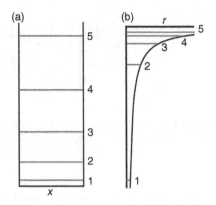

Figure 15.5 Comparison of potential energy functions and energy eigenvalues for (a) particle in a box and (b) the hydrogen atom.

$$\frac{\partial^2}{\partial x^2} + \frac{\partial^2}{\partial y^2} + \frac{\partial^2}{\partial z^2} \tag{15.12}$$

in Cartesian coordinates, but becomes a fear-instilling

$$\frac{1}{r^2}\frac{\partial}{\partial r}\left(r^2\frac{\partial}{\partial r}\right) + \frac{1}{r^2\sin^2\theta}\frac{\partial^2}{\partial \phi^2} + \frac{1}{r^2\sin\theta}\frac{\partial}{\partial \theta}\left(\sin\theta\frac{\partial}{\partial \theta}\right) \tag{15.13}$$

in spherical polar coordinates (see Section 1.7). Thus, the operator, or the mathematical instruction that tells us what to do to the wavefunction, requires us to take (second) derivatives of the wavefunction, either with respect to the Cartesian or to the spherical polar coordinates. This leads to a differential equation, that is, an equation that contains a variable and a derivative of this variable. We encountered differential equations before, in Chapter 13, when rate equations were introduced (Eq. 13.20). These differential equations were solved easily, by straightforward integration, and gave unique solutions.

The differential equation resulting from the total energy Hamiltonian for the hydrogen atom looks intimidating but has been solved, in parts, long before quantum mechanics, by French mathematicians of the nineteenth century. It was E. Schrödinger who realized that the hydrogen atom problem could be cast in the form of these known differential equations. He published these results in 1926 and thereby became the father of quantum mechanics. The first of three classical papers by Schrödinger actually is entitled "Quantisierung als Eigenwertproblem" (quantization as an eigenvalue problem) and introduces the hydrogen atom wavefunctions $\psi_{nlm}(r,\theta,\phi)$ as the product of a radial part, $R_{nl}(r)$, and an angular-dependent part, $Y_l^m(\theta, \phi)$:

$$\psi_{nlm}(r,\theta,\phi) = R_{nl}(r)\, Y_l^m(\theta, \phi) \tag{15.14}$$

(see also Eq. 5.22). Here, the subscripts n, l, and m are three quantum numbers associated with the variables r, θ, and ϕ, respectively. They were introduced in Chapter 5 and given the names of main, angular (orbital), and magnetic quantum numbers, respectively. It should come as no surprise that three quantum numbers are required to describe the problem since we needed one quantum number for the one-dimensional particle-in-a-box and two quantum numbers for either the square or circular two-dimensional particle-in-a-box. As discussed in Chapter 5, the solutions of both the radial and the angular parts are polynomials in their respective variables. For the radial part, these are known as the Laguerre polynomials. The first three Laguerre polynomials in x are (1), $(1-x)$, and $\left(1 - 2x + \frac{x^2}{2}\right)$, respectively. The first three radial parts $R_{nl}(r)$ are as follows:

$$R_{10} = 2\left(\frac{1}{a}\right)^{3/2} e^{-r/a} \tag{15.15}$$

$$R_{20} = \frac{1}{\sqrt{2}}\left(\frac{1}{a}\right)^{3/2}\left(1 - \frac{r}{2a}\right)e^{-r/2a} \tag{15.16}$$

$$R_{30} = \frac{2}{\sqrt{27}} \left(\frac{1}{a}\right)^{3/2} \left(1 - \frac{2r}{3a} + \frac{2r^2}{27a^2}\right) e^{-r/3a} \quad (15.17)$$

In Eq. 15.15, "a" is a constant (the Bohr radius, $a = 5.29 \cdot 10^{-11}$ [m] = 52.9 [pm]), and r is the distance from the nucleus. These functions all contain an exponential decay function (see Section 1.6), $e^{-r/na}$ that indicates a maximum at $r = 0$ as depicted in Figure 5.10a for the R_{10} and R_{20} functions, and an exponentially decrease in all directions from the nucleus. Notice that the R_{20} and R_{30} functions cross the abscissa. This represents a spherical region at a certain distance from the nucleus where the wavefunctions, and their squares, are zero. This spherical "nodal plane" describes a region in space where the electron cannot be found.

Example 15.5 Use EXCEL or a similar graphics program to plot the three radial wavefunctions R_{10}, R_{20}, and R_{30} given in Eqs. 15.15–15.17. Based on these plots, how many spherical nodal planes will be there in each of the 1s, 2s, and 3s hydrogen-like orbitals? How far from the nucleus are the nodal planes?

Answer: See answer booklet

The exponential decrease of these functions, in all directions from the nucleus, can best be visualized by a probability density plot shown in Figure 15.6. This figure represents the probability of finding an electron at a given point in space around the nucleus if you perform thousands of individual experiments. Thus, these plots show the square of the radial wavefunctions. Panel (a) depicts this probability density in three-dimensional space for R_{10}. Panel (b) shows a cut through this probability cloud along the plane indicated in light gray in Panel (a). Any plane that contains the coordinate center would show exactly the same probability density, because of the spherical symmetry. Panels (c) and (d) show the corresponding probability density for R_{20}, and Panels (e) and (f) for the R_{30} radial functions. Notice that Panel (d) in Figure 15.6 represents a three-dimensional analog of the electron density of the particle-in-a-cylindrical box shown in Figure 15.4b.

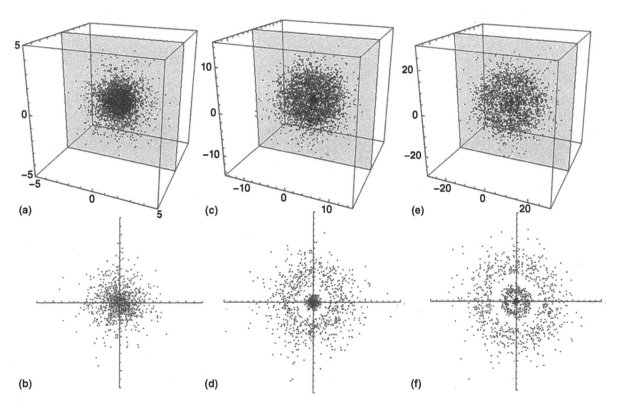

Figure 15.6 Probability density plots for the radial wavefunctions R_{10}, R_{20}, and R_{30}. Notice the size of the volumes: the edges shown are for r/a values from −5 to 5 for the R_{10} function, from −15 to 15 for the R_{20}, and from −30 to 30 for the R_{30} function. *Source:* Probability density plots created by https://demonstrations.wolfram.com.

The angle-dependent solutions of the Schrödinger equation were discussed before in Chapter 5, and are the spherical harmonic functions. We have discussed how the shapes of these spherical harmonics determine the shape of the hydrogen atom orbitals and the shape of the orbitals of elements with more than one electron. The spherical harmonic functions $Y_l^m(\theta, \varphi)$ are polynomials in the polar coordinates θ and φ, and the super- and subscripts l and m determine the power of the variables θ and φ in the polynomials, respectively. The simplest spherical harmonic is given by

$$Y_0^0 = \frac{1}{2}\sqrt{\frac{1}{\pi}} \tag{15.18}$$

Here, both l and m are zero; consequently, θ and φ do not occur explicitly on the right-hand side of Eq. 15.18. Since the right-hand side of Eq. 15.18 is just a constant, this function describes a simple spherical surface, as shown in Figure 5.7, top row (see Section 1.7 for an explanation of the last statement). According to Eq. 15.14

$$\psi_{nlm}(r, \theta, \phi) = R_{nl}(r)\, Y_l^m(\theta, \phi) \tag{15.14}$$

the hydrogen 1s orbital wavefunction is the product of Eqs. 15.15 and 15.18:

$$\psi_{100} = \psi_{1s} = \left(\frac{1}{\pi}\right)^{1/2} \left(\frac{1}{a}\right)^{3/2} e^{-r/a} \tag{15.19}$$

For the hydrogen 2s orbital, the angular part of the wavefunction will be the same as for the 1s orbital, since both l and m are zero

$$\begin{aligned}\psi_{200} = \psi_{2s} &= R_{20}(r)\, Y_0^0(\theta, \phi) = \frac{1}{\sqrt{2}} \left(\frac{1}{a}\right)^{\frac{3}{2}} \left(1 - \frac{r}{2a}\right) e^{-\frac{r}{2a}} \frac{1}{2}\sqrt{\frac{1}{\pi}} \\ &= \left(\frac{1}{8\pi}\right)^{1/2} \left(\frac{1}{a}\right)^{3/2} \left(1 - \frac{r}{2a}\right) e^{-r/2a}\end{aligned} \tag{15.20}$$

For all s-orbitals, the angular part is just a constant (Eq. 15.18); thus, the electron density probability functions are just the square of the radial part, as shown in Figure 15.5. When l is nonzero, such as in the 2p wavefunctions, both the radial and the angular parts change. The radial part for the 2p orbitals is

$$R_{21} = \frac{1}{\sqrt{24}} \left(\frac{1}{a}\right)^{3/2} r\, e^{-r/2a} \tag{15.21}$$

This function has a node at $r = 0$. For $l = 1$, there are three possible values for m, namely 0 and ± 1 (see Chapter 5). For $m = 0$, the spherical harmonic, Y_1^0, is given by

$$Y_1^0 = \frac{1}{2}\sqrt{\frac{3}{\pi}} \cos\theta \tag{15.22}$$

This function is shown in Figure 5.7 and represents the dumbbell shape associated with p orbitals. The $2p_0$ wavefunction, often referred to as the $2p_z$ wavefunction, is just the product of Eqs. 15.21 and 15.22:

$$\psi_{210} = \psi_{2p_z} = R_{21}\, Y_1^0 \tag{15.23}$$

Things get more complicated when $m = \pm 1$, that is, for the other two 2p orbital wavefunctions. While the radial part stays the same (Eq. 15.21), the angular functions are

$$Y_1^{-1} = \frac{1}{2}\sqrt{\frac{3}{2\pi}} \sin\theta\, e^{-i\varphi} \tag{15.24}$$

$$Y_1^1 = -\frac{1}{2}\sqrt{\frac{3}{2\pi}} \sin\theta\, e^{i\varphi} \tag{15.25}$$

In Eqs. 15.24 and 15.25, an expression $e^{\pm i\varphi}$ appears that contains the unit of imaginary numbers, $i = \sqrt{-1}$ in the exponent. $e^{i\varphi}$ is a complex expression (i.e., containing real and imaginary parts) that can be written as

$$e^{i\varphi} = \cos\varphi + i\sin\varphi \text{ (Euler's formula)}. \tag{15.26}$$

Thus, it is a periodic trigonometric function in complex space discussed in Section 1.10. Therefore, the functions Y_1^{-1} and Y_1^1 cannot be plotted in real space, since they are complex functions. Multiplication of Y_1^{-1} and Y_1^1 with the radial part,

R_{21}, gives the wavefunctions ψ_{211} and ψ_{21-1}, which are also complex functions. However, a simple operation can combine the ψ_{211} and ψ_{21-1} to eliminate the imaginary part and give real functions:

$$\psi_{2p_x} = \tfrac{1}{2}\{\psi_{211} + \psi_{21-1}\} \tag{15.27}$$

$$\psi_{2p_y} = \tfrac{1}{2}\{\psi_{211} - \psi_{21-1}\} \tag{15.28}$$

This procedure is legal since ψ_{211} and ψ_{21-1} correspond to degenerate eigenvalues, and – according to Eq. 15.10 – any linear combination of degenerate wavefunctions is also a possible solution to the differential equation describing the problem. The resulting functions have the same shape as the $\psi_{210} = \psi_{2p_z}$ function, but are rotated in space to lie along the x or y axes, rather than the z axis.

Example 15.6 Verify Eq. 15.28, using Euler's formula (15.26).

Answer: See answer booklet

The 3s orbital is again a spherical orbital but has one more nodal plane than the 2s orbital. The 3p orbitals have a shape similar to the 2p orbitals, also with one more nodal plane. Both 3s and 3p orbitals are larger than the 2s and 2p counterparts. In addition, at the $n = 3$ level, d orbitals are allowed, with $l = 2$, and m values of $-2, -1, 0, 1, 2$. These 3d orbitals, i.e. $3d_{-2}, 3d_{-1}, 3d_0, 3d_1$, and $3d_2$, are even more difficult to visualize than the $n = 2$ orbitals discussed above. However, they are degenerate, and combinations can be formed that can be plotted. These hybrid orbitals are referred to as the $3d_{xy}$, $3d_{xz}$, $3d_{yz}$, $3d_{x^2-y^2}$, and $3d_{z^2}$ orbitals. Pictures of them can be found online. They are similar in shape to the spherical harmonic functions $Y_2^{-2}, Y_2^{-1}, Y_2^0, Y_2^1$, and Y_2^2, depicted in Figure 5.7.

15.3 Atomic Energy Levels and Atomic Emission Spectroscopy

In the last section, we explored the fundamental concepts of quantized atomic energy levels in the simplest of all atoms, the hydrogen atom. We showed that the energy levels first introduced in Chapter 5 and depicted in Figure 5.6 can be adequately described by quantum mechanics, and that the transitions that give ride to the hydrogen absorption and emission spectrum can be explained by the energy differences between these energy levels. These transitions are insinuated by the vertical lines in Figure 15.7a, which was presented before as Figure 5.6.

Whether or not a transition between energy levels may occur is determined by another quantum mechanical quantity known as the transition moment, which must be nonzero for a transition to be allowed. For the transition moment to be nonzero, certain selection rules must be fulfilled. These selection rules stipulate, for the case of the hydrogen atom, that the change in the main quantum number, Δn, can be any positive or negative integer:

$$\Delta n = \pm 1, \pm 2, \pm 3 \ldots \tag{15.29}$$

as indicated in Figure 15.7a. However, there is an additional requirement, namely that the orbital quantum number must change by 1 unit:

$$\Delta l = \pm 1 \tag{15.30}$$

This is shown in Figure 15.7b, which indicates that for a transition starting from the ground state (with $n = 1$ and $l = 0$), the electron must end up in orbitals in p orbitals (2p, 3p, 4p, etc.).

The concept of selection rules will be with us for the rest of this chapter, since all spectroscopic transitions are determined by them. As in most cases, it was experimental evidence that led to the formalism to calculate selection rules. In case of the H-atom emission spectrum (see Section 5.3), the evidence was provided by carrying out the hydrogen emission experiment in the presence of a magnetic field. A magnetic field removes the degeneracy of the 2p orbitals and splits them into three sublevels for

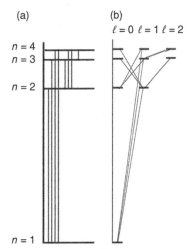

Figure 15.7 (a) Energy level diagram for the H atom (see Figure 5.6). (b) Allowed transitions for the hydrogen atom according to the selection rule $\Delta l = \pm 1$.

the different values of m. Thus, the transition from $n = 1, l = 0$ to $n = 2$, $l = 1$ will appear as three distinct spectral lines. This spectral "fine structure" can only be explained if $\Delta l = \pm 1$ for the transition.

As seen in Figure 15.7, there is a fairly large number of transitions for the hydrogen atom in the visible (400–750 [nm]) and ultraviolet (180–400 [nm]) regions[1] of the spectrum. For atoms with more than one electron, this number increases enormously. Figure 15.8 shows a small fraction of the transitions occurring in the UV-vis spectral region for the sodium atom; in fact, this graph shows only transitions from the ground state of the Na atom ($1s^2\ 2s^2\ 2p^6\ 3s^1$) into 3p, 3d, 4s, and 4p energy levels. Notice that the direct transition from 3s to 4s is forbidden, since $\Delta l = 0$, which violates the rule described by Eq. 15.30. Notice also that the transitions from the 3s to the 3p orbitals have slightly different energies depending on the spin state of the excited atom. This graph does not include transitions into higher orbitals, nor does it show any transitions from electrons in orbitals with lower energy than the 3s orbital. Finally, the lines connecting the orbitals – the transitions – can occur up or down: when a Na atom is excited thermally, by heating, into excited states, it may emit photons with energies corresponding to the difference in energy levels. This emission process is shown in Figure 15.9a, with the transitions being shown as a down arrow and the creation of a photon. On the other hand, if a Na atom in its ground state is illuminated with light, it may absorb the wavelength corresponding to the energy difference. This is known as an absorption and is represented by an up arrow in Figure 15.9b. In this case, the photon is annihilated and its energy is transferred to the atom. We see here that light is created or destroyed by atomic processes. This leads us to the fundamental equation of spectroscopy, which was introduced in Chapter 5.

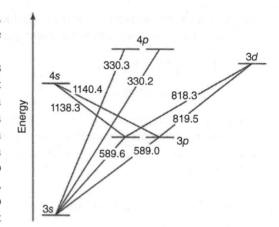

Figure 15.8 Part of the energy level diagram of excited state Na atom and corresponding transitions (in [nm]) between the energy levels.

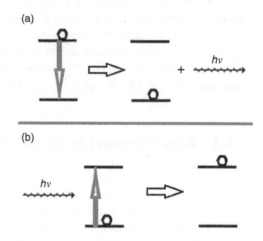

Figure 15.9 Schematic of (a) an emission and (b) an absorption process.

$$\Delta E_{atom} = (E_f - E_i)_{atom} = E_{photon} = h\nu = hc/\lambda \qquad (5.14)$$

This equation indicates that the difference in energy of the final and initial states of the atom, ΔE_{atom}, must be equal to the energy of the photon, E_{photon}. This equation explains why only discrete colors, or wavelength, of light are being emitted or absorbed in a transition from one state to another.

Example 15.7 According to Figure 15.8, excited Na atoms emit light at 589 [nm] when an electron excited into the 3p orbital decays back into the 3s orbital. What is the energy difference between the 3p and 3s orbital in a sodium atom?

Answer: $3.4 \cdot 10^{-19}$ [J]

The distinct colors observed in firework displays result from emissions of excited atoms. Sodium, for example, shows an intense orange color (due to the 589 [nm] emission discussed in Example 15.7. Green color is due to barium, blue due to copper, orange due to calcium, and red due to lithium and strontium. The emission spectral properties of atoms and ions discussed so far are also used in modern analytical methods. Rather than determining the presence and abundance of an element in a sample by classical methods (see below), in modern analytical laboratories, spectral methods are used for this purpose. This reflects an enormous advance in terms of time savings, higher accuracy, and waste generated, as will be discussed next.

The classical methods of determining the presence of metal ions in a sample consisted of dissolving the sample in an acid – generally nitric acid, since most nitrates are soluble in aqueous solutions (see Section 4.5). Subsequently,

---

1 This spectral region is generally referred to as the "UV-vis region."

chloride ions were added to test for and/or precipitate Ag^+, Pb^{2+}, and Hg^+ ions that form insoluble chloride salts. Next, the residual acidic solution was treated with S^{2-} ions. Several metal ions, among them Cu^{2+}, Cd^{2+}, Sn^{2+}, and As^{3+}, form insoluble sulfide precipitates in acidic solutions. These precipitates needed to be redissolved and tested for each of the possible cations using specific reagents. Other cations could be precipitated in neutral or slightly basic solutions with S^{2-} ions. Finally, carbonate ions were added to the remaining analyte to test for alkaline earth ions that all form insoluble carbonates. This qualitative analysis was followed, once the ions in a sample were identified, by quantitative methods. For example, a specific ion may have been precipitated by one of the counterions enumerated above, filtered, dried, and weighted (gravimetric analysis). Alternatively, a solution of a metal ion might have been titrated with a chelating agent and a specific indicator to determine the concentration of the unknown. All these methods were labor-intensive and required a lot of specific reagents.

In contrast, modern spectral methods can perform both qualitative and quantitative analysis in one step. The method to be introduced here is known as inductively coupled plasma atomic emission spectroscopy (ICP-AES). Behind that fancy name is a technique that is quite easy to understand from the knowledge we have gained from atomic emission spectroscopy. The ICP part of the experiment basically consists of utilizing a very hot "flame," or torch (6000–10000 [K]) that is created by subjecting a flow of argon gas to a high (radio) frequency electromagnetic field that ionizes the argon and creates a plasma. The analyte solution is "nebulized" (i.e., injected into the plasma in the form of microscopic droplets). At the temperature of the argon plasma, the solvent immediately vaporizes and leaves the analyte atoms in highly excited states from which the emission spectrum is observed. Since atomic species exhibit abundant emission lines, it is straightforward to identify elements by their multiple emission colors. The intensity of each emission is directly proportional to the abundance of the element; therefore, it is just a matter of calibrating the instrument with standard solutions of accurately known concentrations (see Example 4.5) to arrive at a qualitative and quantitative analytical method.

Over 75 elements can be detected by ICP-AES methods, and the detection limits often are as low as 1 ppb. Multiple elements can be detected simultaneously, thus providing an easy analytical workflow. Many commercial applications of ICP-AES methods exist, for example, testing of drinking water for toxic contaminates such as As, Cu, Cr, or others.

15.4 Molecular Energy Levels, Spectroscopy, and Structural Methods

In the previous section, we explored how *atomic* structure and energy levels can be used in an analytical setting to determine both the presence and the abundance of atomic species in a sample. In the present section, spectroscopic methods will be introduced that answer the question that popped up several times in previous chapters for *molecular* systems, such as: How do we know orbital energies of molecules? How do we know bond angles and distances? As insinuated earlier, the answer is "from experiment." Now the next logical question is: "What experiments give us the desired answers?" This will be the subject of the following section, which is a very superficial introduction to spectral methods to elucidate the structural information of molecules.

15.4.1 Electronic Energy Levels and UV-vis Absorption Spectroscopy

In our discussion of MO theory (Section 6.6), bonding and antibonding orbitals were introduced. These define molecular energy level between which electronic transitions can occur. In other words, just like electrons can be promoted into higher energy atomic energy levels in atomic systems by absorption of light (see previous section), electrons can be promoted into higher – and often antibonding – molecular energy levels by light. Just like in atomic spectroscopy, the energies required for a photon to promote an electron into higher orbitals fall in the UV-vis spectral region (see footnote in Section 15.3).

Objects that appear "colored" to the human eye exhibit these colors because they absorb certain wavelengths. Take a green leave of a tree for example. It appears green since it contains a compound, chlorophyll, that absorbs blue (430 [nm]) and red (660 [nm]) light and reflects green light. Thus, when illuminated by the broad spectral emission from the sun, leaves appear green. In general, the transitions that give rise to absorptions in the UV-vis region occur between π and π^* orbitals of C=C, C=N, or C=O double bonds, such as in ketones (see later), or in aromatic moieties. In addition, metal ions such as Fe, Cu, Co, Cr, etc., can have transitions in the visible spectral region. The green color of a beer bottle glass is due to iron atoms, whereas the deep red color of a ruby crystal is due to chromium atoms. Since these transitions involve the promotion of an electron, they are generally referred to as "electronic transitions" (which are observed in the UV-vis spectral region, as mentioned before).

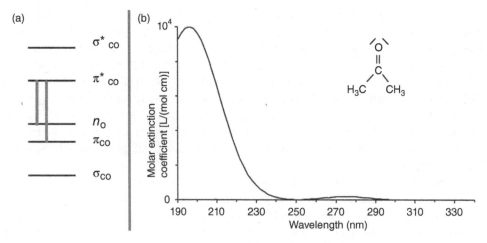

Figure 15.10 (a) Energy level diagram for the two electronic transitions observed in a ketone such as acetone. (b) Schematic UV absorption spectrum of acetone.

Example 15.8 If you wore a red T-shirt into a room that is illuminated by red light only, would it still appear red to you?

Answer: See answer booklet

The UV absorption spectrum of acetone, $(CH_3)_2C=O$ (see Section 3.3), will be discussed next to present details of the energy levels, the method to observe an absorption spectrum, and the quantities derived from the observations. Acetone is a colorless liquid used, for example, in nail polish remover. Since it is colorless, it has no transitions in the visible region of the spectrum, but has two electronic transitions in the UV region. Figure 15.10a shows a part of the MO level diagram of acetone. The MOs are labeled according to a valence bond description: the lowest orbital shown corresponds to the C—O single bond and is marked σ_{CO}. The next lowest orbital shown is the π orbital of the C=O double bond, marked π_{CO}. Each of these orbitals is filled with two electrons. The energy level marked n_O corresponds to the two nonbonding orbitals located on the oxygen atom, accommodating four electrons. The π_{CO} and n_O are the highest occupied molecular orbitals (HOMOs). The higher orbitals, π^*_{CO} and σ^*_{CO} are unoccupied (lowest unoccupied molecular orbitals (LUMOs)). Two transitions are observed for acetone, one from the π_{CO} orbital into the π^*_{CO} orbital (written backward as $\pi^*_{CO} \leftarrow \pi_{CO}$) and one from the n_O to the π^*_{CO} orbital ($\pi^*_{CO} \leftarrow n_O$). Since the $\pi^*_{CO} \leftarrow \pi_{CO}$ transition has a larger energy difference than the $\pi^*_{CO} \leftarrow n_O$ transition, it occurs at a lower wavelength, corresponding to a higher energy photon. Thus, a UV spectrum is observed with two absorption features, one at about 200 [nm] and the other at about 280 [nm], see Figure 15.10b. While the abscissa of this plot is obvious, the y-axis needs some explanation: what is actually plotted along the y-axis?

This leads us to a short discussion of spectrophotometers, the scientific instruments used to obtain an absorption spectrum. A block diagram of such a machine is shown in Figure 15.11. Broadband light from a source – such as a hot wire in an incandescent light bulb or from an electric discharge – is focused into a component known as a monochromator. It employs a prism or grating to disperse white (broadband) light into its components like the rain droplets in a rainbow do. The monochromatic light exiting the monochromator is focused through the sample into a detector that converts the light into an electric signal that can be recorded. The electric signal is proportional to the intensity of the light. The monochromator allows different colors to be selected consecutively by rotating the prism or grating to allow only one color at a time to reach the sample and detector.

In order to measure a UV-vis absorption spectrum, the instrument response – i.e., the light intensity (displayed as the detector signal) as a function of wavelength – is measured without sample. This corresponds to the trace marked "$I_0(\lambda)$" in Figure 15.12.

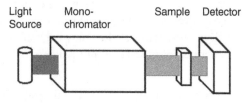

Figure 15.11 Block diagram of a UV-vis spectrophotometer.

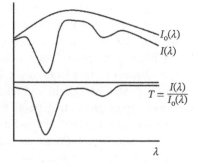

Figure 15.12 Schematic of the measurement process.

Next, the sample contained in a suitable sample cell is inserted into the instrument, and the signal with the sample is collected as a function of wavelength. This is marked "$I(\lambda)$" in Figure 15.12. The ratio $I(\lambda)/I_0(\lambda)$ is called the transmittance of the sample. A plot of the transmittance vs. wavelength is referred to as a "transmission spectrum." The absorbance A, also known as the optical density (OD), is defined as the negative logarithm of the transmittance:

$$A(\lambda) = -\log I(\lambda)/I_0(\lambda) = -\log T \qquad (15.31)$$

1 OD unit corresponds to 10% transmission, while 2 OD units correspond to 1% transmission. The absorbance is used in all quantitative work since it is related to the molar concentration C_M (in [mol/L]) of the sample according to the Beer-Lambert's law:

$$A(\lambda) = \varepsilon(\lambda)\ C_M\ l \qquad (15.32)$$

Here, l (in [cm]) is the path length the light beam travels through the sample (the sample thickness) and $\varepsilon(\lambda)$ is the molar extinction coefficient, expressed in units of [L/(mol cm)]. Since A is a logarithmic quantity, it is unit-less. The molar extinction coefficient is a measure of how "allowed" a transition is and is directly related to the quantum mechanical transition moment (see Section 15.3). The y axis in the schematic absorption spectrum of acetone, Figure 15.9b, is expressed in units of ε. The absorption at a wavelength of ca. 200 [nm], the $\pi^*_{CO} \leftarrow \pi_{CO}$ transition, has an ε value of just under 10 000, which corresponds to an allowed electronic transition. The absorption at ca. 280 [nm], the $\pi^*_{CO} \leftarrow n_O$ transition, is much weaker and actually forbidden by the symmetry-based selection rules. The fact that it is still observable is well understood but beyond the scope of this introductory discussion.

The take-home message from this discussion is two-fold. First, UV-vis spectroscopy allows the experimental determination of differences in orbital energies and thereby provides information for the computational MO methods (see Section 6.6). Second, the Beer–Lambert law is extremely useful for determining concentrations of species in aqueous solutions. For example, if you ever donate blood, a drop of your blood is used to determine the hemoglobin concentration of your blood by measuring its OD. Since hemoglobin has an extremely high extinction coefficient at 420 [nm] ($\varepsilon_{420} \approx 10^6$ [$1/(C_M$ cm)]), just one drop of diluted blood gives a strong absorbance even at a short path length of 0.1 [cm]. In clinical settings, the degree of oxygenation of blood can be determined via spectral measurements as well since oxygenated hemoglobin exhibits its absorption maximum at a larger wavelength (ca. 430 [nm]).

Example 15.9 An absorption band of a sample at a given wavelength has a measured OD of 1.54. What percentage of the photons incident on the sample is transmitted?

Answer: 2.9% of the photons are transmitted

Example 15.10 Using the epsilon value of hemoglobin given above, what concentration of hemoglobin in water would give an absorbance of 0.6 when measured in a 0.1 [cm] path length cuvette?

Answer: $C = A/\varepsilon l = 6 \cdot 10^{-6}$ [M]

15.4.2 Vibrational Energy Levels and Infrared Spectroscopy

In the previous section, electronic spectroscopy in molecular systems was introduced where electrons are promoted into higher molecular orbitals by photons in the UV-vis spectral region. Next, we shall discuss a branch of spectroscopy in which infrared (IR) photons (2.5–25 [μm] wavelengths) are used to illuminate a sample. These IR photons do not possess sufficient energy to promote electrons into higher orbitals, but they have just the right energies to excite molecular vibrational transitions. Thus, this branch of spectroscopy is referred to as infrared or vibrational spectroscopy. How can we visualize what is happening in vibrational spectroscopy?

We have discussed the concept of vibrational motion in molecules before (see Section 11.2.1) in the context of the heat content and specific heat of compounds. Unless you are at absolute zero temperature (0 [K]), this motion of atoms in a solid, liquid, or gas can never be stopped (this is actually a formulation of the third law of thermodynamics: absolute zero temperature cannot be achieved since it would require perfectly still and localized atoms, which is not possible). However, this vibrational motion obeys certain rules that are exploited in vibrational spectroscopy. For large molecules, these rules become complicated and somewhat hard to understand; so, let us start the discussion using a simple, diatomic molecules

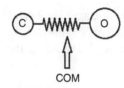

Figure 15.13 Model of a diatomic molecule, CO, as masses connected by a spring. COM: center of mass.

such as carbon monoxide. Although vibrational energy states are quantized and vibrational transitions are governed by quantum mechanics, we can start this discussion with a very plausible classical analog assuming that the C and O atoms are point masses, connected by a spring as shown in Figure 15.13. In fact, chemical bonds very much resemble a mechanical spring in that they obey Hook's law quite nicely, which states that the force F to stretch a spring is given by

$$F = k\,x \tag{15.33}$$

where x is the elongation and k is the spring constant or force constant that describes how stiff a spring is. The units of k are force divided by displacement or [N/m]. Equation 15.33 holds for small elongations only and does not predict that the spring might break if elongated too much. The vibrational frequency of such a system is given by

$$\nu = \frac{1}{2\pi}\sqrt{\frac{k}{m_R}} \tag{15.34}$$

In Eq. 15.34, the frequency is given in units of [Hz = 1/s], and m_R is the reduced mass of the carbon monoxide molecule,

$$m_R = \frac{m_C m_O}{m_C + m_O} \tag{15.35}$$

At this point, a couple of comments are appropriate. First, a short explanation in Section 1.9 demonstrates how to get from Eqs. 15.33 to 15.34. This involves solving a differential equation and meshes nicely with the discussion of differential equations in Section 15.1. Second, the reduced mass introduced in Eq. 15.35 is used to ascertain that the center of mass (COM) indicated in Figure 15.13 does not move during the vibration. This aspect – the separation of translation from the purely vibrational motion of the molecule – is discussed in detail in the literature. Third, an analysis of the magnitudes and units involved in Eq. 15.34 will be the subject of Example 15.11.

Example 15.11 Calculate the classical vibrational frequency of the CO molecule from the value of the force constant, $k = 1870$ [N/m].

Answer: $\nu = 6.42 \cdot 10^{13}$ [Hz]

This vibrational frequency is an inconveniently large number. In vibrational spectroscopy, the energy or frequency of a transition is usually expressed in units of inverse wavelength, known as the wavenumber,

$$\tilde{\nu} = 1/\lambda. \tag{15.36}$$

This unit can be thought of the number of waves per unit length (usually a centimeter). Thus, the wavenumber is expressed in units of [cm^{-1}] and has the advantage that it is directly proportional to energy (as opposed to the wavelength of a photon, which is inversely proportional to energy):

$$E = h\nu = \frac{hc}{\lambda} = hc\,\tilde{\nu} \text{ and} \tag{15.37}$$

$$\nu = \frac{c}{\lambda} = c\,\tilde{\nu} \tag{15.38}$$

Thus, to convert the answer of Example 15.11 into a more common notation, we use Eq. 15.38, $\nu = c\,\tilde{\nu}$ and get

$$\tilde{\nu} = \frac{6.42 \cdot 10^{13}}{3 \cdot 10^{10}} = 2140 \left[\frac{s^{-1}}{\text{cm s}^{-1}} = \text{cm}^{-1}\right] \tag{15.39}$$

Here, we have used the velocity of light, expressed in units of [cm/s]. The vibrational frequency expressed in this new energy unit is 2140 [cm^{-1}]. You will find this energy unit in many spectroscopic applications that involve vibrational or rotational spectroscopic transitions. Notice that there is no amplitude of the vibration given, only its frequency. You may remember that fact from a simple experiment in a physics laboratory: if you suspend a mass from a spring attached to the ceiling, elongate the spring, and then release it. The frequency of the resulting vibration is given by Eq. 15.34, where the reduced mass is just the mass attached to the spring. The amplitude of the oscillation is undefined here as well: if you elongate the spring further, the mass moves up and down more quickly, but the frequency does not change!

Example 15.12 (a) Use the frequencies for carbon–carbon single, double, and triple bond vibrations given in the last paragraph of Section 6.3 to calculate the vibrational frequencies in wavenumber units for ethane, ethylene, and acetylene.

(b) Assuming that these three molecules are "diatomic" molecules X–X, Y=Y, and Z≡Z with atomic masses of 15, 14, and 13 for X, Y, and Z, respectively, calculate the force constants for each of the three species presented in the last paragraph of Section 6.3.

Answer:
(a) Ethane: $\tilde{\nu} = 993$ [cm^{-1}]; ethylene: $\tilde{\nu} = 1623$ [cm^{-1}]; acetylene: $\tilde{\nu} = 1974$ [cm^{-1}]
(b) Ethane: $k = 439$ [N/m]; ethylene: $k = 1094$ [N/m]; acetylene: $k = 1503$ [N/m]

The quantum mechanical treatment for the vibration of a diatomic molecule involves setting up the vibrational Schrödinger equation that contains the potential energy term

$$V = \tfrac{1}{2} k x^2 \tag{15.40}$$

which is the integrated form of Eq. 15.33:

$$V = \int F\, dx = \int kx\, dx = \tfrac{1}{2} k x^2 \tag{15.41}$$

The vibrational Schrödinger equation again is a complicated differential equation that had been solved, long before quantum mechanics, by a French mathematician, C. Hermite. The mathematics, again, is quite complicated and so are the vibrational eigenfunctions. However, the vibrational eigenvalues $E(v)$ for a diatomic molecule are very simple:

$$E(v) = (v + \tfrac{1}{2})\, h\nu \text{ or } E(\tilde{\nu}) = (v + \tfrac{1}{2})\, hc\tilde{\nu} \tag{15.42}$$

where v is the vibrational quantum number, $v = 0, 1, 2, ...$, and ν is the vibrational frequency described by Eq. 15.34. Equation 15.42 implies that the energy of the vibrational ground state $E(0)$, is given by

$$E(0) = \tfrac{1}{2} h\nu \tag{15.43}$$

which implies that the ground state energy is nonzero. The energies of the excited states are given by

$$E(1) = \left(\tfrac{3}{2}\right) h\nu,\ E(2) = \left(\tfrac{5}{2}\right) h\nu,\ E(3) = (7/2)\, h\nu,\ \text{etc.} \tag{15.44}$$

The selection rules for the vibration of a diatomic molecule are

$$\Delta v = \pm 1 \tag{15.45}$$

This is shown for the CO molecules in Figure 15.14. The quantum mechanical result incorporates the fact that there is always a "zero-point" vibrational energy of $\tfrac{1}{2} h\nu$, (Eq. 15.43) just a little over 1000 cm^{-1} for CO. Furthermore, we see that a photon absorbed for a vibrational transition to occur must have the energy $h\nu$ or $hc\tilde{\nu}$. Since the energy difference between $v = 0$ and $v = 1$ is the same as the energy difference between $v = 1$ and $v = 2$, there should be only one absorption observed, at 2140 cm^{-1}, which is true to a first approximation.

Things get a bit more complicated in a polyatomic molecule, such as water. Whereas there was only one energy ladder in a diatomic molecule (shown in Figure 15.14), a polyatomic molecule of N atoms will have 3N-6 such energy ladders, each associated with a "normal mode Q" of vibration, and each with a different spacing between energy levels. This is shown schematically in Figure 15.15 for the water molecule. Since water is a triatomic molecule, it has 3·3−6=3 normal modes that are shown in Panel (a) of Figure 15.15. These three modes are the antisymmetric stretching mode Q_1, where one of the O—H bond contracts, while the other one elongates. In the symmetric stretching mode, Q_2, both O—H bonds elongate and contract in phase, and in the deformation mode, Q_3, the H—O—H angle increases and decreases. Each of the normal modes is associated with a different absorption wavenumber, listed in Table 15.1. The observed infrared absorption spectrum, therefore, contains three major peaks, as shown in Figure 15.15b. Panel (c) shows an approximate energy level diagram (starting for each of the energy ladders at $\tfrac{1}{2} h\nu$); the arrows in this panel correspond to the transitions shown in Panel (b).

As molecules get bigger, the complexity of the infrared spectra increases significantly. Consider the molecule acetone, discussed before (see Figure 15.10) for which two electronic transitions were described. Since acetone consists of 10 atoms, it will have 3·10−6=24 normal modes. Although the theoretical treatment of such a

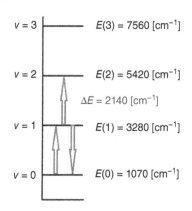

Figure 15.14 Vibrational energy level diagram and allowed transitions in CO.

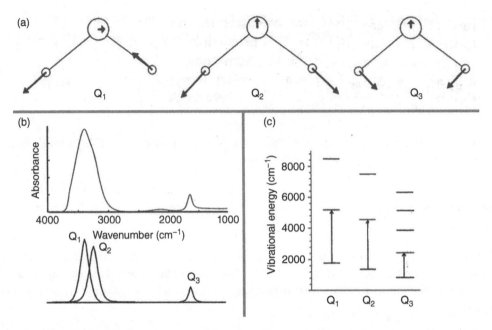

Figure 15.15 (a) Normal modes of vibration of the water molecule. (b) Observed infrared absorption spectrum of water. (c) Vibrational energy level diagram for water.

Table 15.1 Observed vibrational transitions for the water molecule.

Mode	Description	Wavenumber
Q_1	Antisymmetric stretching mode	3750 [cm^{-1}]
Q_2	Symmetric stretching mode	3650 [cm^{-1}]
Q_3	Deformation mode	1620 [cm^{-1}]

molecule is possible (and practical), most interpretation of infrared spectra is being carried out in a qualitative way. This is possible since certain chemical groups exhibit the same transitions, regardless of the group's environment. A carbonyl group, for example, exhibits a strong infrared peak at between 1680 and 1720 cm^{-1}, due to the C=O stretching motion. C—H stretching motions, for example in a methyl group, occur at about 2950 cm^{-1} for the symmetric stretching mode (all three C—H bonds stretching in phase) and about 2980 cm^{-1} for the antisymmetric stretching mode, where one C—H bond stretches while the other two compress. Methyl groups also have deformation modes at about 1380–1450 cm^{-1}. The modes discussed here for carbonyl and methyl groups can readily be identified in the infrared spectrum of acetone, which is shown in Figure 15.16.

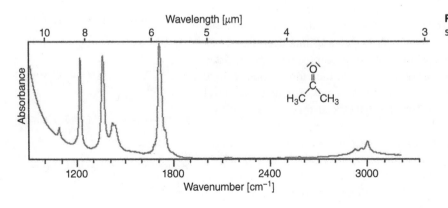

Figure 15.16 Infrared absorption spectrum of liquid acetone.

Vibrational spectroscopy is a widely used technique in modern science. Many of the structural aspects discussed in earlier chapters were discovered by infrared spectroscopy. We revisit here the structure of SF_4 that was introduced in Chapter 6 (see Figure 6.14). If SF_4 were a tetrahedral molecule such as CH_4 or a square planar molecule such as ClF_4^-, it would have an infrared spectrum quite different from what was observed. This is because in highly symmetric molecules, such as CH_4 or ClF_4^-, some vibrational modes are forbidden by symmetry rules. For example, the vibrational mode in CH_4 where all four C—H bonds stretch in phase is forbidden in infrared absorption. Since S—F stretching vibrational modes are observed in SF_4, we conclude that it is not a tetrahedral molecule. Infrared absorption spectra are observed with instruments and methodology similar to those described before for UV-vis spectroscopy, but using semiconductor detectors sensitive to IR light.

15.4.3 Rotational Energy Levels and Microwave Spectroscopy

In earlier chapters, we discussed structural parameters like bond lengths and bond angles in molecules from the viewpoint of the valence shell electron pair repulsion (VSEPR) model, and we stated that the H—O—H bond angle in water, at 104.5°, is smaller than the H—C—H bond angle in methane, although in both cases, the central atom is sp^3 hybridized. Again, you may have asked the question: how do we know the bond angles with such a high degree of accuracy? The answer here is from structural methods such as rotational spectroscopy and/or X-ray diffraction methods. Rotational or microwave spectroscopy (see Table 5.1) utilizes the principle that not only the linear momentum, but also the angular momentum is quantized. Let us take a look at the classical definition of linear and angular momentum and see how the quantization of the rotational energy of a molecule gives us the desired information, namely structural parameters.

Classically, the rotational kinetic energy of a rotating body is described by

$$E_{\text{kin}} = \frac{L^2}{2I} \tag{15.46}$$

where L is the angular momentum and I is the moment of inertia. This definition is analogous to the definition of the (linear) kinetic energy in terms of the linear momentum:

$$E_{\text{kin}} = \frac{p^2}{2m} \; (= \tfrac{1}{2} m v^2) \tag{15.47}$$

The kinetic energy, in both cases, is proportional to the square of the corresponding momentum, divided by a quantity measuring the inertia or "resistance" to this motion. The angular momentum of a particle in a rotatory motion, defined in Eq. 15.46, is the vector product of its linear momentum and the radius of the circular motion.

The moment of inertia of a molecule depends very much on its shape. In the simplest case of a linear, diatomic molecule, the moment of inertia I is given by

$$I = m_1 r_1^2 + m_2 r_2^2 = m_R r_{12}^2 \; [\text{kg m}^2] \tag{15.48}$$

where m_1 and m_2 are the masses of the two atoms, r_1 and r_2 their distances from the COM (see Figure 15.12), m_R is the reduced mass (Eq. 15.35), and r_{12} is the bond distance, $r_{12} = r_1 + r_2$. The quantum mechanical description of rotational energy is similar to that of the radial part of the H-atom Schrödinger equation. The solution of the rotational Schrödinger equation gives us the rotational energy levels for a diatomic molecule as

$$E_{\text{rot}}(J) = B\, J\,(J+1) \tag{15.49}$$

where J is the rotational quantum number, $J = 0, 1, 2, ...$ The rotational constant, B, is related to the moment of inertia I by

$$B = (h/c) \frac{1}{8\pi^2 I} \tag{15.50}$$

The units of B, when c is expressed in units of [cm/s], are

$$\left[\frac{\frac{\text{kg m}^2}{s^2} \text{s}}{\frac{\text{cm}}{\text{s}}} \frac{1}{\text{kg m}^2} = \text{cm}^{-1} \right] \tag{15.51}$$

An energy level diagram for the rotational energy is shown in Figure 15.17a. Notice that the rotational energy increases quadratically with the

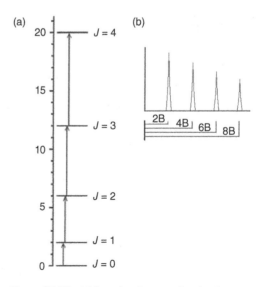

Figure 15.17 (a) Rotational energy levels of a diatomic molecule. (b) Observed rotational spectrum.

quantum number, in contrast to the situation in vibrational spectroscopy, where the energy depends linearly on the vibrational quantum number v, and in the hydrogen atom, where the energy depends on $1/n^2$.

The selection rule for a rotational transition is

$$\Delta J = \pm 1 \qquad (15.52)$$

Therefore, transitions are allowed as shown by the up arrows in Figure 15.17. For the $J = 0$ to $J = 1$ transition, the energy difference between the levels is $2B$. For the $J = 1$ to $J = 2$ transition, the energy difference is $4B$, and for the $J = 2$ to $J = 3$ transition, the energy difference is $6B$. Thus, the rotational spectrum will appear as equidistant absorption features, spaced by $2B$, as shown in Figure 15.17b. Example 15.13 will demonstrate how the observed rotational spectrum can be translated into highly accurate structural data.

Example 15.13 The rotational constant for deuterium fluoride, $^2D-^{19}F$, is 11.007 cm^{-1}. The atomic masses for ^{19}F and ^2D are 18.99840 and 2.01410 [amu], respectively (from mass spectrometry, see Section 15.5). Calculate the bond length of deuterium fluoride to the maximum number of significant figures consistent with this information.

Answer: 91.709 [pm]

In general, the rotational spectra of polyatomic molecules get very complicated and will not be discussed here any further. Suffice it to say that the measurement of the moment of inertia allows the determination of bond lengths and bond angles with high accuracy. By the way, rotational spectra can only be observed for gaseous molecules. However, excitation of a hindered rotational motion can be achieved in liquid and solid states: the kitchen microwave stove achieves very fast and efficient energy absorption by shining microwave radiation into food that contains water molecules. In this process, energy is absorbed although the water molecules cannot rotate freely in the items to be heated.

15.4.4 Nuclear Magnetic Resonance Spectroscopy

No spectroscopic method used in chemical and biochemical research has seen as explosive growth over the past six decades as has nuclear magnetic resonance (NMR) spectroscopy. It has become a cornerstone for researchers and students for the identification of molecules and to establish molecular structure in the solution and solid state. NMR spectroscopy is found even in the undergraduate chemistry laboratory: in an organic chemistry laboratory, you most likely will use NMR data, along with IR spectroscopy discussed in Section 15.4.2, to identify and characterize compounds you have synthesized. In this section, the principles of NMR spectroscopy will be introduced.

NMR spectroscopy is based on the observation of transitions due to reorientation of nuclear spins in a magnetic field when exposed to electromagnetic radiation. First, we need to clarify the concept of nuclear spin. The principle of *electron spin* was introduced in Section 5.5 to explain the results of the Stern-Gerlach experiment. These results state that the spatial orientation of the spin angular momentum of an electron is quantized and produces two opposite magnetic moments. A magnetic moment would result if the electron was "spinning" around an axis; thus, the quantization of the angular momentum was called – somewhat misleadingly – "spin," although the electron is not spinning around an axis. Rather, the quantized angular momentum – the spin – is an inherent property of many subatomic particles, including protons, neutrons, and electrons. The proton's magnetic moment, although much smaller than that of an electron, is the fundamental principle of NMR spectroscopy and magnetic resonance imaging (MRI).

Quantum mechanics reveals that there are two nuclear spin wavefunctions that we call α and β ("spin up" or "spin down") with two energy eigenvalues, $\pm \hbar/2$, where $\hbar = h/2\pi$, in the presence of a magnetic field. We can visualize the two spin energies to be due to the nuclear magnetic momentum either parallel or antiparallel to the external field. The energy difference between the two states is proportional to the applied magnetic field, B_0, expressed in units of Tesla, [T]. This is described by Eq. 15.53

$$E = \pm \tfrac{1}{2} \gamma B_0 \qquad (15.53)$$

where the proportionality constant γ is the magnetogyric constant (see later). The splitting of the two energy states depends linearly on the field strength, as shown in Figure 15.18 (ignore the units along the y axis for the moment). The energy difference between the two states is given by

$$\Delta E = 2\,(\tfrac{1}{2})\,\gamma\, B_0 = \gamma\, B_0 \qquad (15.54)$$

Let us stop here for a moment and look at the impact of Eqs. 15.54. First, the energy difference between the α and β states is exceedingly small. For example, in the magnetic field of the earth (ca. 40 [μT]), this energy difference is negligible (see Example 15.14). The earth's magnetic field deflects the solar wind (charged particles ejected by the sun that would cause enormous damage to animal and plant tissue) and causes a magnet, such as the needle of your compass, to align in the magnetic field (and point North). The little magnets used to attach notes to a refrigerator door may have a field strength of 5 [mT] or about 100-fold stronger than the magnetic field of the earth.

The second point to be made concerns the units of the y-axis in Figure 15.18. As seen from Eq. 15.55, the energy required to convert a proton spin from the α to the β state is given by $\Delta E = h\nu$. Therefore, the magnetogyric constant introduced in Eq. 15.53 is reported in units of [MHz/T]:

$$\gamma = 42.58 \; [\text{MHz/T}] \tag{15.55}$$

such that the energy difference in Eq. 15.54 is obtained in units of photon frequency. Hence the units on the y-axis in Figure 15.18.

Thus, in order to cause splitting of a proton's α and β states that can be measured with radio-frequency photons (for example, 100 MHz electromagnetic radiation), a magnetic field strength of 2.34 [T] is required, a magnetic field that is over 58 000 times stronger than the earth's magnetic field

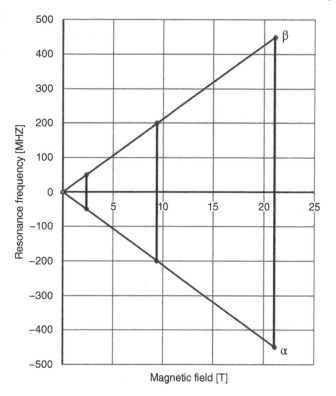

Figure 15.18 Splitting of the nuclear spin states in an external magnetic field. The solid vertical lines indicate proton transition frequencies (100, 400, and 900 MHz) for commercially available NMR machines.

(see Example 15.15). We call the frequency range of the electromagnetic radiation required (between 100 [MHz] and 1 [GHz], depending on the magnetic field strength) the radiofrequency range. Typical FM radio stations have a carrier frequency about 100 [MHz], which you can verify by looking at the dial on your radio.

Example 15.14 Use Eq. 15.54 and numerical values given in the text to calculate the frequency of electromagnetic radiation [in Hz] required to promote a proton from its α to its β states in the earth's magnetic field.

Answer: $\Delta E = 1.7$ [kHz]. This energy is far below the energy corresponding to room temperature.

Example 15.15 What is the energy difference of a proton's α and β state in a magnetic field of 9.39 [T]?

Answer: 400 [MHz]

So far, we have discussed the rather unexciting fact that protons, due to their inherent spin, assume two different energy states when in a magnetic field. That does not explain the enormous amount of structural information contained in an NMR spectrum. The power of NMR spectroscopy results from two physical effects that have not been mentioned so far: the shielding of the external magnetic field that a nuclear magnetic moment experiences by the surrounding electrons and perturbation of the local magnetic field by neighboring proton spins. The first of these effects is referred to as chemical shift and the second as spin–spin coupling.

The chemical shift effect is due to electrons surrounding a proton. These electrons produce their own induced magnetic field B_i, which opposes the externally applied field and thereby "shields" the proton from the external magnetic field, B_o. Thus, a shielded proton experiences the effective magnetic field B that is:

$$B = B_o - B_i \tag{15.56}$$

Consequently, the resonance frequencies of hydrogen nuclei in a molecule vary slightly from the frequencies of naked protons. This shift is called "chemical shift," and depends on the chemical surroundings of hydrogen nuclei: in the presence of strongly electronegative groups, a proton is more "de-shielded" than in an electron-rich environment. The chemical shift is small and only a few hundred Hertz for protons in a different chemical environment.

The chemical shift is expressed in parts per million (ppm) from that of an internal standard, most commonly tetramethyl silane (TMS). Since the Si atom is an electron donor, the proton nuclei in the adjacent methyl groups experience high shielding. Thus, the protons in TMS have one of the lowest proton resonance frequencies of most organic molecules. Let us look at a typical NMR spectrum of an organic liquid, for example, that of pure ethanol, CH_3-CH_2-OH, with a minimal amount of TMS added, recorded on an NMR spectrometer with a field strength of ca. 9.4 [T] (see Example 15.15). This spectrum is shown in Figure 15.19. The signal of all chemically equivalent protons of TMS is observed somewhere around 400 [MHz] resonance frequency; but rather than measuring this frequency accurately (which would be quite difficult), we set this signal arbitrarily to zero as a reference frequency and record the resonance frequencies of the ethanol protons relative to the TMS protons. The methyl protons of ethanol are surrounded by the highest electron density and are, therefore, the most shielded protons and appear closest to the TMS signal at about 1.2 [ppm] chemical shift. Since 1 [ppm] of 400 [MHz] is 400 [Hz], the methyl protons' resonance frequency is shifted by about 480 [Hz] from the TMS signal. The -OH proton, due to the electronegative oxygen and extensive hydrogen bonding, is the most de-shielded or least shielded proton and is furthest away from the TMS signal, at about 4.8 [ppm] or about 1920 [Hz] from the TMS signal. Inspection of Figure 15.19 also reveals that the methylene and the methyl protons exhibit multiplet structure, which is caused by spin–spin coupling, and will be discussed next.

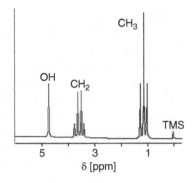

Figure 15.19 Proton NMR spectrum of pure ethanol.

Spin–spin coupling is the interaction between the nuclear spins of neighboring protons. While the chemical shift discussed in the previous paragraph is due to electrons shielding the hydrogen nuclei to different degrees depending on electron densities, the spin–spin coupling is due to neighboring proton spins producing slight variations of the magnetic field. Thus, spin–spin coupling provides information on the number of protons in the vicinity of a given nucleus to which it is covalently bonded. It is responsible for the splitting of spectral features into multiplets in the NMR spectra of simple molecules. The multiplicity – that is, the number of peaks observed for a given spectral feature – is one more than the number of equivalent adjacent protons adjacent to the protons of interest. For ethanol, each of the methyl protons (which are chemically equivalent) couples with the two methylene protons. Therefore, the methyl signal is split into a triplet, or has a multiplicity of three, as shown in Figure 15.19. Each of the methylene protons (which are chemically equivalent) couples with the three methyl protons, so the methylene signal is split into a quartet. The proton of the OH group shows up as a singlet for a number of reasons that are well understood but are beyond the level of discussion here. Also, the chemical shift of this proton can vary enormously, depending on the presence of solvent molecules that may disrupt the intermolecular hydrogen bonding.

15.4.5 X-ray Diffraction

Much of our knowledge of the size of atoms and ions, the lengths of covalent chemical bonds, and the structure of molecules in the solid state is derived from the scattering patterns produced when X-rays interact with or are scattered by the electron clouds around atoms or ions. The basics here are not too hard to grasp: we have discussed before, when we introduced the wave properties of light, that an interference pattern is observed when light passes a pinhole with the size approximately equal to the wavelength of light, see Figure 5.2b. This interference pattern is due to constructive and distractive interference of light wavelets that originate at the boundaries of the pinhole. This interference pattern, shown again in Panel (a) of Figure 15.20, gets more complicated when light passes through more than one pinhole, such as a regular array of pinholes. In this case, the interference pattern appears as shown in Panel (b) of Figure 15.20 and is due to the interference of a light beam that has passed one pinhole with the beams of light passing all other pinholes. This diffraction pattern can be analyzed to reveal the geometric arrangement of the pinholes, as well as their size. These diffraction patterns are readily observed using visible light.

X-rays, as we have discussed before, are a form of electromagnetic radiation with wavelength of ca. 0.1 [nm] = 100 [pm] (see Table 5.1). This wavelength range is comparable to the length of covalent bonds or the distance between the centers of ions in ionic crystalline compounds. Thus, X-rays produce a diffraction pattern similar to that shown in Figure 15.20b when scattered from a single crystal of an ionic compound, such as $CuSO_4$ shown in Figure 15.20c. This panel actually shows the very first observed X-ray diffraction result, published in 1912 by M. von Laue. Subsequent X-ray diffraction pattern of NaCl established the ionic nature and structure of this compound. At this point, chemistry was still in its infancy, and the difference

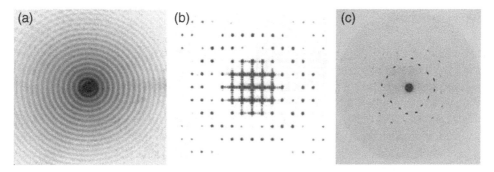

Figure 15.20 (a) Diffraction (interference) pattern of light passing through a single pinhole. (b) Diffraction pattern from a regular array of pinholes. (c) Diffraction from a single crystal of copper (II) sulfate. *Source:* (c) American Museum of Natural History Library/Wikimedia commons/Public domain.

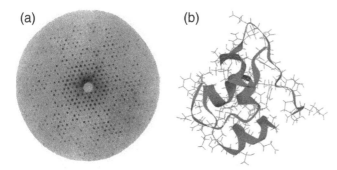

Figure 15.21 (a) X-ray diffraction pattern from a single crystal of insulin. (b) Insulin molecular structure obtained from the diffraction pattern shown in Panel (a). *Source:* Reproduced with the permission from Blundell et al. (1971)/Elsevier.

between ionic and covalent bonding had not been established. However, these early crystallographic studies helped establish our modern view of different types of chemical bonding, in particular, once the structures of diamond and graphite were established in 1913 by W.H. Bragg and W.L. Bragg.

X-ray crystallography is probably the most important structural tool for chemists, physicists, and biologists. Much of our present understanding of the structure and dynamics of molecules and biomolecules is due to crystallographic methods: the structures of DNA, proteins, pharmaceutical molecules, most common chemicals, semiconductors, etc. have been elucidated from X-ray crystallography. In its most sophisticated form, single crystals of the sample, about 1 [mm] in size, are mounted on a goniometer – a device that permits the rotation of an object about all axes – and diffraction patterns are detected for different orientations of the crystal with respect to the incoming X-ray beam using an area detector, basically a large CCD camera similar to the camera in a cell phone, but much larger. Planes of equal electron density in the sample act like an array of pinholes, to produce a diffraction pattern, as shown in Figure 15.21a for different orientations of the crystal. The scattering patterns are converted into electron density maps by a complicated, yet well-understood mathematical procedure (Fourier analysis) and yield structures such as the one shown in Figure 15.21b for insulin, a metabolically important hormone. In the structure shown, the ribbons represent helical and disordered structures of the peptide backbone. When the sample is a single crystal of an ionic compound, such as NaCl, it is straightforward to get from the diffraction pattern, *via* the planes of equal electron density, to the arrangement of atoms in the crystal. In cases like peptides, hormones, DNA segments, or large proteins, it is useful to have more information at hand, such as the sequence of amino acids in a peptide like insulin, to arrive at the desired structure.

While single crystal X-ray crystallography has solved thousands of molecular structures, it is a technique that gives results for molecules in the solid state only. Since biological processes – such as the catalytic action of enzymes – occur in aqueous surroundings, the question arises whether or not solid-state structures obtained using X-ray crystallography really represent biological states. The answer to this question nowadays comes mostly from comparison with NMR spectroscopic results (*cf.* previous section). Modern NMR techniques, not introduced in our preliminary discussion, allow structural determination of very large molecules such as proteins and provide insights into solution structure and dynamics of large biological molecules.

It was found that, in general, X-ray and NMR structures agree well, yet many structural questions are still unanswered. For example, when biological molecules such as insulin (which is needed for the treatment of type-1 diabetes) are synthesized in the laboratory, will the molecules fold the same way naturally produced insulin molecules fold into their native structures? This is a question that cannot be answered by crystallographic methods, since the folding process takes place in solution. This question can be answered using other spectroscopic methods not mentioned in this chapter.

15.5 Mass Spectrometry

The final analytical method to be introduced here is mass spectrometry. It is not a spectral technique like the ones discussed in Section 15.4 where the interaction of matter with electromagnetic radiation is used to gain structural information of atoms or molecules, but rather the mass/charge ratio of ions produced in a mass spectrometer is observed. This allows the determination of the masses of atoms and molecular ions and answers one of our fundamental questions: how do we know the masses of particles, isotopes, and molecules? Knowledge of these masses is fundamental for many aspects discussed in this text, from stoichiometry calculations to the structural methods discussed in the last section. Furthermore, modern mass spectrometry has found its widest applications in forensic science (go watch NCIS!), art conservation, astronomy, biochemistry, pharmacology, and many more. So, how does this technique work?

In a mass spectrometer, gaseous ions of the sample are analyzed, as follows. If the sample has a vapor pressure higher than ca. 10^{-2} [mm Hg], the vapor molecules are sucked into a high vacuum. There, they are bombarded by electrons that knock out some of the molecules' electrons, creating positive ions. Solid samples may be vaporized and ionized by different methods, for example, by hitting a solid sample, embedded in a polymer matrix, with powerful laser pulses, which vaporize and ionize both the sample and the matrix molecules. This method is called matrix-assisted laser desorption/ionization (MALDI) mass spectrometry. Alternatively, samples in solution may be vaporized and ionized using the electrospray ionization technique (ESI). Here, solutions are nebulized in an electric field, which produces positively charged droplets of the solution. In the high vacuum of the instrument, the solvent evaporates, leaving positively charged ions of the analyte.

The positively charged molecular ions are subsequently accelerated in an electric field. The beam of collimated ions is then passed through a magnetic field, which deflects the ion beam based on the mass/charge ratio of the molecular ions. Higher-mass molecular ions are deflected less, and more highly charged ions are deflected more. The deflected molecular ions are detected by an ion detector, basically an electronic device that measures the charge impinging on it. Since the amount of deflection is determined by quantities that can be controlled very accurately, the observed distribution of molecular ions, and therefore the masses of the molecular ions, can be determined with high accuracy.

Originally, the range of the highest mass number that could be detected was a few hundred Daltons [Da], where a Dalton is 1/12 of the mass of a ^{12}C atom, or, in other words, the atomic mass unit (amu). There were, in the early days of mass spectrometry, two major pieces of information available from the experiment. One was the determination of the sample's molecular mass (in units of Da) by the largest mass over charge signal in the mass spectrum, corresponding to a singly positively charged molecular ion. In addition, there are generally a few signals at lower mass-to-charge ratios observed in a mass spectrum. These are due to the molecular ion undergoing fragmentation after ionization. This is the case since the electrons being knocked out in the ionization process can leave the molecular ion in an unstable state, and the molecule can fall apart between the time of ionization and detection. This fragmentation pattern can often help establish the nature and the structure of the sample.

More recently, with the ionization methods described above, much larger molecular ions can be analyzed by mass spectrometry. Many auxiliary techniques have been developed that allow the sequencing of proteins, and mass spectrometry has become one of the most powerful analytical methods.

For a summary of the techniques introduced in Sections 15.4 and 15.5 (except X-ray crystallography), and a few very useful sample exercises, see https://www2.chemistry.msu.edu/faculty/reusch/virttxtjml/Spectrpy/spectro.htm#contnt

The information contained in this website goes beyond the introductory level presented in Chapter 15 and is appropriate for second-year courses in organic chemistry courses.

Further Reading

L. Pauling and E. B. Wilson, "Introduction Quantum Mechanics", McGraw-Hill, New York, (1936).
G. L. Breneman, "The Two-dimensional Particle in a Box", *J. Chem. Ed.* 67(10), 866 (1990).
https://chem.libretexts.org/Bookshelves/Analytical_Chemistry/Instrumental_Analysis_(LibreTexts)/08%3A_An_Introduction_to_Optical_Atomic_Spectroscopy/8.01%3A_Optical_Atomic_Spectra.
M. Diem, "Quantum Mechanical Foundations of Molecular Spectroscopy", Wiley-VCH, Weinheim, Germany (2021).
"Mass Spectrometry", https://en.wikipedia.org/wiki/Mass_spectrometry.

Epilogue

Let us pause and look back to where we started out in this conceptual approach to general chemistry. From a very simplified view of atoms and molecules – just enough to understand the concept of moles and stoichiometry – we have worked our way through a microscopic view of atoms and molecules, which allowed us to determine the shape, bonding, and polarity of molecules, and many of their chemical properties, such as melting and boiling points and reactivity. Subsequently, we have taken a much more detailed look at chemical reactions at the macroscopic scale: we have discussed quantitatively how far a chemical reaction goes, i.e. we have introduced the concept of chemical equilibria and applied them to aqueous solutions of acids, bases, buffers, and solubility. In Chapter 11, we have returned to the question that looms large in science. This question, of course, is: Why do things happen? In the earlier chapters on the microscopic view of chemical reactions, we discussed the energetics of bond formation. However, we have seen in this chapter that energetics alone is not a criterion for a reaction to occur, but that it is the free enthalpy that determines whether or not a reaction goes. This free enthalpy has enthalpic and entropic contributions; thus, probability as well as energetics determine whether or not a reaction occurs.

Subsequently, we also have discussed and answered the question of how fast and by what mechanism a reaction occurs. There are chemical processes that take millennia, while others occur in microseconds. In biochemistry, many reactions that are too slow to be useful are catalyzed by protein enzymes with a 1000-fold increase in reaction rates. In Chapter 14, we saw an aspect of chemistry that is often neglected, namely the transmutation of atoms of one kind into others, such as the change of the isotope ^{14}C into ^{14}N, or uranium to lead. We have also touched upon the energetics of such reactions that exceed the energies of chemical reactions million-fold. Finally, in Chapter 15, two main goals were achieved: we answered the question of how do we know all the properties of atoms and molecules by introducing the spectroscopic methods used in modern chemistry. The other goal was to show how atoms and molecules obey laws of mathematics that were formulated decades before our understanding of chemistry developed. This view suggests that mathematics is the universal language of nature.

Appendix

In the narrative of the text, units are presented in square brackets following a numeric value, such as 3.0 [mmol]. In equations, units are presented in square brackets after the symbolic equation, such as in Example 4.1:

$$n_{Fe} = 100.0/56\,[g/(g\,mol^{-1})] = 1.786\,[mol]$$

List of Constants

c	Velocity of light	$2.998 \cdot 10^8$	[m/s]
e	Electronic (elementary) charge	$1.602 \cdot 10^{-19}$	[C]
F	Faraday constant	96 485	[C/mol]
g	Gravitational acceleration	9.81	[m/s^2]
h	Planck constant	$6.6 \cdot 10^{-34}$	[J s]
k	Coulomb constant	$8.99 \cdot 10^9$	[N m^2/C^2]
k	Boltzmann constant	$1.38 \cdot 10^{-23}$	[J/K]
m_e	Electron mass	$9.1\,094 \cdot 10^{-31}$	[kg]
m_n	Neutron mass	$1.6\,749 \cdot 10^{-27}$	[kg]
m_p	Proton mass	$1.6\,726 \cdot 10^{-27}$	[kg]
N_A	Avogadro constant	$6.022 \cdot 10^{23}$	[mol^{-1}]
R	Gas constant	0.082	[L atm/(K mol)]
		= 8.3	[J/(K mol)]

List of Abbreviations and Symbols

$\mathcal{A}$	Atomic mass (formerly referred to as "gram atomic mass")
C_M	Molar concentration
C_m	Molal concentration
c_p	Heat capacity
χ	Mole fraction
d	Density (general)
ε	Voltage or electric potential (Chapter 12)
$\varepsilon(\lambda)$	Molar extinction coefficient (Chapter 15)
F	Force
K	Equilibrium constant
k	Rate constant
L	Angular momentum
l	Sample path length (Chapter 15)
I	Moment of inertia
i	Current
$\mathcal{m}$	Molecular mass (formerly referred to as "gram molecular mass")
m	Mass
N	Number of atoms or molecules
N	Number of neutrons in an atom
n	Number of moles
n, l, m	Quantum numbers
Q	Quotient of initial concentrations
Q	Electric charge
q	Elementary charge
w	Work
Z	Atomic number = number of protons in an atom

Index

Bold page numbers indicate that the subject is discussed on consecutive pages

a

Acetone   32, **110**, 197, **220**, **223**
Acetylene (ethine)   36, **72**, 222
Acid   **129**, **135**, **140**
Actinides   25
Activity   **126**
Actual yield   43, 48
Adenine   87
Alcohol   32, 45, 156
Alkali (metals)   21, 25, 32, 48, 66, 131, 170
Alkaline earth (metals)   25, 48, 131, 219
Allotrope   2, 80, 86
Amide linkage   87
Amine groups   87
Ammonia   34, 36, 71, 74, 87, **116**, **121**, **129**, **133**, **138**, 196, 209
Ammonium (ion)   24, 71, 130, 134, **137**
Analytical balance   46, 95
Angular momentum quantum number   59
Anion   **20**, **24**, **29**, **34**, 66, 85, **129**, **138**, 143, 170
Antibonding orbitals   **76**, 79, 84, 219
Area   **3**, 7, 13
Aromaticity   80
Arsenic   45
Atmospheric pressure   4, 91, **96**, 110, 112, 119, 154, 175
Atom   17, **21**, **29**, **41**, **51**, **55**, **58**, **61**, **69**, 95, 150, 169, **199**, **210**, 213
Atomic mass (unit)   17, **20**, 26, **37**, 100, 103, 223, 226, 230
Atomic number   **18**, 65, 199, 202
Atomic radius   18, 23, 25, 65
Aufbau principle   23, 51, 62, 64, 66, 69, 74, **199**
Average mass number (of an element)   20
Avogadro's law   42, 99
Avogadro's number   1, 25, **27**, 41, 53, 99, 103, 162, 185
Arrhenius   129, 196

b

Bar   96
Barometer   97
Base (mathematics)   9

Base   35, 88, **129**, **137**
Benzene   80
Black body radiation   53, 55
Bohr radius   62, 215
Boiling   3, 5, 99, 110
Boiling point   38, 76, 88, **91**, 95, 110, 112, 150, 158
Boiling point elevation   38, 45, **90**
Boltzmann   102
Boltzmann constant   104, 162
Bond order   **78**
Bond   **30**, **69**, 87, 129, 157, 197, 209, 211, 220, 223, **225**
Bonding orbital   70, **76**
Borane   34, 129
Boron   **34**, **64**, 73
Boyle's law   **97**
Brass   44
Bromide   21
Butane   110, **155**
Brønsted-Lowry acid/base   129

c

Calcium carbonate   29, 44, 118
Calcium chloride   47, 91
Calcium fluoride (feldspar)   123
Calorie   4, 149
Carbon dioxide   33, 37, 41, 43, **74**, 89, 138, **155**, 170, 190
Carbon tetrachloride   75
Carbonate ion   24, 34, 36, 44, 118, 219
Carbonyl group   224
Cation   20, 33, 35, 66, 129, 131, **137**, 170
Celsius   3
Centigrade   **3**, 7, 98, 113, **149**
Chain rule (of differentiation)   13
Chalcogens   25
Charles' law   98
Chemical equilibrium   107, **113**, 121
Chemical reactions   21, 25, **41**, 85, 107, **116**, 130, 138, **147**, **150**, 158, 170, **185**, 191, 196, 199

Chloride   **20**, 24, 29, 44, 122, 135, 219
Chlorine   20, **24**, 35, 43, 167, 170, 173
Chlorine trifluoride   75
Clathrate   90
Clausius   102
Clausius – Clapeyron equation   111
Collagen   29
Colligative properties   90, 110
Combustion   41, 147, 152, **154**, **169**
Common ion effect   **124**, **135**
Complex numbers   15
Compound (chemical compound)   24, **29**, **33**, **41**, 44, 66, 72, 83, 101, 141, 170
Concentration (solution concentration)   38, **45**, **90**, 105, 107
Concentration quotient   119, 174
Condensation   83, **100**, **108**, 113, 127
Conductivity   26, **83**, 122, 129
Conductivity band   84, 210
Conductor   80, **85**, 178
Copper   44, 83, 126, 169, 176
Coulomb   174
Coulomb's law   85
Covalent bond   **24**, **30**, 36, 41, 66, **69**, 71, **75**, 228
Covalent solids   83, **86**
Crystal lattice   29, 37, **84**
Cubic equations   8
Cyanide ion   33, 37
Cytosine   87

d

Davisson / Germer experiment   54
de Broglie   **53**
Decadic logarithm   111
Degeneracy (of orbitals)   **22**, 61, 64, **78**, 84, 212, 217
Density   3, 18, 45, 90, **95**, 100, 105, 113
Definite integral   **13**
Derivative   **12**, 210, 214
Delocalized electrons   34, 76, 80, 84, 86, 180, 211
Deoxyribonucleic acid (DNA)   87
Diamagnetism   66, 79
Diatomic molecule   **25**, **30**, 33, 70, 75, 78, 211, 223, 225
Diamond   33, 86, 156, 229
Differential calculus   **12**, 154, 189
Differential equations   14, 59, 61, 214, 222
Diffraction   **52**, 62, 209, 225, **228**
Diffusion   **104**, 185
Dilithium   78
Dilution   **45**, 90
Dimer   35
Dimerization   35
Dinitrogen monoxide   37
Dinitrogen tetroxide   35

Dipole moment   66, 69, **75**, **88**, 90
Dipole-dipole interaction   **88**, 108
Dissociation   **122**, 126, **129**, 143
Divalent   24
Double bond   **32**, 73, 80, 89, 211, **219**
Dynamic equilibrium   90, **107**, **113**, **122**, 193

e

Effusion   **104**, 185, 207
Einstein   **52**, 201, 205
Elastic collision   102
Electrolyte solution   122
Electromagnetic radiation (light)   62, **226**
Electromagnetic wave   **51**
Electron   **17**, **29**, 51, **54**, **58**, **69**, **84**, 129, **169**, **199**, **209**, **226**
Electron affinity   23, **65**, 157, 170
Electron correlation   62
Electron diffraction   54, 62
Electron microscopy   54
Electron shell   32, 57
Electronegativity   32, 66, 75
Electronic configuration   21, 29, 59, 64, 66, 78
Electrostatic forces   24, 29, 85
Element   **17**, **29**, **41**, **63**, 75, 86, **155**, 170, 173, **199**, 206, 209
Elementary charge   17, 57
Elementary step   11, 185, 190
Emission spectrum   **55**, 217
Empirical formula   38, 42
Energy level   **56**, 77, 84, 211, **217**, **224**
Enthalpy of vaporization   108, 111, 196
Equation (algebraic)   **5**, **7**, 12, 14
Equilibrium   90, 104, **108**, **112**, **131**, **138**, 158, 163, **165**, 179, 181, 185, **192**, 196
Equilibrium calculations   115
Equilibrium concentrations   109, **113**, 121, 134
Equilibrium constant   **113**, 116, 120, 131
Erythrocyte   92
Ethane   31, **36**, **72**, 222
Ethene (ethylene)   36, 72
Ethine (acetylene)   36, 72
Euler's formula   15, 216
Euler's number   9
Evaporation   83, 91, **108**, 113, 119
Exponential expressions   1, **3**, 8, 10, 18
Exponential functions   3, 9, 14, 110

f

Fahrenheit scale   3, **4**
Feldspar (calcium fluoride)   **123**
Ferromagnetism   66
Fluoride   21, 29, 36, 87, 124, 126
Fluorine   19, 26, **30**, 33, **35**, 70, 75, 78, 87, **107**, 119

Forever chemicals   33
Formal charges   34, 36
Formaldehyde   32, **37**, 72, 79
Freezing point depression   38, 45, **90**
Frequency (of light)   **51**, **72**, 222, **227**
Fugacity   127

g
Gallium   83
Gas   3, **21**, **95**, **108**, 127, 148, 185
Gas constant   4, 99, 104, 148, 162
Gay-Lussac law   98
Glass   86, 171
Glucose   **37**, 202
Graham's law   105
Gram atomic mass   27
Gram molecular mass   **37**
Graphene   80, 86
Graphite   33, 80, 86, 156, 180, 207
Graphs   7, 98, 155, 157
Gravitational acceleration   56, 148
Growth function   **8**
Guanine   87
Gypsum   47

h
Heat   4, 83, **85**, 108, 112, **147**, **160**, 169, 181, 207, 221
Heat capacity   149, 151, 158
Heat of vaporization   108, 113
Heisenberg   58, 210
Helicase   88
Helium   18, **22**, 26, 62, 95, 101, 108, 200, 202, 206
Hemoglobin   92, 221
Homogeneous mixtures   43
Hund's rule   23, 64, 66, 70
Huygens   52
Hybridization   **70**, 80, 211
Hydrochloric acid   130
Hydrogen atom   **1**, 26, **55**, **61**, **69**, 76, 109, 209, **213**, 218
Hydrogen atom energy levels   57, 61
Hydrogen atom orbitals   216
Hydrogen bonding   **87**, 95, 108, 228
Hydrogen fluoride   32, 87, 107
Hyperbola   6, 97
Hydroxyapatite   29
Hydronium ion   29, 130
Hydrobromic acid   130
Hydroiodic acid   130

i
Ideal gas   3, 6, **98**, 109, 127, **163**
Ideal gas law   92, **100**

Imperial system   **2**
Indefinite integral   13
Insulators   84
Integral calculus   12
Integration   10, 13, 85, 111, 189
Intermolecular forces   45, 69, **75**, **87**, 90, 95, 101, 105, **108**, 158
Ion   **20**, **29**, **34**, 66, **124**, **135**, 170, 180
Ionic compounds (solids)   24, 29, 33, 37, 44, 85, 90, 170, 228
Ionic structure   85
Ionization energy (potential)   23, 29, **65**, 170, 201, 230
Iron   18, 29, **42**, 64, 66, 149, 191, 205
Isoelectronic   **33**, **71**, 79
Isotonic solution   92
Isotope   **20**, 26, 190, 199, **202**

j
Joule   3, 99, 148

k
Kelvin scale   3, **6**, **98**, **110**, **149**
Ketone   32, 79, **219**
Kinetic theory (of gases)   99, **101**, 111

l
Lattice energy   44, 85, 123, **156**
Laughing gas   37
Le Chatelier's principle   **120**, 133
Length   2
Lewis acid   34, 129
Lewis base   35, 129
Lewis structure   **30**, **70**, **78**
Light (electromagnetic radiation)   33, **51**, **55**, 76, 84, 211, **218**
Light emitting diode (LED)   84
Limiting reagent   **42**
Linear combination of atomic orbitals (LCAO)   76
Liquid   **83**, **86**, **90**, **95**, **107**, **118**, **159**
Liter   3
Logarithmic functions   9
Logarithms   1, **8**, 13, **110**, 139, 182, 196, 221
London dispersion forces (induced dipole forces)   85, 89
Luster   83

m
Magnesium   24, 85, **156**
Magnetic moment   **63**, **226**
Magnetic quantum number   21, 23, 61, 214
Magnetic resonance imaging (MRI)   63, 226
Main group element   25, 65
Main quantum number   21, 25, 30, 57, **59**, 65, 217
Malleability   83
Manometer   96
Mantissa   1, 4

Mass 1, 3, **17**
Mass number (of atoms) **19**, **26**, 37, **200**, 230
Mass percentage **45**
Mass spectrometry 39, 209, 226, 230
Mathematical Tools 1
Maxwell 102
Maxwell's equations **51**
Mean square velocity **102**, 105
Mercury 3, 83, 96, 110
Metal 3, 43, 45, 53, 66, 84, 151, **162**, 174, 218
Metallic bonding 26, 83, 95
Metalloid 26
Meter **2**, 52
Methane **31**, 37, **70**, 88, 209
Methanol (methyl alcohol) **32**, 37, 90, 171
Methyl fluoride 33
Metric System **1**, 18, 96
Molar composition 41
Molar concentration (molarity) 45, 92, 221
Molality 45, 91
Molarity 45, 118
Mole **26**, **41**, 99, 177, 179
Mole fraction 45, 76, 91
Molecular dipole moment 69, **75**
Molecular formula 36, **38**, 180
Molecular ion 69, 72, 76, 78, 230
Molecular mass **37**, 87, 100, **103**, 110, 230
Molecular orbital 30, 33, 69, 77, 79
Molecular orbital (MO) theory 33, 69
Molecular polarity 44, 69, **75**
Molecular solids 86
Molecule 17, **25**, **29**, **42**, 51, **69**, **219**
Momentum **53**, **102**, 225
Monatomic gas 25
Monovalent 24
Multiple bond 32

n

Natural gas 31, 207
Natural decay function 9
Natural growth function 9
Natural logarithm 1, 10, 111
Net ionic equation 48, 169, 172
Neutron 17, **200**, 207
Newton **3**, 73, 96, 148
Nitrate ion 34
Nitric acid 37, 130, 218
Nitrite ion 35, 139
Nitrogen 23, **33**, 64, **71**, 74, **78**, 87, **100**, **117**
Nitrogen dioxide **35**, 72
Nitrogen monoxide 35
Nitrous acid 37, **131**, **139**

Noble gas 21, **25**, 66, 100
Nodal plane 59, 77, **215**
Non-polar covalent bond 30
Nuclear chemistry 17, 67
Nuclear magnetic resonance (NMR) 63, **226**
Nuclear reactions **199**
Nuclear shell model 67, **199**
Nucleon 18, **205**
Nucleus **17**, 25, 30, 57, 62, **65**, 76, 129, **199**, 210, **213**

o

Octane 110, 117, 152
Octet (configuration) **24**, 29, **32**, 186
Orbital 21, **30**, 51, **58**, 64, 69, 74, **76**, 84, 195, **212**
Orbital energy **61**, 64
Orbital (angular momentum) quantum number 21, 62, 217
Organic chemistry 36, 129, 226
Osmosis 92
Osmotic pressure 90, **92**
Oxygen **18**, 23, 33, 38, 41, 75, 78, 92, 170, 205
Ozone **33**, **72**, 186

p

Paramagnetism 66, 79
Partial pressure 101, 109
Particle-wave duality 18
Pascal 4, 96
Pauli (exclusion) principle **63**
Pauling 32
Percent composition 29, 38
Percent yield **42**
Periodic chart (of elements) **19**, 29, 33, 51, 59, **61**, 67, 78, 84, 199, 209
Periodic properties (of elements) 21
pH **9**, 126, **130**, 173
Phase 29, 43, 83, 172, 188
Phase diagram 113
Phase transitions 83, 89, 101, **157**
Phosphate 30, 87, 140, 171
Phosphorus 33, 43, 89
Phosphorous pentachloride 74, 167
Phosphorous trichloride 167
Photocathode 53
Photoelectric effect 52
Photon **53**, 80, 84, 186, 201, 211, **218**, 222, 227
Pipette **46**
Planck's constant 53, 211
Platonic solid 71
Polar coordinates 11
Polar covalent bond **30**, 75
Polarity 32, 44, 66, **75**, 87
Polonium 25

Polynomial   8, **58**
ppb (parts per billion)   45, 219
ppm (parts per million)   45, 228
Precipitation reactions   47
Pressure   **3, 95, 100**
Product rule (of differentiation)   13
Proportionality   **5**, 57, **98**, 150, 163, 186, 205
Protein   29, 88
Proton   18, 25, 63, 66, 76, **127**, 200
Proton spin   63, 227

q
Quadratic equations   **7**, 14, **116**, 122, 133
Quantization of light   53, 58
Quantized states   63
Quantum mechanics   21, 32, 55, 58, **61, 209**, 212, 217, 226
Quantum numbers   21, 25, 30, **56**, 59, **63**, 210, **212**, 217, **223**
Quark   17
Quartic equations   **7, 121**
Quartz   25, 86
Quotient rule (of differentiation)   13

r
Radial distribution function   62
Radial part (of wavefunction)   **59, 214**, 225
Radial polar coordinates   11
Radical   35, 72, **186**
Raoult's law   91
Rare earth metals   25
Real gas   89, 98, **100**
Resonance (structures)   **32**, 74, 80
Rest mass   54
Root mean square velocity   **103**
Rydberg constant   57

s
Saturated solution   122, 124
Schrödinger   58, 214
Schrödinger equation   **58**, 62, **69**, 210, **214, 223**
Scientific calculators   1
Scientific method   17, 95
Scientific notation   1
Sea of electrons   **84**
Sea water   45, **124**
Semiconductor   25, 84, 210, 225
Semipermeable membrane   92
Second derivative   **12**, 210, 214
Significant figures   **1**, 4, 20, 26, 42, 112, 116, 124, 133, 226
Silicate   29, 34
Silicon   25, 84, 86, 210
Silicon dioxide   86
Silver chloride   122, 125, 135
Single bond   **32**, 70, 78, 220

Sodium   20, 24, 29, 38, 44, 85, 124, 170, 218
Sodium chloride   20, **44**, 85, **91**, 170
Solubility   85, 90, **122**, 135
Solubility calculations   123
Solubility product   **123**
Solubility product constant   **123**
Solution stoichiometry   43
Solutions   38, **43**, 85, **90**, 118, **122, 129**, 169, 172, 219
Solvation   69, 90, **122**
Solvation energy   85, 123
Spatial quantization (of angular momentum)   63
Spherical harmonic functions   12, **58**, 70, **216**
Spherical polar coordinates   11
Spectator ions   47, 169, 171
Spectral lines   55, 218
Spectroscopy   51, 63, 76, 89, 209, **217, 225**
Spin   **22, 63**, 78, 199, 209, 218, **226**
Spin multiplicity   23
Spin quantum number   **63**
Spin-spin coupling   **227**
sp hybrid orbital   **71**, 73
sp^2 hybrid orbital   **71**, 75, 80, 87
sp^3 hybrid orbital   **71**, 86, 225
Square planar (structure)   71, 75, 225
Stationary states   55
Stern-Gerlach experiment   63, 226
Stoichiometry   17, 25, 30, **41**, 48, 107, **117**, 126, 172, 194
Subshell   21
Sublimation   83, 91, 113
Sucrose   **90**
Sulfate   29, 36, 47, **178**
Sulfur   25, 36, 38,
Sulfur dioxide   36, 75
Sulfur hexafluoride   36, 75
Sulfur tetrafluoride   74
Sulfur trioxide   36, 76

t
Temperature   **3, 5**, 26, 29, 55, **83, 87, 95, 107**, 117, 120, **149**, 155, **157, 162**, 181, **195**, 221
Tetrahedral (structure)   **71**, 86, 209, 225
Theoretical yield   **42**
Thermite reaction   **42**, 107
Thymine   87
Three-body problem   69
Time   102, 148, 173, 185, **187, 192**, 204
Transitions (spectral transitions)   80, **217, 221**, 226
Transition metal   19, 25, **65**
Trigonal pyramidal (structure)   71, **73**
Trigonometric functions   12
Triple bond   33, 36, **72**, 78
Tungsten   95

u

Ultraviolet (UV) light 33, 52, 80, 186, 218

v

Vacuum pump 96
Valence band 84
Valence bond 220
Valence bond (VB) theory **69**, 76
Valence electron 32, 34, 73
Valence shell electron pair repulsion (VSEPR) model 34, 69, 72, **73**, 225
Vapor pressure **107**, **109**, **118**, 122, 127, 196, 230
Vaporization 83, 91, **107**, 118, **157**, 196
van der Waals equation 101
van der Waals interactions 87
Velocity of light **54**, 200, 205, 222

Volume 3, 6, **45**, 90, **95**, 113, 152, 165
Volume work 148, 152
Volume contraction 45
Volumetric flask **46**

w

Wavefunction **58**, 78, **209**, 214, 216
Wavelength (of light) **51**, 57, 80, 201, 211, 218, **220**, 228
Wavenumber 222
Water molecule 31, 44, **70**, 74, 76, 90, 108, 130, 135, **223**

x

X-ray diffraction 209, 225, **228**

z

Zinc 44, 176, 178